D0384629

MIND

MIND

A JOURNEY TO THE HEART OF BEING HUMAN

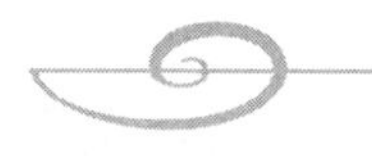

Daniel J. Siegel, MD

W. W. NORTON & COMPANY
INDEPENDENT PUBLISHERS SINCE 1923
NEW YORK LONDON

MIND
A Journey to the Heart of Being Human
Daniel J. Siegel

Printed in the United States of America
First Edition

For information about permission to reproduce selections from this book,
write to Permissions, W. W. Norton & Company, Inc.,
500 Fifth Avenue, New York, NY 10110

For information about special discounts for bulk purchases, please contact
W. W. Norton Special Sales at specialsales@wwnorton.com or 800-233-4830

Manufacturing by Maple Press
Book design by Molly Heron
Production manager: Christine Critelli

Library of Congress Cataloging-in-Publication Data

Siegel, Daniel J., 1957 author.
Title: Mind : a journey to the heart of being human / Daniel J. Siegel, MD.
Description: First edition. | New York : W. W. Norton & Company, [2017] |
Includes biographical references and index.
Identifiers: LCCN 2015049567 | ISBN 9780393710533 (hardcover)
Subjects: LCSH: Psychology. | Brain. | Intellect. | Self. | Consciousness.
Classification: LCC BF131 .S46 2017 | DDC 150--dc23 LC record available at
http://lccn.loc.gov/2015049567
ISBN: 978-0-393-71053-3

W. W. Norton & Company, Inc.,
500 Fifth Avenue, New York, N.Y. 10110
www.wwnorton.com

W. W. Norton & Company Ltd., Castle House,
75/76 Wells Street, London W1T 3QT

1 2 3 4 5 6 7 8 9 0

Important Note: *Mind* is intended to provide general information on the subject of health and well-being; it is not a substitute for medical or psychological treatment and may not be relied upon for purposes of diagnosing or treating any illness. Please seek out the care of a professional healthcare provider if you are pregnant, nursing, or experiencing symptoms of any potentially serious condition.

For
Caroline

CONTENTS

ACKNOWLEDGMENTS xi

1 Welcome 1

2 What Is the Mind? 26

3 How Does the Mind Work in Ease and Dis-Ease? 62

4 Is the Mind's Subjective Reality Real? 99

5 Who Are We? 123

6 Where Is Mind? 145

7 A Why of Mind? 188

8 When Is Mind? 212

9 A Continuum Connecting Consciousness, Cognition, and Community? 257

10 Humankind: Can We Be Both? 302

REFERENCES 333

INDEX 345

EXPANDED CONTENTS

1. Welcome 1

The Mind's Curiosity About Itself 5

A Common View: The Mind Is What the Brain Does 7

Our Identity and the Internal and Relational Origin of Mind 12

Why this Book About the Mind? 13

An Invitation 16

The Approach of Our Journey 17

Words Reflecting on Reflecting Words 20

2. What Is the Mind? 26

Working on a Working Definition of Mind (1990-1995) 26

The System of Mind: Complex Systems, Emergence, and Causality 42

Reflections and Invitations: Self-Organization of Energy and Information Flow 51

3. How Does the Mind Work in Ease and Dis-Ease? 62

Self-Organization, Lost and Found (1995-2000) 63

Differentiation and Linkage: The Integration of Healthy Minds 75

Reflections and Invitations: Integration and Well-Being? 85

4. Is the Mind's Subjective Reality Real? 99

Adapting to a Medical World that had Lost Its Mind (1980-1985) 100

Mindsight in Health and Healing 110

Reflections and Invitations: The Centrality of Subjectivity 115

5. Who Are We? 123

Exploring the Layers of Experience Beneath Identity (1975-1980) 124

Top-Down and Bottom-Up 127

Reflections and Invitations: Identity, Self, and Mind 140

6. Where Is Mind? 145

Could Mind Be Distributed Beyond the Individual? (1985-1990) 145

Neuroplasticity and Cultural Systems 177

Reflections and Invitations: Within and Between 182

7. A Why of Mind? 188

Meaning and Mind, Science and Spirituality (2000-2005) 188

Integration as the "Purpose of Life?" 200

Reflections and Invitations: Purpose and Meaning 206

8. When Is Mind? 212

Exploring Presence in Mind and Moment (2005-2010) 212

Attunement, Integration, and Time 227

Reflections and Invitations: Awareness and Time 235

9. A Continuum Connecting Consciousness, Cognition, and Community? 257

Integrating Consciousness, Illuminating Mind (2010-2015) 258

Consciousness, Non-Consciousness, and Presence 277

Reflections and Invitations: Cultivating Presence 287

10. Humankind: Can We Be Both? 302

Being, Doing, and Integrating Mind (2015-eternal present) 302

The Systems of a Plural Self and Integration of Identity 316

Reflections and Invitations: MWe, an Integrating Self, and a Kind Mind 323

ACKNOWLEDGMENTS

IMAGINE TRYING TO ARTICULATE A FEELING FULL OF GRATITUDE for this gift of being here, for being human, for being alive. These words pale in contrast to the sensation of appreciation, of love, of awe I feel for the many individuals with whom I've been given the privilege of sharing this life's journey. In many ways, this is also a feeling of connection to each of us here, now, who live on this planet, our collective home, this place we've named Earth. The sun is setting now as I write these words, the planet doing a fine job making its daily turn, thank you Earth, the crimson collage adorning the sky above this crescent bay where, from my earliest days, I grew to love sea and sand. Even though now long beyond those days of youth, life still feels vibrant and young. And this feeling, too, is of deep respect for those who've lived in the years, decades, centuries, and millennia that we call "past," and for those to come in our "future." We are all a part of one unbroken chain, one interconnected whole of life, one common human family.

In this experience here, now, there have been many individuals who've travelled the journey of discovery that was the path to the birth of this book. A wonderful group of interns worked hard to explore the scientific, philosophical, clinical, contemplative, spiritual, and popular literature to see how the notion of "mind" has been treated across a wide range of pursuits. These helpful and thoughtful people include Megan Gaumond, Carly Goldblatt, Rachel Kiekhofer, Deena Margolin, Darrell Walters, and Amanda Weise. Thank you for your spirit of adventure in taking this deep dive.

The book has also had the good fortune of having readers of the initial manuscript who offered useful feedback and insights that were woven into subsequent revisions. These individuals include

Diane Ackerman, Ed Bacon, Aldrich Chan, Adriana Copeland, M. Lee Freedman, Lisa Freinkel, Don Hebert, Nathaniel Hinerman, Lynn Kutler, Maria LeRose, Jenny Lorant, Sally Maslansky, Ronald Rabin, and Caroline Welch. Thank you all for your time and focused energy in reading and reflecting on *Mind*.

Each year we have our Interpersonal Neurobiology conference at UCLA, organized by my dear friends and colleagues, Marion and Matt Solomon and Bonnie Goldstein. Lou Cozolino, our present Norton Interpersonal Neurobiology Series editor, and Allan Schore, the prior series editor, are also wonderful collaborators in this effort to bring our interdisciplinary field out into the world. It is an honor to be working alongside them and the many other authors, scientists, and practicing professionals, who have courageously joined forces to construct the bridges to link independent sources of knowledge.

I am also grateful for the bridges that the Garrison Institute has been constructing to connect our collective mind's inner wisdom with social action through its visionary initiatives and educational programs that link personal reflection with planetary responsibility. It is an honor to serve on the board with co-founders Diana and Jonathan Rose, Lisette Cooper, Rachel Cowan, Ruth Cummings, Rachel Gutter, Paul Hawken, Will Rogers, Sharon Salzberg, Bennet Shapiro, Monica Winsor, and Andrew Zolli.

Over these years of being an educator, Rich Simon of the Psychotherapy Networker has been a wonderful collaborator in imagining how to bring cutting edge science and its applications to the wide range of professionals in the field of mental health. Rich took the initiative early on to create a multi-disciplinary approach, embracing poetry, mindfulness, and neuroscience, in his celebrated annual conference and award-winning magazine.

I would also like to thank the wonderful team at Norton, including Kevin Olson for his commitment to connecting with our readership, and Elizabeth Baird for her careful oversight of the copyediting of the manuscript. Deborah Malmud, a Vice President at Norton, has been a fabulous collaborator in creating the Norton Series on Interpersonal Neurobiology. I deeply appreciate her taking a chance on this new field. Thanks for all your support, and for the fun along the way. Thanks as well to the rest of the Norton group: Julia Gardiner, Natasha Senn, and Mariah Eppes.

This journey has taken me around our world to engage in educational programs with people from many walks of life. I am grateful for all the learning I absorb from workshop and conference participants as we build on these ideas about mind, brain, and relationships—and the central importance of kindness and compassion in our interconnected world. On one of those trips, my host in Norway, a wonderful psychologist and painter, Lars Ohlickers, took me out for a daylong hiking adventure along the fjords of Norway. As I was stepping out onto the plateau of a cliff, Lars snapped a photo with his camera that he later shared with me during my teaching at the conference he had organized. When I saw the image, I felt right away that it ought to be the cover for this book. I thank Lars for the insight of seizing the moment, and for the invitation to join with him and the magic of nature on that day. I'd also like to thank the wonderful individuals who graciously offered to let their photographs be a part of the story of this journey, including Lee Freedman, Alexander Siegel, Madeleine Siegel, Kenji Suzaki, and Caroline Welch. I am so grateful for each of these individuals and the feelings their images have helped create.

These days I am primarily an educator, providing various ways of learning about the mind and how to develop well-being in our lives to a diverse collection of groups, professions, and scientific disciplines. Journeying inward and journeying around the world is made possible by our wonderful staff at the Mindsight Institute, the center for Interpersonal Neurobiology. Jessica Dreyer, Diana Berberian and Deena Margolin, along with Eric Bergemann, Adriana Copeland, Liezel Manalo, Mark Seraydarian, Andrew Schulman, Ashish Soni, Alta Tseng, and Priscilla Vega, are a fabulous team that makes the work a pleasure.

These decades of exploring the nature of mind from a vast array of disciplines would not have been possible without the support of my family. My mother, Sue Siegel, has been a life-long inspiration in how to think deeply about the world. My son and daughter, Alexander and Madeleine Siegel, have each been engaging conversation partners in exploring the nature of reality and of our interpersonal experiences. And my life partner, and more recently educational partner as head of the Mindsight Institute, and guide in all things mindful, Caroline Welch, is a continual source of inspiration, support, and lifelong learning. My life would not be what it has become without her. I thank you all for traveling together along this journey to the heart of being human.

MIND

CHAPTER 1

Welcome

HELLO.

A simple communication offered from me to you.

But who is it that knows I greeted you with 'hello'?

And how do you know?

And what does knowing really mean?

In this book we'll explore the nature of the who, how, what, why, where, and when of the mind, of your mind, of your self, the experience you have that knows I am welcoming you with *hello.*

Some use the term *mind* to mean intellect and logic, thought and reasoning, contrasting mind to heart, or mind to emotion. This is not how I use the broad term mind here, or in other writings. By *mind,* I mean all that relates to our subjective felt experience of being alive, from feelings to thoughts, from intellectual ideas to inner sensory immersions before and beneath words, to our felt connections to other people and our planet. And mind also refers to our consciousness, the experience we have of being aware of this felt sense of life, the experience of knowing within awareness.

Mind is the essence of our fundamental nature, our deepest sense of being alive, here, right now, in this moment.

Yet beyond consciousness and its knowing within awareness of our subjective felt sense of being alive, mind may also involve a larger process, one that connects us to each other and our world. This important process is a facet of mind that may be hard to measure, but is nevertheless a crucial aspect of our lives we'll explore in great depth in the journey ahead.

Though we may not be able to quantify in numerical terms these facets of our mind at the heart of the experience of being here in this life, this internally felt subjective phenomenon of living, and the ways we can feel our connections to one another and the world, are subjective phenomena that are real. These non-measureable facets of the reality of life have many names. Some call this our essence. Some call this our core, soul, spirit, or true nature.

I simply call this *mind*.

Is *mind* just some synonym for subjectivity—the feeling of our emotions and thoughts, memories and dreams, inner awareness and interconnectedness? If mind also includes our way of being aware of this inner sense of moment-to-moment living, then mind would additionally involve the experience called consciousness, our way of being aware, of knowing what these aspects of our subjective life are as they unfold. So at a minimum, mind is a term that includes consciousness and the way we are aware of our felt experience, our subjective lives.

But something also happens beneath awareness that involves what we usually refer to as *mind* as well. These are our non-conscious mental processes, such as thoughts, memories, emotions, beliefs, hopes, dreams, longings, attitudes, and intentions. Sometimes we are aware of these, and sometimes not. Though we are not aware of these at times, perhaps even the majority of the time, these mental activities happening without consciousness are real and influence our behaviors. These activities can be seen as a part of our thinking and reasoning, as some process that enables "information" to flow and transform. And without awareness, it may be that these flows of information do not evoke subjective feelings, as they are not a part of conscious experience. So we can see that beyond consciousness and its awareness of subjective experience, the term *mind* also includes the fundamental process of information processing that does not depend upon awareness.

But what does mind-as-information-processor really mean? What is information? If information drives how we make decisions and initiate behavior, how does mind, conscious or not, enable us to make willful choices on what to do? Do we have free will? If the term *mind* includes aspects of subjectivity, consciousness, and information processing, including its problem-solving and behavioral control, what makes up the essence of what mind is? What is this "mind stuff" that is a part of this spectrum of mental processes from felt sense to executive control?

With these common descriptions of the mind involving consciousness, subjective experience, and information processing, and how these are manifested in ways that you may be familiar with, including memory and perception, thought and emotion, reasoning and belief, decision-making and behavior, what can we say ties each of these well-known mental activities together? If mind is the source of everything from felt sensations and feelings to thought and the initiation of action, why are these all subsumed under the word *mind*? What can we say the mind is?

Mind as a term, and mind as an entity or process, can be seen as a noun or verb. As a noun, mind has the sense of being an object, something stable, of something you ought to be able to hold in your hands, something you can possess. You have a mind, and it's yours. But what is that noun-like stuff of mind actually made of? As a verb, mind is a dynamic, ever-emerging process. Mind is full of activity, unfolding with ceaseless change. And if the verb-like mind is indeed a process, what is this "dynamic stuff," this activity of our mental lives? What, really, is this mind, verb or noun, all about?

Sometimes we hear a description of the mind as an "information processor." (Gazzaniga, 2004). This generally indicates how we have representations of ideas or things and then transform them, remember events by encoding, storing, and retrieving memory, and move from perception to reasoning to enacting behavior. Each of these forms of mind activity is part of the information processing of the mind. What has intrigued me, as a scientist, educator, and physician working with the mind for more than thirty-five years now, is how common these descriptions of the mind are, yet how a definition of what the mind actually is, a clear view of the mind's essence beyond lists of its functions, is missing from a wide range of fields that deal with the mind, from clinical practice and education to scientific research and philosophy.

As a mental health professional (psychiatrist and psychotherapist), I've also wondered how this lack of at least a working definition of what the mind might actually be could be limiting our effectiveness as clinicians. A *working* definition would mean we could work with it and change it as needed to fit the data and our personal experience. A *definition* would mean we could clearly state what the essence of mind means. We so often hear the word *mind* yet rarely do we notice it lacks a clear definition. Without even a working definition of mind in scientific, educational, and clinical professional worlds, and without one in our personal and family lives, something seems missing, at least in my own mind, from our understanding and conversations about the mind.

With only descriptions and no attempt at even a working definition of what mind is, could we even define what a healthy mind is?

If we stay at the level of description, of mind as being made of thoughts, feelings, and memories, of consciousness and subjective experience, let's see where it takes us. For example, if you reflect for a moment on your thoughts, what is your thinking truly made of? What is a thought? You might say, "Well, Dan, I know I am thinking when I sense words in my head." And I could then ask you, what does it mean to say "I know" and to "sense words?" If these are processes, a dynamic, verb-like aspect of information processing, what is being processed? You may say, "Well, we know that it is simply brain activity." And you may be surprised to find that no one knows, if this brain-view is indeed true, how the subjective sense of your own thinking somehow arises from neurons in your head. Processes as familiar and basic as thought or thinking are still without clear understanding by our, well, our minds.

When we consider the mind as a verb-like, unfolding, emerging process, not being, or at least not only being, a noun-like thing, a static, fixed entity, we perhaps get closer to understanding what your thoughts may be, and in fact, what mind itself might be. This is what we mean by the description of the mind as an information processor, a verb-like process. But in either case, mind-as-noun indicating the process*or* or mind-as-verb indicating the process*ing*, we are still in the dark about what this information transformation involves. If we could offer a definition of the mind beyond these commonly used, important, and accurate descriptive elements, perhaps we'd be in a better position to clarify not just what the mind is, but also what mental well-being might be.

These have been the questions that have occupied my mind over these past four decades. I've felt them, they've filled my consciousness, they've influenced my non-conscious information processing in dreams and drawings, and they've even shaped how I relate to others. My friends and family, teachers and students, colleagues and patients, all know firsthand how obsessed I've been with these basic questions regarding the mind and mental health. And now you do too. But like them, perhaps you'll also come to see how attempting to answer these questions is not only a fascinating process in itself, but also results in useful perspectives that can offer us new ways of living well and creating a stronger, more resilient mind.

This book is all about a journey to define the mind beyond its common descriptions. And once we can do that, we can be in a more empowered position to see the scientific basis for how we might cultivate healthy minds more effectively.

The Mind's Curiosity About Itself

This interest in the mind has been with human beings for as long as we have recorded history of our thoughts. If you, too, are curious about what the mind might be, you are not alone. For thousands of years, philosophers and religious leaders, poets and storytellers, have wrestled with descriptions of our mental lives. The mind seems to be quite curious about itself. Perhaps this is why we've even named our own species, *homo sapiens sapiens*: the ones who know, and know we know.

But what do we know? And how do we know it? We can explore our subjective mental lives with reflection and contemplative practices, and we can set up scientific studies to explore the nature of the mind itself. But what can we truly know about the mind using our minds?

In the last few centuries to present day, the empirical study of the nature of reality, our human mental activity called science, has attempted to systematically study the characteristics of mind (Mesquita, Barrett, & Smith, 2010; Erneling & Johnson, 2005). But as we'll see, even the various scientific disciplines interested in the nature of the mind have not established a common definition of what the mind is. There are many descriptions of mental activities, including emotion, memory, and perception, but no definitions. Odd, you may think, but true. You may wonder why the term, *mind*, is even used if it is not defined. As an important academic "placeholder for the unknown," the word *mind* is a reference term without a definition. And some say that the mind *should not* even be defined, as I've been personally told by several philosophy and psychology colleagues, as it will "limit our understanding" once we use words to delineate a definition. So in academia, amazingly, the mind is studied and discussed in wonderful detail, but not defined.

In practical fields that focus on helping the mind develop, such as education and mental health, the mind is rarely defined. In workshops over the last 15 years, I have repeatedly asked mental health professionals or educators if they have ever been offered a definition of the mind. The results are quite startling, and surprisingly consistent. Of over 100,000 psychotherapists of all persuasions from around the globe, only 2 to 5 percent have ever been offered even one lecture that defined the mind. Not only are over 95 percent of mental health professionals without a definition of the *mental*, but they are also without a definition of the *health*. The same small percentage of over 19,000 educators I've asked, teachers of kindergarten through twelfth grade, have been offered a definition of the mind.

So why attempt to define something that seems to be so elusive in so many fields? Why try to put words to something that may simply be beyond words, beyond definition? Why not stick with a placeholder for the unknown, embracing the mystery? Why limit our understanding with words?

Here is my suggestion to you about why it may be important try to define the mind.

If we could offer a specific answer to the question of what the essence of mind is, provide a definition of mind that takes us beyond descriptions of its features and characteristics, such as consciousness, thought, and emotion, we might be able to more productively support the development of a healthy mind in our personal lives as much as we might cultivate mental health in families, schools, places of work, and society at large. If we could find a useful working definition of mind, we'd then become empowered to illuminate the core elements of a healthy mind. And if we could do that, perhaps we might be better able to support the way we conduct our human activities, not only in our personal lives, but with one another, and with our ways of living on this planet we share with all other living beings.

Other animals have minds too, with feelings and information processing such as perception and memory. But our human mind has come to a place of shaping the planet so much now that we—yes, we with language who can name things—have come to call this epoch the "Human Age" (Ackerman, 2014). Coming to define the mind in this new planetary Human Age might just enable us to find a more constructive and collaborative way of living together, with other people and all living beings, on this precarious and precious planet.

And so from the personal to the planetary, defining the mind might be an important thing to do.

The mind is the source of our capacity for choice and change. If we are to change the course of our planet's global status, we can propose that we'll need to transform our human mind. On a more personal level, if we have acquired compromises to our brain's functioning, through experiences or genes, knowing what the mind is could enable us to more effectively change the brain, as many studies now reveal that the mind can change the brain in a positive way. That's right: your mind can transform your brain. And so mind can influence our basic physiology and our broadest ecology. How can your mind do that? This is what we'll explore in this book.

Finding an accurate definition of mind is more than just an academic exercise; defining the mind may empower each of us to create more health in our individual lives as well as our collective life so we hopefully might create more well-being in our world. To approach these pressing issues, this book, *Mind*, will attempt to address the simple but challenging question, what is the mind?

A Common View: The Mind Is What the Brain Does

A view commonly stated by many contemporary scientists from a range of academic disciplines such as biology, psychology, and medicine, is that the mind is solely an outcome of the activity of the neurons in the brain. This frequently stated belief is actually not new, as it has been held for hundreds and even thousands of years. This perspective, so often stated in academic circles, is concretely expressed this way: "The mind is what the brain does."

If so many esteemed and thoughtful academicians hold this view, and hold it with energized conviction, it would be natural to think that perhaps this idea is the simple and complete truth. If this is indeed the case, then your inner, subjective, mental experience of my hello to you is simply the brain's neural firing. How that might happen—to move from neural firing to subjective experience within knowing—no one on the planet understands. But the assumption within academic discussions is that one day we will figure out how matter becomes mind. We just don't know right now.

So much in science and in medicine, as I learned in medical school and in my research training, points to the brain's central role in shaping our experience of thoughts, feelings, and memories, what are often referred to as the contents—or activities—of mind. The state of being aware, the experience of consciousness itself, is considered by many scientists a byproduct of neural processing. Therefore, if *mind=brain activity* turns out to be the simple and complete equation for the origin of mind, then the scientific search for the neural basis of mind, for how the brain gives rise to our feelings and thoughts, and what are called the "neural correlates of consciousness," may be long and arduous pursuits, but ones that are on the right track.

William James, a physician whom many consider to be the father of modern psychology, in his textbook, *The Principles of Psychology*, published in 1890, stated, "The fact that the brain is the one immediate bodily condition of the mental operations is indeed so universally admitted nowadays that I need spend no more time in illustrating it, but will simply postulate it and pass on. The whole remainder of the book will be more or less of a proof that the postulate was correct" (p. 2). Clearly, James considered the brain central to understanding the mind.

James stated, too, that introspection was a "difficult and fallible" source of information about the mind (p. 131). This view, along with the difficulty researchers faced in quantifying subjective mental experience, an important measuring process many scientists engage in to apply crucial statistical analyses, made studying neural processes and externally visible behaviors more appealing and useful as the academic fields of psychology and psychiatry evolved.

But is the stuff in your head, the brain, truly the *sole* source of mind? What about the body as a whole? James stated, "Bodily experiences, therefore, and more particularly brain-experiences, must take place amongst those conditions of mental life of which Psychology need take account" (p. 9). James, along with physiologists of his day, knew that the brain lives in a body. To emphasize that, I sometimes use the term, "embodied brain," which my adolescent daughter emphatically reminds me is ridiculous to say. Why? Her response to me: "Dad, have you ever seen a brain not living in a body?" My daughter has a wonderful way of making me think about all sorts of things I might otherwise not consider. While she's right, of course, in modern times we often forget that the brain in the head is a part of not just the nervous system, but also part of a whole bodily system. James said, "Mental states occasion also changes in the calibre (sic) of blood-vessels, or alteration in the heartbeats, or processes more subtle still, in glands and viscera. If these are taken into account, as well as acts which follow at some remote period because the mental state was once there, it will be safe to lay down the general law that no mental modification ever occurs which is not accompanied or followed by a bodily change" (p. 3).

Here we can see that James knew that the mind wasn't merely enskulled, it was fully embodied. Nevertheless, his emphasis was on bodily states being associated with mind, or even following mental states, but not causing or creating mental activities. Brain was seen, from long ago, to be the source of mental life. Mind in academic circles is a synonym for *brain activity*—events in the head and not the full body. As one illustrative but commonly stated example, a modern psychological text offers this view as the full glossary definition of mind: "The brain and its activities, including thoughts, emotion and behavior" (Cacioppo & Freberg, 2013).

These views of mind coming from brain are at least 2500 years old. As the neuroscientist Michael Graziano states: "The first known scientific account relating consciousness to the brain dates back to Hippocrates in the fifth century B.C...He realized that mind is something created by the brain and that it dies piece by piece as the brain dies." He then goes on to quote Hippocrates' *On the Sacred Disease*: "'Men ought to know that from the brain, and from the brain alone, arise our pleasures, joys, laughter and jests, as well as our sorrows, pains, griefs and tears'... The importance of Hippocrates's insight that the brain is the source of the mind cannot be overstated." (Graziano, 2014, p. 4).

Focusing on the brain in the head as a source of mind has been profoundly important in our lives for understanding challenges to mental health. For example, viewing those individuals with schizophrenia or bipolar disorder, as well those with other serious psychiatric conditions, such as autism, as experiencing some innate atypical functioning emanating from a brain with structural differences, rather than from something caused by what parents have done, or some weakness in a person's character, has been a crucial shift in the field of mental health to look for more effective means of helping people and families in need.

Turning to the brain has enabled us to diminish the shaming and blaming of individuals and their families, a sad and unfortunately all-too-common aspect of past encounters with clinicians, in years not so long ago. Many individuals, too, have been helped with psychiatric medications, molecules considered to act at the level of brain activity. I say "considered" because of the finding that the mental belief a person holds may be an equally powerful factor in some cases, known as a placebo effect, for a percentage of individuals with certain conditions where their beliefs have led to measureable improvements in external behavior and also in brain functioning. And when we remember that the mind can sometimes change the brain, even this view should be coupled with an understanding that training the mind might be of help even in the face of brain differences for some individuals.

Further support for this brain-centric view of mind comes from studies of individuals with lesions in specific areas of the brain. Neurology for centuries has known that specific lesions in specific areas lead to predictable changes in mental processes, such as thought, emotion, memory, language, and behavior. Seeing mind as related to brain has been extremely helpful, even life saving, for many people over this last century. Focusing on the brain and its impact on the mind has been an important part of advancing our understanding and interventions.

Yet these findings do not logically or scientifically mean that only the brain creates the mind, as is often stated. Brain and mind may in fact not be the same. Each may mutually influence the other as science is beginning to quantitatively reveal, for example, in studies of the impact of mental training on brain function and structure (Davidson & Begley, 2012). In other words, just because brain shapes mind, it doesn't mean mind cannot shape brain. To understand this, it is actually helpful to take a step back from the predominant view that "mind is brain activity" and open our minds to a bigger picture.

While understanding the brain is important for understanding mind, why would whatever creates, or causes, or constitutes, the mind be limited to what goes on above our shoulders? This dominant *brain-activity=mind* perspective, what philosopher Andy Clark calls a "brainbound" model (2011, page xxv), can also be called a "single skull" or "enskulled" view of mind, a view that, while common, does not take several elements of our mental life into account. One is that our mental activities, such as emotions, thoughts, and memories, are directly shaped by, if not outright created by, our body's whole state. So the mind can be seen as embodied, not just enskulled. Another fundamental issue is that our relationships with others, the social environment in which we live, directly influence our mental life. And here, too, perhaps our relationships create our mental life, not only influencing it, but also being one of the sources of its very origins, not just what shapes it, but what gives rise to it. And so the mind in this way may also be seen as relational, as well as embodied.

Linguistics professor Christina Erneling (Erneling & Johnson, 2005) offers this perspective:

> To learn to utter something meaningful—that is, to acquire semantically communicative skills—is not just to acquire the specific configuration of specific brain processes. It also involves having other people consider what one says as a piece of linguistic communication. If I promise you something verbally, it does not matter what the state of my brain is. The important thing, rather, is that my promise is taken as such by other people. This depends not just on my and your behavior and brain processes, but also on a social network of meaning and rules. To explain typically human mental phenomena only in terms of the brain is like trying to explain tennis as a competitive game by referring to the physics of ballistic trajectories...[I]n addition to analyzing mental capacities in terms of individual performances or brain structure, or computational architecture, one also has to take account of the social network that makes them possible. (p. 250)

So at a minimum we can see that beyond the head, the body and our relational world may be more than contextual factors influencing the mind—they perhaps may be fundamental to what the mind is. In other words, whatever mind is may be originating in our whole body and relationships, and not limited to what goes on between our ears. Wouldn't it be scientifically sound, then, to consider the possibility that mind is more than only brain activity? Couldn't we include the brain as part of something more, part of some larger process that involves the body as a whole as well as our relationships from which the mind emerges? Might this be a more complete, fuller view than simply stating mind is limited to activity in the head?

While the mind is certainly related in fundamental ways to brain activity, our mental life may not be limited to, or solely originating from, what goes on inside our skulls alone. Could the mind be something more than simply an outcome of the firing of neurons in the brain? And if this larger picture turns out to be true, what would that *something more* actually be?

Our Identity and the Internal and Relational Origin of Mind

If who we are—both in our personal identity and felt experience of life— emerges as a mental process, a mental product, a function of mind, then who we are is who our mind is. In the journey ahead, we'll explore everything about the mind—not only the who, but also the what, where, when, why, and how of you, of your mind, of the mind.

We begin with this shared position as a starting place: The mind is shaped by, and perhaps even fully dependent upon, the brain in the head's function and structure. There is no argument against this as a point for us to begin. And so we fully embrace what the majority of mind/brain researchers state—and then propose that we extend the notion of mind further than the skull. The brain in the head concept is just the beginning and may not be the end point of our journey of exploration. We may ultimately choose to abandon this attempt at a larger view as we move forward, and perhaps we will eventually come to the commonly stated conclusion that "mind is only what brain does," but for now let's accept the brain's importance in mental life and open our minds to the possibility that the mind may be something more than simply what goes on in the head. What I am suggesting to you is that we consider that the brain is an important component of a yet fuller story, a broader and more intricate story worth exploring for the benefit of all. That fuller story is what we are going to immerse ourselves in as we move along this exploration. Finding a fuller definition of mind is what our journey is all about.

Some academicians view mind independently of the brain. Philosophers, educators, and anthropologists have long described the mind as a socially constructed process. Written before much of our modern understanding of the brain was known as it is today, these socially-oriented academics see our identity, from our internal sense of self to the language we use, as being made from the fabric of social interactions embedded in the families and culture in which we live. Language, thought, feelings, and our sense of identity are woven from the interactions we have with other people. For example, the Russian psychologist Lev Vygotsky considered thought to be internalized dialogue we've had with others (Vygotsky, 1986). The anthropologist Gregory Bateson saw mind as an emergent process of society (Bateson, 1972). And my own teacher of narrative, the cognitive psychologist

Jerry Bruner, considered stories as arising within relationships people have with each other (Bruner, 2003). Who we are, in these views, is the outcome of our social lives.

And so we have two ways of viewing mind that rarely find common ground: mind as a social function and mind as a neural function (Erneling & Johnson, 2005). Each perspective offers an important window into the nature of mind. But keeping them separate, while perhaps useful for carrying out research studies, and perhaps an understandable and often unavoidable outcome of the nature of a scientist's particular interests or proclivities for ways of perceiving reality, may not be useful for seeing the true nature of mind, one that is *both* embodied and relational.

But how can mind be both embodied and relational? How might one thing be in two seemingly distinct places at once?

How can we reconcile these two descriptive stances of the mind that come from thoughtful reflection and study by dedicated academicians over so many years, that the mind in one view is a social product, and the mind from another distinct view is a neural product? What is going on here? These two views represent what are usually seen as separate views of mental life. Could they actually be part of one essence? Is there a way to identify one system from which the mind might emerge, one system that could be embodied and relational, a view that embraces the internally neural and interpersonally social?

Why this Book About the Mind?

In sum, something does not quite feel right about the notion that the statement "mind is what brain does" is the complete truth. We need to keep an open, well, mind about what the mind is in all its rich complexity. Subjectivity is not synonymous with brain activity. Consciousness is not synonymous with brain activity. Our profoundly relational mental lives are not synonymous with brain activity. The reality of consciousness and its inner subjective texture and the interpersonally social nature of mind, at a minimum, invite us to think beyond the buzzing of neural activity within the skull as the totality of the story of what the mind is.

I understand that this approach to mind may be different from the prevailing views expressed by a majority of modern academics in psychology, psychiatry, and neuroscience, and held by many contemporary clinicians in fields of medicine and mental health. My own doubting mind makes me concerned about these proposals.

My scientific training, however, obligates me to keep an open mind about these questions, to not shut down options prematurely. My training as a physician and psychiatrist, and experience as a psychotherapist for over 30 years, has shown me the minds of those I work with seem to extend beyond the skull, beyond the skin. The mind is within us—within the whole body—and between us. It is within our connections to one another, and even to our larger environment, our planet. The question of what the essence of our mental lives truly may be is open for exploration. The nature of mind remains, from a scientific point of view, still a very open issue.

The purpose of this book, *Mind*, is to address this larger story of what the mind is in a direct and immersive way.

My invitation to you is to try to keep an open mind about these questions as we move along. This journey into the nature of mind may require that we re-examine our own beliefs about the mind as we dive deeper into these ideas. Will we come up with new views that have merit in your own life? I hope so, but you'll see what emerges as we move into the journey ahead. As we travel on this exploratory trip together, we may end up with more questions than answers. But hopefully the experience of inquiry into the nature of mind will be illuminating, even if we don't agree upon or even come to final answers.

For these and many other reasons we'll explore, we may wish to keep an open mind—whatever and wherever that mind is ultimately revealed to be—about this question of what the mind is. This sense that there may be something more to the mind than simply enskulled brain activity is not instead of brain, but rather in addition to it. We are not discarding the achievements of modern science; we are exploring them deeply, respecting them fully, and potentially expanding them to reveal a larger truth of what the mind is. We are opening the dialogue in a scientific way, inviting inquiry into mind for all, including academicians, clinicians, educators, students, parents, and anyone with an interest in the mind and mental health. The purpose of this journey is to hopefully broaden discussions, deepen insights, and widen understanding.

Opening the discussion about mind and mental health will hopefully enable us to more effectively pursue research, conceptualize and conduct clinical work, organize educational programs, inform family life, deepen how we understand and live our individual life paths, and even shape society. This exploration holds the potential to deeply empower our personal lives, illuminating the nature of our minds and how we might cultivate more well-being in our day-to-day world.

Our modern life is often flooding us with information, digitally bombarding us yet also linking us across the globe; while at the same time we as a modern human species are more and more isolated and despairing, overwhelmed and alone. Who are we? And what are we to become if we don't conscientiously consider the consequences of how energy and information are flooding our lives? Now, more than ever, it is crucial that we clearly identify what the core of human life is, what the mind is, and learn how to cultivate the essence of mental health—to know what is essential to create a healthy mind.

One possible strategy would be to simply create a new word instead of *mind*, and then use that new term to clarify from where and how our interpersonal connections and embodied lives, subjective experience, inner essence, sense of purpose and meaning, and consciousness each arise. What would you call these essential features of our lives if you were not going to use the term *mind*?

Finding a different term that symbolizes a process that is distinct from "mind is equivalent to brain activity" is one approach. And maybe that's a fair solution. But this exploration is more than just a semantic discussion about terms, definitions, and interpretations. If the mind is a term for the centrality of our essence, for the heart of who we are, let's see if we can preserve those meanings of the term "mind" and see what this mind, this heart of being human, is truly all about. How about this suggestion: We use the term, "brain activity" for referencing neural firings. In this way, we are stating what it is, neuronal activations taking place within the skull, within the brain inside the head. Then we can freely explore the reality of mind in its fullness without evoking the common arguments I've heard, among them that this attempt at exploring a wider view "reverses science," as some have said to me, since it says mind is more than brain activity. Even if mind fully depends on brain activity, it does not make mind the same as brain activity.

For now, for this beginning of our journey, let's stick with *mind* as our term and see how it goes. We can come back to new linguistic representations later if we choose. In our everyday language, between you and me along this path we are about to embark upon, let's simply agree, for the moment, that *mind* will have the broad meaning of something that at times has an awareness with a subjective quality, and that is filled with information flow, with and without awareness.

For now, we don't need another term, but let's keep an open mind about it. And let's explore how we can clarify the nature of mind so we can know it deeply and support its function and development toward health fully.

An Invitation

After an extensive review of a range of published academic, clinical, and popular texts, it has become clear that this combined *inner* and *inter* nature of mind is something rarely discussed in scientific, professional, or public circles. Sometimes inner is the focus, sometimes inter, but rarely both. But couldn't mind be both inner and inter? If we can define the essence of mind clearly we could more robustly help one another individually, in families, schools, and our larger human communities and societies. For these reasons, the time seems ripe to offer something that may help move the conversation forward about a broader view of mind.

Though I've written extensively about the mind academically (in *The Developing Mind, The Mindful Brain,* and *Pocket Guide to Interpersonal Neurobiology*), discussed its applications in clinical practice (*Mindsight* and *The Mindful Therapist*), and explored everyday applications in various books for the general public including for adolescents and parents (*Brainstorm, Parenting from the Inside Out* [with Mary Hartzell], *The Whole-Brain Child* and *No-Drama Discipline* [both with Tina Payne Bryson]), a book that focuses deeply on this specific proposal of what the mind may in fact be seems needed, one that does so in a more direct and integrated manner.

By integrated what I mean is this: As the mind, at the very least, includes our inner subjective experience of being alive, our felt, embodied sense within conscious awareness, then a book focusing on the question of what the mind actually is may perhaps best be structured by inviting the reader and writer, you and me, to be present fully, feeling and reflecting on our own subjective mental experiences, as we move along in discussing the fundamental concepts. We need to become aware of our inner experiences beyond merely discussing facts, concepts, and ideas, devoid of inner felt awareness and subjective textures. This is a way to invite your conscious mind to explore your personal experience as we move along. Ideas are able to have their greatest impact when they are combined with a fully felt experience. This is a choice I can offer to you as the author in the form of an invitation, one you can participate in, if you choose, as a reader. In this way, this book can be a conversation between you and me. I'll offer ideas, science, and experiences, and you can empower your own mind to receive and respond to these communications. As the pages and chapters of this journey unfold, your own mind will become a fundamental part of the exploration of what the mind is.

If the mind is truly relational, then this book needs to be as relational as possible as well as encouraging of your reflections on your inner felt experience. You may be reading the words these fingers of my body have typed, but the intention is for this to be a collaborative journey of discovery, one that invites your mind and my mind to be as present as possible.

In other words, the process of reading *Mind* ought to reflect the content of the book itself, the journey to explore what the mind may be.

If we leave out either the embodied or relational side of our mental lives, the inner and inter, we may miss the heart of what the mind truly is during our explorations. How can we do this? Here's an idea. If I, as the writer, can be present both personally and intellectually, perhaps you, as the reader, can too. This is how we can blend the scientific and personal as they become deeply interwoven in seeing the mind clearly.

Being scientific about the mind requires that we not only respect empirical findings, but also honor the subjective and interpersonal. Not a typical approach, perhaps, but it seems necessary to truly explore what the mind is.

That's my hope for this book, that this be a journey, for you and me, to openly explore the nature of our human mind.

The Approach of Our Journey

We live our lives in each moment. Whether we are feeling our bodily sensations now, reflecting on the present with a filter of our experiences in the past, or becoming lost in memory, these all happen now. We anticipate and plan for the future in this moment as well. In many ways— especially if time is not actually some unitary thing that flows—all we have is this moment, all we have is now. The mind emerges within memory as well as the moment-to-moment experiences that unfold in the present as sensory immersions, along with the mental images we have of future experiences—how we anticipate and imagine what is to come next. This is how we link past, present, and future, all in the present moment. Yet even if time is not really what we imagine it to be, as some physicists propose in ways we'll explore in-depth in the journey ahead, change is in fact real—and this linkage across time that is our mind's own construct is a way of interconnecting experience across change. The mind is filled with an ever-flowing experience of change. The reading of *Mind* will therefore involve these mental experiences of change, what in time terminology we refer to as the future, present, and past. The memory researcher, Endel Tulving (2005), calls this "mental time travel" as we link past, present, and future. If time is a mental construct, mental time travel is what our minds do—it's how we organize our mental experience of life, our representations of change.

To respect this central change source of mind, the way our minds construct our selves across time, I've elected to structure this book in frames of mental time travel. We'll be exploring the ideas of mind in ways that will involve this past-present-future orientation of mind. To achieve this, I'll be using a chronological structure, one that unfolds as a narrative, reflecting on past and present times as we open our minds to the future.

The entries of the book have both conceptual discussions and non-fiction narratives that help communicate the material and hopefully make it more memorable in the reading. Stories are how our minds recall information best, and the ways we feel as we immerse ourselves in those stories is what appears to impact how an experience stays with us. I'll also invite you to consider aspects of your own experiences related to particular discussions of mind as we move along. In this way you'll be reading some of my stories, and perhaps reflecting on and even writing down some of your own.

To embrace this integration of past-present-future, and of the personal and conceptual, I've divided these narratives into five-year periods or demi-decade epochs, called simply "epoch entries," that help temporally structure and conceptually organize our dive into the exploration of mind. Please bear in mind that these entries are not always in chronological order. We'll be exploring autobiographical reflections, the subjective experience of mind in day-to-day experiences and moment-by-moment reflective immersions, along with relevant conceptual views inspired by science. These are the empirical findings of studies from a range of disciplines that will be compared and contrasted, their confluent insights synthesized and extended through scientific reasoning. These will be further woven with practical applications and mental reflections.

As these entries unfold, I invite you to explore your own reflections in the here-and-now of your subjective reality. You may find that your autobiographical reflections of how your mind has developed across periods of time in your life begin to emerge and become a focus of your attention. Even your sense of future possibilities may become opened in new ways. This is an invitation for you to open your own mind to its innate mental time travel orientation. Reflecting on these past-present-future experiences as they arise in you, and perhaps even writing down your reflections if you are so inclined, may deepen the experience. We live as sensory beings as well as autobiographical ones as our minds emerge in each moment, within reflections on the past and imaginings of the future. Sensory input, reflections on memory, and imagination are fundamental parts of mental time travel that can be fun to explore—hence, they are *fun-da-mental.*

Between the initial epoch entry and the final reflective invitation section, you'll find a central section of each chapter that focuses primarily on scientific concepts that extend and deepen the discussion. In these middle sections, we will pause from the more autobiographical narrative reflections and focus specifically on some core concept or question related to the narrative notions of mind just presented, only this time the discourse will be primarily exploring an intellectual, conceptual framework. You may feel within you, as you read these more predominantly scientific sections, that this way of communicating from me to you evokes a different mental experience, perhaps one that is a bit more abstract, has a more distant feeling to it, and may even feel less engaging. If this, or anything else, is what arises in you, this is what arises in you. I apologize now for the shift, but let the shift itself offer an experience that can possibly teach something. Each moment is important, and whatever arises can be an emergence that has something to offer. Let every experience be an opportunity inviting us to learn. Ansel Adams is often quoted as stating, "In wisdom gathered over time, I have found that every experience is a form of exploration."

If these more conceptual middle sections don't initially work for you, you can skip over them if you choose. This is your journey. But I urge you to try, at first at least, to simply let the experience of reading them be a source of learning about the mind, about your mind, and about the nature of how we connect with each other through facts or stories. So let's see how you feel, SIFTing your own mind as you go: checking in with your *s*ensations, *i*mages, *f*eelings, and *t*houghts. Let every experience be an invitation to reflect and an opportunity to deepen our learning about ourselves and the mind. This is your journey of exploration.

If instead you are looking for more of this conceptual discussion, seeking a purely theory-based, subjectively-distant discussion about the mind, you'll need to explore other more standard books rather than this one. The strategy and structure of this book, *Mind*, focuses on defining what the mind may be, embracing the reality of subjectivity and inviting you to explore the nature of your own experience as you go, attempting to illuminate the mind's nature in both scientific discussions and experiential reflections. This interdisciplinary approach to exploring the nature of our minds is, I believe, true to science, even if this is not a typical science book. This book can be of benefit to anyone curious about the mind and interested in creating a healthier mind. Exploring the mind deeply needs more than simply focusing on fascinating conceptual discussions and scientific findings; it means combining these with subjectively felt life.

Words Reflecting on Reflecting Words

Even in these words we use to connect, me with you, you with yourself in your own inner thoughts, you in sharing with another person in reflective conversations, or as written words within self-reflections in a journal, we have actually begun to shape, and also limit, our comprehension of mind. Once a word is "out there" for us to share, and even when it is "in here," inside ourselves, shaping our thoughts and ideas and notions, it limits our understanding. This may be why some scholars, as I mentioned earlier, have urged me to not define the mind as it will limit our understanding. For this reason alone, they probably would not be happy with this book. Yet without words, without language inside us or between us, it's challenging if not impossible to share ideas, let alone explore them, either conceptually in our communication or empirically in science. As a clinician, educator, and parent, trying to find a truth-based definition using words is worth the effort and the potentially useful outcome it might create, as long as we crucially acknowledge these limitations of words.

But let's take a moment to respect and inspect words and their limitations—these ways you and I will be connecting initially in this book as we move forward. No matter what we do once we speak or write, even if the words we carefully choose are accurate, they are innately limiting and limited. This is a big challenge for any project that is word based, and perhaps for living a talking life itself, not just when we focus on the mind. If I were a musician or painter, perhaps I'd perform a piece without words or craft a canvas with only color and contrast. If I were a dancer or choreographer, perhaps I'd create a movement that more directly revealed the nature of mind. But I am a word-person and this is a word-format, so for now that's all I've got to connect with you. I am so driven to explore this notion of this mind that connects us to each other that words are what we'll use, as limited and limiting as they may be. Let's be patient with each other, and with ourselves, as we share these words with one another. We need to remember that words both create and constrain. Keeping this in mind will help us deepen our understanding of the process of our exploration, and the conceptual notions that arise. Let's make some music, paint a picture, and share in a dance of the mind as best we can with these words that connect us to each other.

If we bear in mind the meaning of linguistic symbols as a form of information we'll share, then the nature of words themselves can be used to reveal aspects of the nature of the mind.

For example, if I were to say how we were "grasping" the notion of the mind, we'd see, too, how embodied our views are, the words based on the embodied language we choose: We reach out with our hands to grasp something; we reach out with our minds to comprehend something. We comprehend, "with-grasp." We even understand one another, as we "stand-under." That's the embodied linguistic nature of mind. Words are information, as they are symbols for something other than the energy pattern of which they are composed. But even as representations, as symbols of sound or light, the terms such as *grasping* and *understanding* don't fully capture the essence of deep comprehension, of sitting with truth, of seeing clearly, and perhaps nothing short of the inner sensation of clarity will.

And using the term *share*, even with words within yourself, there is a *betweenness*, a relational side to mind, reflected even in our *languaging*, our putting into words, the inner nature of mind itself. As blind and deaf Helen Keller noted in her autobiography, she felt as if her mind was born at the moment she shared a word for water with her teacher, Anne Sullivan (Keller, 1903). Why does sharing give birth to the mind? And is this why we talk to ourselves with the internal privacy of our own inner voice? These words we share in mind become the words we bear in mind when we learn about ourselves and reflect on our lives. We indeed have a relationship with ourselves as well as with others. We need to remember, throughout this journey, that the language we use, and the language that surrounds us, interconnects, illuminates and imprisons all at once, and we need to be and remain aware, as best we can, of this linkage, liberation, and limitation words create in our lives.

Once the word train starts to leave its station of non-worded reality, though, we can stray from our original effort to reveal truth and unveil deep meaning, and wander away from the way things actually are. This is just one part, important as it is, of the journey into mind to keep in mind. For this setting and journey, sharing language that helps us grasp and share the nature of mind is how we'll best traverse the path that lies ahead. While we'll refer to science and concepts, we'll also be communicating directly about the experience happening right now inside of us. Words will begin to get at some of that experience, but they will likely not to be quite enough, not exactly what we mean.

Let's acknowledge that we can always say something like, "Well, it's more complicated than that" or, "It isn't exactly that way." These statements are certainly true, no matter what we actually put into words, so yes, it's not exactly like this. And yes, it is more complicated than that. Absolutely. Sometimes, the best way to be accurate is not to speak. Just stay silent. And that is certainly really important to do, regularly. Let's perhaps see, though, that beyond these inherent linguistic limitations, we may in fact find words, and the ideas and experiences they attempt to describe, that get close to something we can simply call *truth*. Something that is real. Something that has predictive value, something that helps us live our lives more fully, more truthfully. Silence is a good place to start. And words can be a powerful way to continue that journey into illuminating the nature of mental reality. Perhaps words can even help us connect in deeper ways to not only others whom receive these worded sentences, but even to ourselves, as we are invited to attend to what our minds are experiencing, even without words, with the truth illuminated in silence.

For every train of words you and I share, you'll also have your own non-worded mental life that arises. We sometimes tune in to that non-worded world best with silence, as we take a "time-in" to attend to the sea inside. You'll have sensations, images, feelings, and both worded and non-worded thoughts, so I invite you to let yourself silently SIFT your mind as these words evoke different elements of your own mental life.

I've also included a few photographs to try to access some non-worded ways in which visual images may evoke sensations closer to what I have in mind, even though what happens deep in your mind and deep in my mind may not be the same as we view the same photo. In fact, the concern I have about using these photographic images is they may evoke in you something that might be quite different from what they evoked in me when I chose the picture. But alas, we can never know. So enjoy the images, and if they make you wonder what was going on in my mind when I chose this one or that one, wonderful. I may not even know why, it may have been simply a bodily sensation inside of me that had a "yes" reaction when I saw the image and thought of the entry. Or maybe it was the cascade of images it evoked in me that felt right. Or perhaps the emotions I felt with the image matched how I felt writing that entry. And maybe even my thoughts evoked with the photo were just the ones I hoped would come up in you. You can SIFT my imagined mind in your mind, and you can SIFT your own mind and see what those pictures inspire to arise. You will have your own experience, and being open to whatever arises is a stance we can take along this journey. There is no right or wrong, just your experience. I'm simply inviting you to be aware of the fullness of your mind beyond merely the literal linguistic statements made with the words in this book.

We can only do our best to connect in our communication, remaining open to the journey and not worried too much about the endpoints. It is this traveling across moments as they unfold, like mind itself, which continually emerges. This, too, is why we'll be exploring the very nature of time, of what it really means to be present in life.

These questions are intended to not only evoke exploration, but also to ignite the illumination that arises from the questioning. As my old mentor, Robert Stoller, MD once wrote, "Still, yearning for clarity contains a pleasure of which I am only now fully aware. Sometimes, on paring a sentence down to its barest minimum, I find it transforms into a question, paradox, or joke (all three being different states of the same thing, like ice, water and steam). That is a relief: clarity asks, it does not answer" (Stoller, 1985, p. x).

You'll see here a focus on fundamental questions—investigations to have fun with—related to various elements of mind that attempt to be woven into one tapestry. We'll navigate the path along the way by examining aspects of the who, what, where, when, how, and why of the mind. This will be our common ground, a six-part compass we'll use to navigate through our journey with two lenses. One will be the lens of personal, felt experience: mine in the descriptions, yours in the reflections on your experiences as they emerge. The other lens is one of scientific and conceptual reasoning, explorations of research findings and their implications.

One reason I've chosen this particular way of creating the journey of *Mind* is to invite you, as well as me, to blend the personal experience of your own mind with your own evolving understanding of the scientific ideas underpinning this exploration. My hope is for this to be "active reading" involving your own curiosity and imagination, as well as your own personal reflections on mental life, combined with the construction of a scientific foundation of mind. This is a book of questioning we can create together as we explore the fundamental nature of mind. The words are only a starting place, perhaps even an initial meeting place for us to connect. The journey ahead is beneath, before, and beyond the words themselves.

I'm not so good at telling jokes, as my kids have often reminded me, but I think we'll find plenty of paradoxes and a quenching quantity of questions that emerge during our expedition. Sometimes reflecting on the deep nature of mind is mind-boggling, and mind-blowing. Sometimes it's outright hysterical. There are many books that offer you proposed answers from serious science or personal reflection. This book offers you both personal reflections and scientific knowledge in an integrated format, filled with questioning that directs our journey ahead in a way I hope will be engaging and illuminating.

One challenge of discussing the mind is that we need to consider the mind as both a personal experience and a scientifically understandable process, entity, object, or thing. This tension between the personally knowable, non-externally observable, and unquantifiable and the objectively knowable, externally observable, and quantifiable is an inherent conflict that has led our major academic pursuits over the last century to broadly turn away from insight and reflection on subjective experience in formal studies of mind. Yet whoever we are, whatever we are, and whenever we are, where the mind is, how it functions, and why we are here, are each aspects of our mental life that, I believe, can be best grasped when we honor both the subjective and objective nature of mind at the heart of each of these facets of our lives.

My deepest hope is to join with you to elucidate the nature of our minds, illuminate our beliefs and uncover our disbeliefs, demonstrate the mind's central importance in our lives, and offer some basic ways to define the mind so we can then explore what a healthy mind might actually be. The natural next step, once we've explored these issues, is to suggest the various ways we might choose to empower ourselves to cultivate a healthy mind, personally and in others.

And so to discover, explore, reclaim, and cultivate our minds, I invite you to join me on this journey as we dive deeply into the heart of being human.

Ready to dive in? Let's begin—and I hope you enjoy our journey ahead.

CHAPTER 2

What Is the Mind?

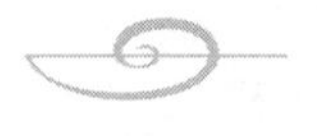

In this entry, we'll dive into a proposed working definition of one aspect of mind as being a function of a system comprised of energy and information flow. This system is both within the body and between ourselves and other entities—other people and the larger environment in which we live. This is a useful place to start our journey into the nature of the *what* of mind.

Working on a Working Definition of Mind (1990-1995)

The 1990s were called "The Decade of the Brain."

I felt like a kid in a candy store, loving to weave what I was experiencing with my patients as a practicing psychiatrist with the explorations of memory and narrative emerging with research subjects, continually striving to link these with what we were now learning in brain science. I had completed my clinical training with my internship year in pediatrics followed by a residency in first adult, then child and adolescent psychiatry. After a National Institute of Mental Health research fellowship at the University of California—Los Angeles, studying how parent-child relationships shape the growth of the mind, I was asked to direct the clinical training program for child and adolescent psychiatry at the university. I took that educational role very seriously, thinking about how a comprehensive view of the developing mind, the new understandings of the brain, and the science of relationships I had been learning might all come together to form some kind of core curriculum for the new generation of clinicians there. At the same time, I started a study group with my former teachers and colleagues on campus to address the pressing question: What is the relationship between the mind and brain?

Photo by Lars Ohlckers

Forty people came to our group, mostly researchers from academia and a few clinicians. Many fields were represented, including those of physics, philosophy, computer science, biology, psychology, sociology, linguistics, and anthropology. The one question that brought us together initially was this: What is the connection between mind and brain? The group could define the brain—a collection of interconnected neurons and other cells in the head that interact with the whole body and the environment. But there were no definitions of the mind short of the familiar "brain activity" neuroscientists in the room would state, which was not an acceptable view for the anthropologists or linguists in the room, who focused on the social nature of mental processes, like culture and language.

My own professor of narrative I mentioned earlier, Jerome Bruner, had said during my graduate course with him as a research fellow that narrative doesn't happen within a person, it happens between people. Even in my course paper where I wondered how narratives were mediated in the brain of traumatized individuals, he urged me not to make such "an error" and to realize the social nature of narrative. These stories we tell—the narratives of our lives revealing our memories and life's meanings—are core mental processes. I was now studying how the findings of attachment research revealed that the narrative of a child's parent was the best predictor of that child's attachment to that parent. We knew from careful empirical studies that what seems like a solo act of your own life story is somehow related to the interpersonal interactions between parent and child that are facilitating the child's growth and development, a process we call "secure attachment."

I had learned that narrative was a social process, something between people. These stories are what connected us in one-on-one relationships, families, and communities. I wondered what other elements of mind beyond narratives—our feelings, thoughts, intentions, hopes, dreams, and memories—were also deeply relational.

At the time, I was meeting people with whom I would have ongoing conversations and connections that would shape who I was becoming. The psychologists Louis Cozolino, Bonnie Goldstein, Allan Schore, and Marion Solomon became close colleagues and friends, and little did I know that our lives would remain intertwined even up to the present day in deeply stimulating and rewarding ways, now a quarter of a century later. My relationships with them, and many other individuals along this journey, became a part of the narrative of who I was. Little did I know that this decade would also bring the end of life for three of my main teachers who had shaped my professional development: Robert Stoller, Tom Whitfield, and Dennis Cantwell. With teachers and colleagues, friends and family, we find connections that deeply transform us. Relationships are the crucible in which our lives unfold as they shape our life story, molding our identity and giving birth to the experience of who we are, and liberating—or constraining—who we can become.

Even though I had been taught in medical school a decade earlier that a person's body was the source of disease and target of our interventions, somehow the human mind seemed to be broader than the body. These deep lessons of the primacy and social nature of narratives affirmed that some profoundly important source of meaning in our lives—the stories that bind us to each other, help us make sense of experience, and enable us to learn from each other—was located deeply in a *between* domain of our relational lives.

Certainly these elements of mind would likely also be related to brain function—this relationship was something we'd known in neurology for over a century, but thanks to recent advances in brain scanning technologies it was now more illuminated and refined. Still, being dependent on the brain does not mean being limited to the brain alone, nor does it mean that the mind is the same as brain activity, as we've seen.

So I responded back to Professor Bruner during my final presentation for the course that I was interested to know how the neural processes in both brains of people within a relationship contributed to the social nature of narrative. He just waved his hands at me, with a look of frustration and perhaps confusion. I understood then that the bridging of disciplines—neural and social—was not so easy to do.

Later I'd learn that the term *consilience* could be used to identify a process where we discover the universal findings across often-independent disciplines (Wilson, 1998). I seemed to be, without my knowledge of that term, on a quest to find consilience in understanding the mind.

But even if the disciplines and their human proponents couldn't find overlap, perhaps reality itself was filled with such consilience. Perhaps neural and social were part of one fundamental process—not merely social stimuli influencing the brain like light stimuli influencing the optic nerve, but one fundamental flow of something. But what could that something really be, something that would facilitate, for example, a collaborative, connecting conversation between a neuroscientist and an anthropologist?

In our newly formed fellowship of 40, there was no consensus. Without a definition of what *mind* actually was, short of saying it was only "brain activity," it was hard to come to some shared understanding of the link between brain and mind, let alone some way to communicate effectively and respectfully with one another.

The group seemed on the verge of dissolution.

With all the focus on disease models of psychiatric disorders in those days of the *Diagnostic and Statistical Manual of Mental Disorders*, the *DSM*, along with the increasing prominence of pharmaceutical interventions, and the scientific statements that mind was just an output of the brain, discussing the issue in our study group became quite intense: Was mind just brain activity, or was it something more?

The group was at a standstill given this lack of a shared view of the mind. As the facilitator of the group, having relationships with each individual in the room I had personally invited, I felt an urgent need to do something that might enable these thoughtful people to better communicate and collaborate. If the group was to continue meeting, something had to be done.

As a college student 15 years earlier, I worked in a biochemistry lab searching for the enzyme that could enable salmon to transition from fresh to salt water. At night I worked on a suicide prevention crisis assistance phone line. I learned as a biology student that enzymes were necessary for survival; and as a mental health volunteer, I learned that the nature of emotional communication between two people during a crisis could mean the difference between life and death.

I wondered if enzymes and emotions shared some common ground, some common mechanism of salmons' survival and suicide, couldn't the brain and relationships have some common element as well? In other words, if the molecular processes of energy activation that enzymes enabled permitted fish to survive, and if emotional communication between two people could keep hope alive, could life itself depend on some fundamental transformations that were shared by enzymatic energy processes and the energy of emotional connections? Couldn't the brain and relationships share some consilient grounding of their essence? Couldn't they be two aspects of one system? And could this essence that linked brain and relationships reveal the nature of the mind? Could there be something in this essence that each group member might embrace to keep the group from imploding from tension and lack of mutual understanding and respect?

I took myself for a long, long walk on the beach that week after the first meeting of the group, focusing on the waves of the shore where I'd grown up, wandering and wondering up and down the coastline of the Santa Monica Bay. Reflecting on that place where sea meets land, and on the life I'd lived there on that sandy shore, filled me with some sense of continuity, something that linked then and now, water and land. It seemed to me that a shared element between brain and relationships was waves, waves of energy. Waves are ever changing, each moment unfolding in new and emergent ways, creating patterns that are dynamic—meaning they arise and fall, changing and unfolding, influencing each other.

Energy waves arise as patterns, as changes of energy flow emerging moment-by-moment. Energy comes in various forms, like light or sound, as a range of frequencies, and a distribution of amplitudes. Even time can be related to the emergence of energy patterns, as modern physicists are now exploring in their emerging views of the nature of energy and reality. In these new views, the fixed energy waves of the past influence the emergence of waves in the present and shape the unfolding of the potential waves to come. Fixed, emergent, and open, time may involve the changing of energy along a spectrum between possibility and actuality.

Energy, physicists say, is best described as a potential to do something. This potential is measured as the movement between possibility and actuality along a spectrum of probabilities, what is sometimes called a wave function or probability distribution curve. We experience this flow of energy not as some magical mysterious non-scientific thing, but as fundamental to the world in which we all live. We may not see the energy fields that surround us, as famed scientist Michael Faraday described two centuries ago in his discovery of electrolysis and electromagnetism, but they are real. We also may not often sense the origins of energy as a sea of potential, but we experience in our awareness the emergence of possible into the actual. That is the flow of energy, the change of this probability function. The light is off, now the light is on. The room is silent, now you speak. You see someone coming toward you, a dear friend, and you receive a warm welcoming hug. That is the transformation of possibility into actuality. It is the flow of energy we experience each moment of our lives.

Some of this emergent flow of energy has symbolic value with meaning beyond the pattern of energy itself. I knew from the field of cognitive science that such symbolic meaning could be called "information." I write or speak gibberish, and there may be no meaning. But I write or say, "Golden Gate Bridge," and voilà, energy has information—it stands for something other than the pure form of energy that manifested from a sea of possibilities into this one actuality. Now I say "Eiffel Tower," and arising from that vast sea of nearly infinite potential arises this one energy pattern, information manifesting as the linguistic symbol of that architectural structure in Paris.

Yet not all energy patterns have information. So the common element shared by the brain and relationships might be energy itself; or, to be complete, that common element might simply be called "energy and information." When questioned, many scientists state that all information is carried along energy waves, or energy patterns. Other scientists view the universe as fundamentally comprised of information, and energy patterns arise from that basis of reality, a universe constructed of information. So in each view, information expresses itself in the world by way of energy transformations, the unfolding of the potential to do something into an actual something. That's energy in a nutshell. With either of these perspectives, the two terms, energy and information, might be a useful foundation to consider, especially when paired as a concept into a single unit.

These patterns or waves emerge as energy changes across time, as it flows, each moment unfolding in the present. For our experience of mental life, continually emerging and changing, the notion of flow seems to fit well. Even if the proposal from some physicists that time is not a unitary process that is as we may imagine it to be turns out to be true, that time is not its own distinct entity in the world that flows, but instead is a mental construction of our awareness of change, all scientists concur that reality is filled with change, if not across time then across space or the probability curve. Change across the probability curve means the movement of energy along the range between open potential to realization as an actuality. Therefore we can use the term *flow* to signify the change across time or space or probability, or perhaps even some other aspect of reality. Flow means change. We can use the phrase "across time," as in "flow across time," to simply signify ways of tracking this flow, the various dimensions of change in our lived reality. And so the fundamental phrase for this proposed central element of mind might be called "energy and information flow."

It seemed to me back then, as it does now, that one could propose that *energy and information flow* is the central element of a system that is the origin of the mind.

But what is this system from which the mind emerges? What is it, what are its boundaries, and what are its characteristics? The basic element of this system might be energy and information flow—but where does that occur?

Walking along the beach, watching the waves, it seemed that the shore was created by both sand and sea. The emergent coastline arose from the sand and the sea, not from either alone. The coast was both shore and sea.

Could the mind, somehow, be *both* within and between?

Energy and information flow within the whole body, not just the brain. Energy and information also flow between a person and other people in patterns of communication, and in connections with the larger environment in which that person lives—like these words from me carried through this book to you. We can say that energy and information flow occurs between our body and the non-body components of the world—the world of "others" and our environment—as well as within us—within our body, including its brain. I put quotations around the word *others* to remind us that this is just a word—the notion of self versus other needs to be kept in the front of our minds as we move along this exploration.

But if energy and information flow, within and between, turns out to be the system that gives rise to the mind, what could the mind actually be? Feelings, thoughts, and memories, you may say. Yes, those are great, accurate descriptions of the contents or activities of mind. These are the ways we describe the subjective reality of mental life. Many fields offer such important descriptions of mental processes. But what are those actually? Amazingly, no one really knows. On the level of neuroscience, as we've mentioned, no one understands how neural firing might create the subjectively felt experience of a thought, memory, or emotion. We just don't know.

Years later, the philosopher and physicist Michel Bitbol and I went for a long stroll during a week-long gathering of a group of about 150 physicists and agreed that subjectivity may just be a "prime" of the mind—one not reducible to anything else. I could see then that perhaps subjective experience as a prime might arise from energy and information flow. How that occurs, we just don't know. But as a prime, it simply cannot be reduced to something else, or possibly even reduced to just one location, such as the brain's firing. But at least identifying a possible link between subjective experience and energy and information flow gives us a place to begin to deepen our understanding of mind. Seeing energy and information flow as a fundamental part of a system that gives rise to the mind, including its subjective textures of life, seems to be a fair starting place for deepening our understanding.

While we also don't understand how being aware of subjective experience might arise from neural firing, could this experience of consciousness also be a prime of energy and information flow? In other words, to have subjective experience we need to be aware, so perhaps both awareness and the subjective experiences awareness facilitates are primes of the flow of energy and information. This doesn't explain in any way, really, how these important aspects of mind actually arise, but at least it may point us in the right direction for our journey.

We can also move beyond this prime of subjectivity and perhaps consciousness itself and ask about the information processing of our thinking, remembering, or evaluative emotional lives. What constitute these mental activities?

If I ask you to say what a thought is, for example, you may find it hard to articulate exactly what this common mental activity consists of. The same might happen if you consider a feeling and try to say what that is all about. What an emotion truly is, no one really knows. There are many descriptions of what goes into a thought or feeling published in an abundance of books and articles, but even when you take these sophisticated scientific, philosophical, and contemplative views into account, or discuss this with their authors directly, the core essence of thoughts and feelings remains, in my mind, quite elusive.

We could say something at least a bit more specific about the mind as being subjectively experienced energy flow patterns that sometimes contain information. That's a great start, as we can begin to trace energy and information flow as the origin of mind, and its location as both within the brain and other locations.

We have a brain in the body, an embodied brain. We also have relationships with other people and the planet, our relational reality. Energy and information flow within us (through the mechanisms of the body including its brain) and between us (in our communication within relationships).

Great. So we are clarifying the basic element (energy and information flow) and the location (within and between) of a possible system of mind. We are beginning to illuminate more fully aspects of the *what,* and *where,* of mind.

This is not the way people often write about or speak about our lives, I know. The notion that something is both within us and between us, two places at once, may seem strange, counterintuitive, and even flat out wrong. When I prepared to present this view to the group of 40 in the autumn of 1992, I felt nervous that such a view would seem odd and unfounded. But let's explore some of the implications of these ideas and see where they take us.

If this embodied and relational system of energy and information flow is the source of mind, what exactly might the mind be within that system? Yes, we are suggesting the system is made of energy and information, and these change over time, space, probability distribution, or in some other fundamental way. That change is called flow. And we're suggesting that this flow is both within and between.

So we've moved closer to shedding light on the possible basic *what* and *where* of mind.

But what might the mind actually be within that system? Perhaps our mental activities simply are primes of energy and information flow as they unfold within and between us. In this way, the system is the source of mind itself. But beyond activities of the mind such as feelings, thoughts, and behaviors, beyond information processing, and beyond consciousness and its prime of subjective felt textures, could the mind also include something more? Could a definition of the mind as having something to do with energy and information flow beyond these common descriptions be formulated?

To address these fundamental questions, we need to examine the nature of this system we are proposing that may give rise to mind.

The system of energy and information flow within and between us has three characteristics: 1) It is open to influences from outside of itself; 2) It is capable of being chaotic, meaning, roughly, it can become random in its unfolding; and 3) It is non-linear, meaning that small inputs lead to large, not easily predictable results. These three criteria for mathematicians, especially the third for some, enable them to define a system as complex: open, chaos-capable, and non-linear.

Some people hear the term *complex* and become nervous. They understandably want more simplicity in their lives. But being complicated is not the same as complexity. Complexity is elegantly simple in many ways.

If you think of your own life, your inner experience and relational worlds, do you notice that these three characteristics are present? On that walk on the beach, I reflected on my own life, on the experience of mind, and imagined how it had been open, chaos-capable, and non-linear. If you feel that way too, then you may understand the emerging reasoning, and may even feel the excitement, to be able to say that the mind is some aspect of a complex system.

Big deal. Why would anyone care?

Well, the importance of this view rests on the implications that arise with the following facts and inductive reasoning. A system is comprised of interacting fundamental elements. One feature of complex systems is that they have emergent properties—aspects of the system that arise simply from the interaction of elements of the system. In the case of the system of mind, the elements we are proposing as the core features, the essence of this system, are energy and information. The ways these elements interact are revealed in the flow of energy and information. That's the *what* of the mind, in part. And the *where*? Within us—within the body as a whole, not just in the head—and between us—in our relationships with other people, and with our environment, with the world.

Okay, that's the *where*, and the partial *what*—something that emerges naturally from energy and information flow within and between.

Fine. Next we see that one emergent property of complex systems has an intriguing name: *self-organization*. Straight out of math, the process of self-organization is the way a complex system regulates its own becoming. In other words, arising from the system (the emergent aspect) is some process that, in a recursive, self-reinforcing way, organizes its own unfolding (self-organization).

If that feels counterintuitive, you are not alone. What this means is that something arises that turns back and regulates that from which it arose. That is the emergent, self-organizing aspect of a complex system.

I wondered, what if the mind were the self-organizing property of energy and information flow as it unfolds within and between us? Others had described the brain as a self-organizing system, but what if the mind were not merely limited to the brain? Some thinkers had described the mind as embodied (Varela, Thompson, & Rosch, 1991). But what if the mind were not only fully embodied, but also fully relational? What if the full system was not limited by skull, nor even by skin? Couldn't that system have completeness to it, and be an open, chaos-capable, non-linear system of energy and information flow that is both within and between? And if so, wouldn't there be a mathematically supported notion of an emergent self-organizing process that arose from both within and between? Instead of being in two places at once, the system of energy and information flow is one system, one place, which is not bounded by our brains or bodies.

Skull and skin are not limiting boundaries of energy and information flow.

Embracing both the withinness and betweenness of this one process of mind was, and remains, somehow, not how academicians or clinicians would speak about mental life. One process distributed inside us and between us? On the surface, it just doesn't make sense. But this was the fundamental idea stirring inside me in the midst of the sea of notions of mind as brain activity alone in this Decade of the Brain.

Beyond the important subjective quality of mind, beyond even our awareness of this subjectivity, and perhaps even distinct from information processing, the idea was this: Could one aspect of the mind be seen as a self-organizing emergent property of this complex system of embodied and relational energy and information flow?

I returned from that walk on the beach, and read about this further to prepare for the next week's meeting, and it blew my mind. I could find nothing in the literature to support linking the embodied and relational, but it seemed to be a logical inference from the math of complex systems and from reflecting on the mind as part of an open, chaos-capable, non-linear system of our lives. If one saw the fundamental element as energy and information flow, then perhaps a bridge linking the life's work of both neuroscientists and anthropologists, and everyone else in the room, might be collaboratively established. That next week, I proposed to the 40 academics assembled that we might consider a working definition of one aspect of the mind as this: *an embodied and relational, self-organizing emergent process that regulates the flow of energy and information both within and between.*

In condensed terms, this self-organizing aspect of the mind can be briefly defined as *an embodied and relational process that regulates the flow of energy and information.*

Where is this occurring? Within you and between you. *What* is it? At least one aspect of mind—not the totality of mind but one important feature—can be seen as a self-organizing process that emerges from, and regulates, energy and information flow within and between us.

This proposal of one aspect of the mind as an embodied and

relational self-organizing emergent process of energy and information flow does not explain the prime of subjective experience, but it might end up being related to it, in ways we don't yet understand. Or it might be that the subjective experience of lived life, while perhaps an emergent property of energy flow, is something distinct from self-organization. We'll track that question as we go along our journey.

This view does not explain consciousness, our ability to be aware and have a sense of knowing. Within this consciousness, too, we have an awareness of the known, and even a sense of the knower. But these aspects of consciousness, like the subjective experience we feel in awareness, may also arise from energy flow but ultimately be distinct from the self-organizing aspect of mind.

Information processing, too, may or may not be a part of self-organization though the notion of regulating energy and information flow seems, of each of these facets of mind, the most likely to be linked to self-organization. We'll keep an open mind about the interrelationships of these four named facets of mind: subjectivity, consciousness, information processing, and self-organization. Each may be embodied and relational, but the exact interrelationships of these facets we'll keep as an active focus of our questioning along the journey.

It is important to note, too, that while subjective reality, consciousness, and even information processing may in the end be located within our body, perhaps even dominant in the brain, this self-organizing aspect of mind may be distributed in both body and relationships. However, the more we see cloud computing and the ways interconnected computers can collaboratively contribute to information processing, processing that is driven, however, at least in part by the intentions of human beings, it is likely that within and between are a fundamental part of the information processing facet of mind. We'll explore these issues, including consciousness and its felt texture of subjectivity, much more as we journey onward.

It is this differentiation among these facets of mind that can help us more freely and fully dive in to our exploratory attempt to define what the mind is. These careful distinctions, too, may help reduce some of the tension among mind researchers who may be studying different facets of mental experience without realizing these may be differentiated aspects of one reality, the reality of mind. Language and careful reflection may cultivate clarity that can promote collaborative connections.

To be abundantly clear: This working definition of mind as an embodied and relational self-organizing process makes no presumptions about explaining the origins of subjective reality, consciousness, or information processing. But what it does do is offer a clear working place from which we can dive more deeply into other important aspects of mind. It suggests that this self-organizing facet of mind naturally arises from and also regulates energy and information flow—within and between us. This view clarifies not only the *what*, but also the *where* of this aspect of mind.

Relationships are how we share energy and information. The terms *brain*, or *embodied brain* refer to the embodied mechanism of energy and information flow. This proposal suggests that at least one facet of mind is the embodied and relational self-organizing emergent process that arises from and regulates energy and information flow. In other words, energy and information flow is embodied (the embodied brain or simply, brain), shared (relationships), and regulated (mind).

Some academics, upon hearing this definition, have become distressed, and as one professor said to me personally, "Energy is not a scientific concept and should never be used to describe the mind." But if physics is science, energy is fair game for a scientific proposal. Another researcher said that this view "divides the mind from the brain" and "pushes us backward in science." But while we can appreciate these concerns, the proposal, in my view, actually does just the opposite. It brings the various fields of science together, rather than dividing them, as is the all-too-frequent result of contemporary approaches (see Mesquita, Barrett, & Smith, 2010). This proposal actually doesn't separate brain from mind; it suggests their deep inter-dependence. In fact, it brings into view an important often scientifically disregarded but fundamental element of human life and the human mind—our relationships with each other and the world in which we live.

Brain, relationships, and mind are three aspects of one reality: energy and information flow. This perspective can be viewed as a triangle of human experience.

This view does not divide reality into separate, independent pieces; it recognizes its interconnected nature.

Relationships, embodied brain, and mind are three aspects of one reality, like the two sides and edge of one coin. The mind is a part of a complex system with the fundamental element of energy and

MIND

RELATIONSHIPS

EMBODIED BRAIN

The triangle of human experience: energy and information flow.

information flow. This one system's reality is energy and information flow—shared, embodied, and regulated.

This definition also sheds light on some of the fundamental notions of the *who, what, where, how,* and *why* of the mind. Who we are is shaped by energy and information flow. What we are is the sharing, embodiment, and regulation of that flow. Where we are is both within the body we are born into and the relationships that connect this body to other people and other places, other entities beyond the body itself. How this all unfolds we'll explore in the next entry in depth—but from this view we can see that mind is an emergent property of our withinness and betweenness. The why is a large philosophical question, but from a complex systems view, the *why* may be simply an outcome of complexity's emergence, the property of self-organization.

And what of the *when* of mind? Our sense of *when* unfolds as energy emerges, moment by moment—even as we reflect on the past or imagine the future. Emergence happens now, and now, and now. On one level of experience, flow is the unfolding of now from open, to emergent, to fixed, as we've seen as a way of reframing the notion of future, present, and past. If time does not exist as an entity that itself is flowing, as we've mentioned, the term *flow* in our definition can be viewed simply as signifying change. Yes, something can change over time, but change unfolds in space, and can even unfold over other aspects of energy and information like movement of the position along a probability distribution curve. Transformations in patterns, changes in probability distributions, and shifts in many aspects of energy such as density, amplitude, frequency and even form, are what energy flow entails.

So now is now. And change is mostly inevitable. This change can emerge across what we call time, if it exists, and it can happen across space or in the range of features of energy itself. Change can happen as well in the nature of the information being symbolized. For example, in cognitive science it is often said that information itself gives rise to further information processing. The term *mental representation* is itself more like a verb than a noun—the representation of say, a memory, its *re-presentation*, gives rise to more re-presentations, more remembering, and the emergence of more memories, reflection, thought, and feeling. We are an ever-emergent process of energy and information flow as events unfold now. Probabilities change as potentialities transform into actualities.

We'll dive more deeply into the mystery and magic of the *when* of the mind in the journey ahead. But for now, we can consider that the *when* of mind, arising moment-by-moment, is given a sense of immediacy in the emergence of mental life from energy and information flow in all these myriad potential manifestations—all changes continually happening in this moment. This emergence is unfolding as we speak and reflect, even when we reflect on the past of fixed moments in recollections of now, imagine the future of open moments, and experience life as an emergence of now, and now, and now. Shifts and transformations are forever emerging, now.

Back then, at the beginning of the Decade of the Brain, the emergence of a relational mind in that collection of independent minded individuals in our group was exhilarating. You could feel it in the room, the excitement, the unfolding of understanding.

I will never forget, in the ever-emerging nows of my mind even in reflection, of what happened that day when I came to the group.

The group was unanimous in accepting this working definition—all 40 professionals from this wide sampling of disciplines. We went on to meet regularly and discuss wild and wonderful emerging ideas of the mind and brain for four and a half years.

Photo by Madeleine Siegel

The System of Mind: Complex Systems, Emergence, and Causality

If we consider that our minds are a part of an interacting, interconnected system that involves our bodies and brains, as well as the environment in which we live, including our social relationships, we may be able to reconcile how the mind is part of one system that seems to be in two places at once. To understand this possible system of the mind, here we'll explore the science of systems.

Let's start with systems in general. A system is composed of basic elements. These elements change and transform, interacting with themselves, and, if they are open systems, with the world around them. For example, one such open system is a cloud. The basic elements of clouds are water and air molecules. These molecules interact with each other and change, altering their shape, moving in space. Clouds are called open systems because they are influenced by things outside themselves, such as evaporating water from streams, lakes, and oceans below, wind, and sunlight. The shapes of clouds, influenced by these external factors, and the internal air and water molecules, are forever changing as they emerge across the sky.

Systems come in various types and sizes—some are closed and large, like the universe, others are open and more limited in size, like those

clouds in the sky. In the body we have many systems, such as the cardiovascular, respiratory, immune, and digestive systems. There is also the body's nervous system, another example of an open system influenced by elements of the larger body and even outside the body—like these words you are reading right now. In fact, the cells of the nervous system are derived from the ectoderm of the fetus—the outer layer—and so our neurons share the fundamental ways our skin acts as an interface between the inner and the outer worlds. As a part of the body itself, the nervous system exists within the broader system of the whole body. As an open system, the body, too, interacts with the larger world. The terms *inner* and *outer* simply refer to the spatial designations of aspects of our one open system that continually unfolds in our day-to-day lives.

We can open our minds to the notion that the system of mind is not just the inner aspect of the nervous system inside our heads. This mind-system may be something more, something we'll explore, something rarely discussed, yet something we can illuminate, and perhaps even define.

The nervous system has one aspect that sits inside the skull—what we simply call *the brain*. The distribution of neural activity within the skull is interconnected through linkages among the widely separated areas within the brain. The individual cells, the neurons, and supportive glial cells, are themselves micro-systems encased within membranes. But even these cell systems are open, interconnected, and interdependent with the other cells of the body, near and far. Clusters of cells called nuclei link up to form centers, and centers can be part of larger regions. Some neurons serve to link distinct nuclei, centers, and regions to one another, forming circuits. And these various sized clusters of neurons can be interconnected within the divided two halves of the brain as they form hemispheres.

On and on, from micro to macro, the nervous system is composed of layers of interacting components that are themselves open sub-systems embedding their structures and functions into a larger, open system. Such interconnectivity, now termed the *connectome*, reveals how the brain in the head is itself a system, comprised of many interconnected parts and how they function with each other. This head-brain is connected to the remainder of the nervous system and body as a whole. We are even learning how the bacterial cells in our intestines, our biome, directly impact how the neurons in the head, our brain, function in our day-to-day lives.

But whatever varied things shape this neural firing, what is brain activity actually about? When we get down to the cellular level, what is going on when neurons fire—the basis of what some believe is the sole origin of mind? What is this cellular subset of the nervous system, a subset itself of the physiological system of the body, the *body system*, really doing? Neural activity, you say. Fine. But what does it really mean to have *neural firing*?

What we understand at this moment to be the essential nature of neural activity is that the basic cells, the neurons, are active and link to each other through the flow of energy in the form of electro-chemical energy transformations. Whether this is at the membrane level with something called an action potential, or some energy process within the microtubules deep within the neurons themselves, some shift in energy happens at the cellular and sub-cellular levels. An action potential is the movement of charged particles, called ions, in and out of the membrane of the neuron. When this flow, the equivalent of an electric charge, reaches the long axon's end, a chemical called a neurotransmitter is released into the synapse, the space between two connected neurons. This molecule acts like a key and is received by the lock of a receptor at the downstream neuron's membrane, at the dendrite or cell body, to activate or inhibit the initiation of an action potential of this receiving, postsynaptic neuron. There are likely many, many other yet-to-be-studied processes as well, both at the membrane level and in the constituents of the neurons and other cells themselves. But at this moment, our general sense is that brain activity is some form of the flow of something we can simply call electrochemical energy. We can measure this brain activity with magnets and electrical devices; and we can influence this activity with magnets and electrical stimulation. This energy flow is real and measureable.

At a minimum we can say brain activity is related to energy flow.

When those patterns of energy flow symbolize something, we call that *information*. In brain terms, scientists use the words *neural representation* to indicate a pattern of neural firing that stands for something other than itself. As we've seen, this is a "re-presentation" of something other than what was originally presented. For the mind, we use the term *mental representation*. The simplest way of defining what the brain's activity is made of is simply this: the flow of energy and information.

How this brain activity, this neural firing, becomes your subjective mental experience, no one knows. As we've mentioned, this is the big unknown for us humans, an unknown not often discussed. The presumption is that one day we may figure out how brain activity gives rise to mind, but for now this should be seen as only a guess. Still, evidence is strong that something about that brain firing is connected, somehow, to consciousness and our subjective experiences of emotions and thoughts, to non-conscious information processing beneath awareness, and even to our objective output of language and other externally visible behaviors.

Given that brain activity is really energy flow, as we've just described, let's see if we can start with this scientific finding and logically reason our way to the proposal for this wider view of mind. Let's assume that there is something about energy flow that gives rise to, causes, permits, or facilitates the emergence of mental life. That's a big assumption, yet a common belief, but let's try it out for a while and see how it goes. It's the stance of modern science. Neural firing leads to mind. Let's be clear even at the risk of being redundant: no one has proven how the physical action of energy flow and mental subjective experience of lived life are related to each other. No one. Many researchers believe they are in the process of doing so, and it may be fully true, but no one knows for sure how this happens. What modern science seems to do is limit that process of neural-activity-gives-rise-to-mind to what goes on in the head. Let's take a look at that assumption that the head has a monopoly on giving rise to mind.

As part of the larger nervous system, energy and information flows not just in the head, but also throughout the whole body. The parallel-distributed-processor (PDP) system of a spider web-like interconnected neural set of circuits that exists in the brain is connected to neural networks that are distributed throughout the body, within the complex autonomic nervous system and its sympathetic and parasympathetic branches, the intrinsic nervous system of the heart, and perhaps even the complex neural system of the intestines (Mayer, 2011). For example, studies are revealing how our intestines have neurotransmitters, like serotonin, that, along with the biome of organisms that inhabit this inner digestive tube of ours, directly influence our health and mental states—our thoughts, feelings, intentions, and even behaviors—like what we reach for to eat (Bauer et al., 2015; Bharwani et al., 2016; Dinan et al., 2015; Moloney et al., 2015; Perlmutter, 2015).

A question we can then naturally consider is this: If the system that gives rise to mind, as proposed by many modern scientists, is related (somehow) to this distributed energy flow of neural activity of the brain in the head, why couldn't a broader and more fundamental process of energy-flow-giving-rise-to-mind involve the whole nervous system? If mind is, in ways yet to be determined, a product, property, or aspect of energy flow, why would the process of mind as emerging from this flow be limited to the skull or even the nervous system if this flow happens in places beyond the head and even beyond our neural connections? What would make the head the only source of mind? Couldn't this energy-flow-to-mind include the whole of the nervous system? And couldn't, and wouldn't, this flow also involve various other regions of the body? Why would—or how could—this system of energy and information flow be restricted to the inside of the skull?

In other words, if mind somehow emerges from energy flow, it certainly could and would arise from the brain in the head. Absolutely. But how could it be and why would it be limited to the inside of the skull? If energy and information flow somehow is the source of mind, that flow is not limited to the skull.

With this wider perspective, we'd now say that the mind is fully embodied, not just enskulled.

So at a minimum we are proposing that the system that gives rise to the mind, the system that has mind as some aspect of itself, has as its basic element the flow of energy. Sometimes that energy stands for or symbolizes something other than itself. In this case we say that the energy has information. So there is something about energy and information flow that may be fundamental to mind.

Though not commonly viewed this way, we can consider the perspective that the mind may be fundamentally related to energy and information flow. As we've seen, if we further envision that this system of mind extends beyond the boundaries of the skin, beyond a single skull and even a single body, to some kind of distributed process in which mind also arises from our social connections of energy and information flow shared among us, we get a much broader sense of what the mind's essence may be. Couldn't the mind be considered embedded in our connections with others and the environment, a mind that is not only embodied, but also relational? This stance views mind as both a fully embodied and relationally embedded process. It is not that the brain is simply responding to social signals from others; we are suggesting that the mind emerges within those connections as well as from the connections within the body itself. It is these social and neural connections that are both the source of and the shaper of energy and information flow.

In systems terms, energy flow is not limited by skull or skin.

Embodied and embedded energy and information flow—not merely enskulled—reveals the larger system we are proposing as the origin of mind. If this view is accurate, we could then simply state that mind is both embodied and relational.

Seeing the mind as relational is not new, as sociologists and anthropologists as well as linguists and philosophers will attest. Yet how can we combine the social view with the neural view of mind? In contemporary times, modern social neuroscientists appreciate the power of relationships, too. Yet even in this division of neurobiology, itself a branch of biology, the mind is often viewed as brain activity, and the social brain is simply responding to social stimuli—just like the brain responds to light or sound from the physical world, enabling us to see or hear. In this commonly stated view, the brain is simply reacting to stimuli from outside itself, whether it is of physical or social origin. From that contemporary neuroscience perspective, brain activity remains the origin of mind.

Here's the proposal I'm suggesting you and I at least consider: Mind is not just what the brain does, not even the social brain. The mind may be something emerging from a higher level of systems functioning than simply what happens inside the skull. This system's basic elements are energy and information flow—and that flow happens inside of us, and between ourselves and others and the world.

And so we've come to a notion that the location of the system of mind that shapes our identity seems to be not limited to the skull, and not limited to the boundaries of the skin. Mind, in this view, is both fully embodied and relationally embedded.

The system of mind is comprised of energy flow within a complex system. But how does such emergence from complex systems relate to a sense of causality and the notions of free will, choice, and change?

Studying cause-effect relationships enables us to create new understanding, and opens the door to new ways of functioning in the world, including flying around this planet like I am doing right now on this airplane, or leaving our planet for other destinations. The branch of physics dealing directly with the property of energy, the field of quantum mechanics, reveals how reality is comprised of a range of probabilities rather than absolute certainties as in Classical or Newtonian physics. Even with these quantum physics findings, studies of a process called non-locality, or entanglement, reveal ways we can spin electrons differently here, and get them to shift almost instantaneously at a distance way over there. Is that causal? Some would say yes, others would simply say that this finding, now affirmed even for mass, reveals how things are deeply interconnected even though we can't see their connections. Things can be seen to have causal influences if they are all interconnected. Influence one element here, and an element there is influenced as well.

The mind can be seen as causing the brain to fire in certain ways. We can also see that the brain can cause the mind to unfold in a particular pattern. Mind and brain may be interconnected, and can influence the other. At a minimum, we can see that keeping an open mind about the direction of causality, and the fact that this direction of influence can change, is vitally important. Interconnected things influence each other.

In the example of flying, we can see that gravity—itself not fully understood—is the cause of forces that create a motion of smaller objects to move toward larger bodies, like the earth. I'm up in this airplane because of forces and structural features of the wings' shape that enable the jets to propel the plane forward while simultaneously lifting it up by causing more pressure to be created beneath the wings than above them, fortunately for us. This is how differential pressures of air above and beneath the wings have causal influences on this plane flying. Amazingly, it was the human mind that figured all this out.

Human minds have even discovered that gravity, like velocity, changes relational processes called time. No kidding. Both gravitational forces and speed alter the relative nature of time. That by itself is amazing. The fact that the curiosity and creative thinking of the human mind are capable of such non-intuitive discoveries is breathtaking. What minds we have.

The same is true of how our ever-curious intellects created our understanding of complex systems. But in the case of complex systems, using causality in a linear fashion as with gravity and speed may not really apply. For example, a cloud in the sky is a complex system composed of its basic elements, air and water molecules. A cloud is complex because it meets the three criteria of being open to influences from outside itself, chaos-capable, and non-linear. Clouds are shaped in an open way by wind and sun and evaporating water; the water molecules could become randomly distributed; small inputs lead to large and hard-to-predict results.

The magnificent and ever-changing shapes of a cloud are the result of the self-organizing emergent property of this complex system of air and water molecules. There is no programmer such as a cloud-creator, there is no one force, such as gravity, leading to particular shapes at a particular time. The unfolding of clouds is an emergent property, continually arising. The molecules of water are not completely random, nor are they cued up in a straight line.

The unfolding complexity of clouds is a result of one of the complex system's properties: self-organization. This self-organization is not dependent upon a programmer or a program. In other words, it is not caused by a specific something; it simply emerges. Self-organization is an emergent property of complex systems that simply arises as a function of complexity. As a self-organizing process, it is recursively shaping that from which it arises.

Now if you have a penchant for the linear notion of causality, you might push back a bit and say, "Well, Dan, isn't emergence *causing* self-organization?" If you conceptualize emergence as simply arising from the reality that a system is complex, then it isn't necessary to use the notion of causation. Emergence is simply what arises, naturally, from this kind of system. The system isn't really causing emergence, it simply arises from the system.

Self-organization is the natural process of complex systems that tends to maximize complexity, creating ever more intricate unfoldings of the system as it emerges over time, recursively shaping itself, as open moments become emergent and then fixed.

Naturally, if you prefer thinking in linear ways involving views such as *A caused B* thinking, you could even go on to say, "Well, the complexity of the system caused its emergent property of self-organization to shape those clouds in this way." But if you are open to non-linear thinking, then you wouldn't really speak or think like that, but would instead embrace the sense that self-organizing emerges from the system. It is not really caused by the system in a linear notion of *A caused B*. It is simply a property of the reality of complexity.

With the emergent property of self-organization, the non-intuitive recursive feature makes a case for us to not use the notion you may be longing for, the notion of causality. Why? Because energy and information flow, mind may emerge from that flow as a prime, and then, as a self-organizing process, turn back and regulate that from which it arose. Then it arises further, self-organizes further, and on, and on, and on. What is causing what? Self-organization is shaped by the very process that it shapes. This is a recursive feature of being, as we emerge and shape that very experience of emergence.

This is how the mind may have a mind of its own. You can experience and direct the mind, but you cannot always control it. The mind is, we are proposing, a self-organizing emergent property of energy and information flow happening within you and between you, in your body, and in your connections with others and the world in which you live.

As we'll explore, there is a fascinating set of implications to self-organization. But the first stage of taking these ideas in, I'll invite you to consider, is to relax the search for causality. Self-organization simply emerges. We can impair it or we can facilitate it, but self-organization is a natural process that arises from complex systems as they flow.

As we track this one aspect of mind, the self-organizing facet, bear in mind that sometimes we need to get out of our own way in order to not cause impediments to self-organization. In this sense, when we let things happen, as they say, there is a natural arising of this self-organization that needs no conductor, no programmer, no chairperson of the board to direct the show. There is no need to evoke a causal agent, one in charge, the director. There is no need to cause self-organization to happen—if we get out of the way, the system will naturally, emergently,

self-organize. That's why it can be helpful to reflect on our longing for identifying causal relationships and let them go, at least at certain times, and allow the natural essence of self-organization to unfold.

Reflections and Invitations: Self-Organization of Energy and Information Flow

With this working definition of one aspect of the multi-faceted mind, we could not only collaborate deeply as a group, but as a clinician I could experience and look into the lives of my patients through a new lens. This view is not meant to replace the central importance of subjective experience and how we share it within close relationships, it simply offers an additional aspect of mind that may or may not be connected to subjectivity. While we sense our subjective textures of lived life within consciousness, the fullness of the experience of being aware, as we'll see, is larger than the felt sense itself. Mind includes subjective experience, the whole of consciousness that enables us to know that subjective sense, and an information processing, a flow of information that can be within or beneath consciousness.

Self-organization may or may not be related to consciousness and its subjective experience. As we've seen, how we think and remember, conceptualize the world and solve problems, these and more are part of information processing. Would information processing be a part of self-organization, or something distinct? Self-organization, at least on its surface, seems most aligned with this informational flow facet of mind.

I invite you to reflect on this fundamental place we've come to. Mind may be an emergent property of energy and information flow. How does that feel for you? Can you sense the subjective textures that emerge within your lived experience? When energy flows inside your body, can you sense its movement, how it changes moment by moment? This subjective sense of being alive may be one emergent aspect of energy flow. When that flow symbolizes something, when it becomes information, can you feel how that energy pattern is re-presenting some other something in your subjective experience? Energy, and energy-as-information, can be felt in your mental experience as it emerges moment by moment.

Emergent properties of energy flow may include subjective experience—that's simply our proposal—but they also involve the mathematically established process of self-organization. If you consider your own life, can you sense how something seems to organize energy and information flow as you move through a day? "You" don't need to be in charge all the time, even if you feel you do. If one facet of your mind is self-organizing, then it will naturally arise in your life. Self-organization does not need a conductor. Sometimes things unfold best when we get out of the way.

So at a basic level, we are identifying this essence of a system, energy and information flow, as the possible source of mind. That's a proposal we've made, and we are now unpeeling some of the foundational layers of that proposal.

Subjectivity may arise as a prime of this flow of energy and information. Perhaps consciousness has something to do with this flow as well, as we'll explore in-depth soon. Information processing is innate to the notion of energy and information flow. So these three facets of mind—information flow, consciousness, and subjective felt lived life—may each emerge from the flow of energy and information.

Seeing these many facets of mind as emergent properties of energy and information flow helps link the inner and inter aspect of mind seamlessly. Energy and information are within and between, and so the emergent processes arising from them would be within and between as well. This view of the mind as both an embodied and relational process moved us beyond perhaps overly simple, restrictive views of mind-as-brain-activity and enabled anthropologists studying culture, sociologists studying groups, and even psychologists and a psychiatrist like myself studying family interactions and how they shape a child's development, to all have a shared view of how mind emerges as much in relationships as it does from physiological, embodied processes including brain activity. In other words, mind seen this way could be in what seems like two places at once as inner and inter are part of one interconnected, undivided system. In reality, these are not two places, but one system of energy and its flow.

This leads one to consider that the boundaries between synapse and soma, self and society, don't have to be as artificial as they seemed in previous models such as the "biopsychosocial" views that I had been taught in medical school. Mind as emergent was a powerful model; and one aspect of mind as the emergent, self-organizing process that regulated that flow was profoundly helpful to enable us to collaborate as a group coming from such distinct backgrounds. This view of emergent self-organization was not of three different interacting realities as commonly presented in those models, but one reality of energy and information flow.

This flow arises both within us and between us.

Energy and information flow happens in relationships as energy and information is shared; it happens inside of us as the physiological processes, especially of the nervous system including the brain, mediate the embodied mechanism of energy and information flow within us; and the mind is that embodied and relational emergent process of self-organization that regulates that flow.

As we discussed, this working definition of one aspect of the mind as self-organization did not explain away mental experiences such as consciousness and its felt texture of the subjectivity of lived life, or the experience of thought or memory as a part of information processing. Perhaps one day those aspects of mental life will be seen as a part of self-organization, perhaps not. But for the time being, the fact that 40 scientists from a wide range of disciplines could rally around that one single statement, defining at least this one aspect of mind, was a powerful convergence. The collaboration that arose from having a shared statement of what the mind might be helped us collaborate fruitfully for many years.

Does imaging your mind as an emergent aspect of the inner physiology of your body, including your brain, and the inter-connections you have with the world, especially the social world of other people, fit with your reflections on your experience? The notion of emergence can feel, for some, not intuitive, almost nonsensical, perhaps even bizarre. The idea that something simply arises from the interaction of elements of a system—like patterns emerging as water molecules move around in a cloud—may feel odd, or seem to not even apply to living systems, especially to your own life. You may wonder, "Who is in charge here?" Are we just emerging without a sense of free will? Can't we generate intention that drives the system of our self, not simply emerge from it?

These questions, and many, many more, will likely fill our minds as we move ahead. For now, if you focus on our exploration of the emergent aspect of mind—including your conscious experiences and your non-conscious elements of information flow, of which you may only see their shadows, of thoughts, memory, and emotion that later enter awareness, can you sense a quality of emergence, of something arising without you, or perhaps anything, "in charge?"

I invite you to imagine times when your mind seems to have "a mind of its own." For example, if information processing mental activities, like thoughts or emotions, are revealed in fact to be part of the self-organizing aspect of mind, then as emergent processes they can feel as if they are simply arising on their own, without a director or someone, like your "self" in charge. Sounds familiar? This is what an emergent process feels like—it just happens without a conductor in control. In other words, there is no linear causation. The self-organizing facet of mind arises from and then regulates itself. That is the recursive property as it self-reinforces its own becoming. That's the self-organizational aspect of mind. You may feel that as the experience of simply watching life unfold, within you and in your relationships, without having to be the symphony conductor or computer programmer. That's how self-organization functions. You may feel observe, notice, and recognize it, even if at times you are not trying to control it. You just get out of your own way and things naturally organize themselves.

But at other times do you notice that things get so out of sorts that you need to assert some kind of volitional control? That's likely where our conscious intention comes in as we bring consciousness and intention to influence our own experience, as we'll explore in future entries.

Intention and free will can influence our mental lives, but perhaps not fully control them. For me, this blend of active participation as an influencer, along with innate emergence, fits well with the subjective sense of my own mental life. But how does that fit with your experience?

The self-organizational aspect of emergence means that your mind, in addition to emerging from energy and information flow, is turning back and regulating that flow. Now you may be wondering, what does this really mean? Is this some metaphysical proposal of energy patterns that are hard to grasp? Well, not really. Energy is a scientific concept, a process that exists in the physical world, not beyond it—it is not meta-physical.

To address this important issue here, I invite you to explore, both in the conceptual framework and in your own personal reflections, how your mind unfolds. I invite you to consider some fascinating views of energy offered by the science of physics. As we move into these points of view, you may try weaving the scientific concepts with your subjective experience of life, and even of how reading these ideas feels to you in the moment. This gets a little wild for some people, so you may want to put on your seat belt and hold on for this segment of our journey together.

Let's review in more depth and make more personal what some physicists say about this process of energy flow. The physical property of energy, as mentioned before, can be summarized according to many physicists as the potential to do something (Arthur Zajonc & Menas Kefatos, personal communication). Energy can take a variety of forms, from light to sound, electricity to chemical transformations. It comes in various frequencies, like the range of sound waves from high to low pitch, or the spectrum of colors found in visible light. Light we see as red or yellow are both in the form of light, only with different frequencies. Energy can have a range of amplitudes, from quiet sounds and subtle light, to blaring noise and intense light. Amplitude and even density are ways of putting words to the notion of the quantity and quality of intensity. And energy, like light or sound, has a shape and texture to it, such as pulses, colors, and contrasts, which we can simply call its contour.

So on one level, we can see that energy has a range of characteristics: frequency, form, amplitude, density, shape or contour, and even location. We can have energy flowing through our brain, certain parts of our body, and between our bodily selves and others; and we can have that flow happen between our bodily selves and the larger world in which we live.

Energy changes over time and in its various dimensions—intensity and contours, for example—as it influences the world. When I wrote these words to you, energy was being transformed in my nervous system, activated these fingers, typed these words, placed them in a document, and then, ultimately, was sent to you as words on a page of paper, digital screen, or as sounds in the air, depending on how you were receiving the energy from me to you. That's flow. It involves change—change in location, me to you, and even change in the various features, such as form or frequency.

One view of information, as we've seen, is that it involves energy patterns with symbolic value. In many ways, the information processor facet of the mind extracts from energy's profile of change, its patterns of flow, something that symbolizes something other than that profile. We call this information. But information seems, from an energy-as-fundamental perspective, to itself be emerging from mental life. Energy has a profile, an array of features, with or without informational value.

Patterns of energy flow can involve changes in contour, location, intensity, frequency, and form. Here's a new acronym to help us remember this: CLIFF. And so when we say we can regulate energy and information flow we are saying we can monitor and modulate the CLIFF of energy, sensing and shaping its contour, location, intensity, frequency, and form.

You can regulate energy within you, between you and other people, and between you and the larger world. Regulation involves both the sensing process and the shaping process, like when you ride a bicycle or drive a car. You watch where you are going and you change the speed and direction of the vehicle. That's regulating your movement through space. When you regulate energy and information flow, you are monitoring and modifying energy, within your body, and between you and the world. Regulation of energy—a fundamental facet of the self-organizational function of mind—happens within and between.

This CLIFF set of variables is an accessible way of conceptualizing how your mind might sense and shape the flow of energy in each moment of your life.

Still, there is another aspect to energy that is a bit more abstract, but equally relevant for considering how your mind might emerge from and regulate energy flow.

Energy, as we've discussed, can also be viewed as a distribution of potentials. These potentialities are what some quantum physicists see as the fundamental nature of the universe. These potentials can be described as spanning a range from infinite potential to specific actualization of one of those potentialities. In this way, the reality of energy flow—how energy changes—can be proposed, as briefly mentioned earlier, to be the movement of energy from possibility to actuality, the movement from potential to the realization of one of that wide range of possibilities. The energy can continue to flow as it transforms back into potentiality. Abstract and odd, I know, (this is what we may need seat belts for), but this is what many physicists see as the true nature of our universe. When we explore the experience of consciousness in detail later on, we'll return to this view to discuss exciting new possibilities of what consciousness itself may reveal about this view of a sea of potential and the arising of actualities.

Often we live in the classical, Newtonian level of analysis, seeing large objects and how overt forces, like a car driving down a highway or this plane flying in the sky, shape our world. But at another level, quantum mechanics enables us to view the world as filled not with absolutes, but possibilities and probabilities. In fact, much of our modern financial world and advanced computing are based on quantum theory. I bring all this up because if we are to deeply embrace the proposal that the mind is some sort of process that emerges from and regulates energy flow, we need to consider what this proposed idea of energy flow really means.

The basic elements of mind, energy and information, can be seen as smaller than a plane or truck, smaller even than a brain, smaller even than a neuron. So though I am reassured that this plane I am flying in now lives in a dominant Newtonian classical physics set of laws and we can rely, with fair certainty, on the properties of gravity and flow to keep us afloat, the mind doesn't quite work that way. For example, in preparing for the plane's departure earlier this afternoon, a mechanic pressed the wrong button and the emergency evacuation slide was released. Besides the fear generated with the large sound of its deployment, the delay in our flight was another source of distress. The plane's large size made the outer structures and internal mechanisms have high degrees of certainty. We are now up in the air and it can be relied upon that that button will not spontaneously push itself and eject the door and slide mid-air.

But the mechanic's mind is not the same as the plane's structure. His mind could have become distracted, perhaps thinking of a disagreement he had with his co-worker, a distressing concern about one of his kids, or any of an infinite number of thoughts or feelings that, with a few moments distraction, might have led to his compromised attention. Attention—that process which directs energy and information flow—is fundamental to mind.

And so the mechanic's sense of knowing, within his awareness, of what he was doing at that moment, may have no longer been filled with the task of properly checking the plane's status. His attention diverted, his awareness filled with some other energy and information, his hand moves a button automatically, without thinking, and mindlessly the slide is released, we are startled, and now, hours later, we are on a different plane. That is the quantum notion of a range of probabilities. The mind may have quantum probabilities as its dominant mode rather than some Newtonian rules of pressure. The application of classical physics to the mind would evoke the notion of one part of the mind pushing on another, and predictable outcomes that have the certainty we hope this plane will have up here at five miles of altitude. We want the plane to be a Newtonian machine—reliable and predictable in following known laws of action. But the mind may not work according to such classical physics notions.

The quantum or probability nature of reality becomes more readily apparent the smaller the object, even though we are beginning to discover quantum aspects of larger objects, meaning those larger than an atom. The elements of the mechanic's mind are smaller than the fuselage of the plane, and so the unlikely becomes possible and out the slide pops. I suppose we could now nickname him the "Quantum Mechanic."

Energy is small, even though its effects are large. Rather than seeing energy as a force that only creates a pressure in the classic Newtonian view, like the air lifting up this airplane, energy may also function as arising from a plane of potential into a set of plateaus of increased probability and peaks of certainty, and then melting back down again into plateaus and then into a plane of infinite possibility—which is a plane of very low, or near-zero, probability. In other words, when any of a trillion number of things is possible, the likelihood of a particular one of them arising is low. That's a sea of potential, an open plane of possibility.

Later in our ninth chapter, we'll explore how this view can be used to understand consciousness. When we dive more deeply into how we experience a "Wheel of Awareness," we'll be able to explore, firsthand, how the quantum probability view of energy may help us deeply understand the nature of mind. This practice may also further our discussion into the potential overlaps of self-organization and the experience of consciousness. At that time, too, we'll also explore how the experience of mind depicted in the top half of the figure of the Plane and the neural processes of the brain represented in the bottom half of the figure may relate to one another. For now, we will examine the mental side of this proposal, the top half of the graph, and let this guide us to simply consider the notion that the mind does not function like trucks on the road or planes in the sky. Newtonian forces may not be the most useful view of energy when it comes to mind processes. The mind may be more like something small, something that, when we view our large scale world, we just can't see in front of

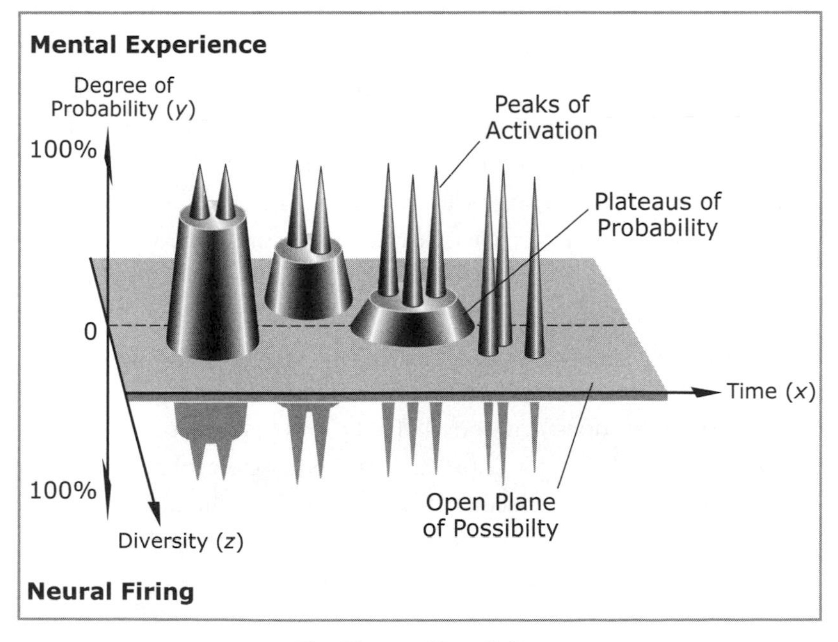

The Plane of Possibility

our eyes or even at times imagine with our conceptual mind. Eyesight helps us see the object world. But seeing the mind may take a very different kind of vision.

In this entry we have been exploring the notion of mind as emerging from energy and information flow. We've seen that neither skull nor skin is a limiting boundary of that flow, so that mind is both fully embodied and relational. At a minimum, the self-organizing aspect of mind would have this emergent embodied and relational property. As we've seen, information processing may be fundamental to that flow, attention being the process that detects and directs its movement within and between us. Consciousness and its subjective sense may or may not be an emergent property as well, perhaps linked to self-organization, perhaps not. We leave that question wide open at this moment.

But if energy and information flow is the source of mind, the source of self, and that flow is both within and between, then how do we know where "I" begin and end? Earlier in this journey we reflected on this issue of the boundaries of self.

As I walked this morning at sunrise, and strolled along a cold beach facing the Atlantic on this wintery day and felt the wind wash over my face. I realized that the sensation of the wind shaped my experience of being alive, and began to hear questions in my mind about where this energy flow would stop . . . Was the wind a part of my mind? If I allowed the flow of sensations of the wind to simply fill me, couldn't that be seen as the sensory experience of my "self?" Was this an aspect of energy flow of my mind, enabling the sensations arising in my body to flow through me, my mind? If so, then perhaps the qualifier "my" of "my mind" needs to be more clearly defined, more clearly delineated with some boundary or another, or else are we saying, that "my mind" might embrace everything? Where does the "self" end? What are the boundaries of this open system?

Do my learned concepts, an outcome of the information processing aspect of my mind that constructs ideas and filters energy into information, does this sense of who I believe myself to be, the constructor of who I am, constrain the experience of my identity? It must, in some ways, become my own self-fulfilling, self-defining sense of, well, sense of self. Now that's a recursive self-organizing process. Does such learning recursively self-organize my sensory flow into generated perceptions and beliefs about the "self," making information of the energy flow into symbols of "me" and who I am, making "me" perceive and believe that I am separate from the wind, separate from the world?

Can I examine my information flow filters that conceptualize and constrain in such a fashion as to widen my sense of who I am, expand my mind, literally, and open my self-organizational emergence to embrace a much larger sense of belonging in this world?

On our journey, this issue of energy and its boundaries has profound implications for understanding the mind and what mental health may be about. So many of these constraints remain hidden from conscious reflection, automatic filters that influence who we think we are. But we may not be what our thoughts tell us we are. We limit our well-being if we limit our sense of self to a completely separate identity from others and the world around us. We need to connect to something "bigger than the self" as so many studies and wisdom traditions have revealed (Vieten & Scammell, 2015). In a recent meeting of representatives from over two dozen nations, there was a deep discussion about the nature of self and the need to expand our sense of self beyond the body for the sake of our personal and planetary well-being.

Perhaps the self is in reality bigger, and that we—our internal, personal, private sense of our minds—just make it smaller. We'll explore how embedding time into our questions of the *who* and *when* of mind expands this discussion even further when we consider that time itself may not actually be what it seems to our minds. The mind's creation of the illusions of the self as limited to the body and the concept of time as something that flows allows us to become preoccupied with the personal past and fret about the uncertain personal future. It is these illusions of self and time that might also limit our freedom in the present. Understanding this focuses us deeply on the present moment and what we can do to embrace the fullness of its potential.

Realizing that potential, facilitating that movement from the possible into the actual, may be what mind as emerging from energy flow is all about. But what then would make for a healthy mind? If one aspect of the mind is indeed self-organization, from both within and between, what optimizes self-organization?

CHAPTER 3

How Does the Mind Work in Ease and Dis-Ease?

IN THIS ENTRY WE'LL FURTHER EXPLORE HOW THE MIND WORKS BY building on the notion that mind emerges from energy and information flow, within and between us. We'll dive deeply into the implications of our working definition of one aspect of the multi-faceted mind as being an embodied and relational, emergent self-organizing process that regulates energy and information flow. Could regulation go well and create well-being, and not go so well and create dysregulation? We will examine ideas about healthy minds and how we may cultivate them by asking the natural question, how is self-organization optimized to create health?

Self-Organization, Lost and Found (1995-2000)

It is the middle of the Decade of the Brain.

Our group of 40 academics is meeting regularly, discussing the connections among the mind, brain, relationships, and life. There is collaboration and respectful conflict, connection and conversation, all focused on the effort to bring clarity to these many issues of being human. In the midst of such academic pursuits, I am also working as a psychotherapist, seeing people of all ages and backgrounds for a wide range of issues that bring suffering to their lives, including serious psychiatric disturbances such as bipolar and obsessive compulsive disorder, relational conflicts, and the aftermath of trauma and loss. My wife and I now have two young children, and life is filled around the clock.

One evening I receive a phone call from Tom Whitfield, a mentor who'd become an important person in my life following my first year of medical school. Tom sounds weak on the phone. He tells me that he's been diagnosed with cancer. He is dying.

I put the phone down and stare out the window.

Tom had been like a father to me, serving as a sanctuary away from Boston as I worked with him that first summer after school in his community pediatric nursing program in the Berkshires of Western Massachusetts. Tom took me on as a student, but I became more like a son. I took him on not only as a teacher and guide, but also a father.

As the months of school during my second year unfolded, I repeatedly encountered various faculty members who seemed to treat their patients, and students, as if we had no minds. What I mean by this is that there was no attention paid to feelings or thoughts, memories or meaning. These inner aspects of mind seemed to not cross the minds of my attending physicians. Later I'd come to understand that these dedicated academic doctors were only attending to the physical aspect of their patients' care, not to the subjective core of the mind at the heart of their patients' lives.

Though I was trained as a biochemistry researcher in college and knew how to think about and measure molecules and their interactions, I never felt that a human being should be considered a bag of chemicals. As we'll discuss later on, during an earlier epoch of that time in the late 1970s and early 1980s when I was in medical school, the approach of the socialization process of medical training seemed to push us to see people as objects rather than centers of internal subjective experience—beings with minds.

After numerous confrontations during my first two years when I was told, emphatically, that asking patients about their feelings or what the illness meant in their lives was "not something doctors do," I made the difficult decision to stop being part of that educational process. I dropped out of school.

After I left, I thought I had disappointed Tom. I had, in my mind, the notion that he expected me to become a pediatrician, like him, perhaps move to the small town where he lived with his wife, Peg, settle there and join him in practice. Big ideas of my own imagination about their expectations imprisoned my mind, and made our relationship distant that year away. When I later returned to school and ultimately chose pediatrics after completing my medical training in Boston, he and Peg seemed pleased. But when I moved to California and chose to transfer to psychiatry training a year later, I again felt like I would be seen as a failure in their eyes—giving up "real medicine" as people used to say at school, for some soft medical specialty of the mind, whatever that was. As I was told by fellow students and some faculty back then, "Only the worst students choose psychiatry."

After I hung up the phone that morning with Tom, all of this flashed rapidly inside me as images flooding my awareness. The story flowed out of the recesses of my mind into the theater of my consciousness, a narrative driven by the shock of Tom's news.

After that first year of pediatrics in the early 1980s, finding myself drawn to working directly with the mind, I transferred into psychiatry training. In the second month of my residency program, Tom and Peg came to stay with me during their trip to watch the summer Olympics in Los Angeles in 1984. I was nervous about their visit, projecting all sorts of worries about their impending disapproval of my leaving Tom's field of pediatrics and finding my own way. But at our first dinner, I was surprised and relieved to discover that Tom had also "dropped out" of pediatrics after 30 years of practice to become a therapist, using hypnosis to help people with medical challenges such as obesity and addiction. It was clear my mind had imagined all sorts of scenarios that I had projected onto my view of what seemed real and accurate about Tom, but in fact were unfounded projections of my fears. I worried myself into a dread of their visit.

My thinking mind was fabricating all sorts of concerns, from memory and imagination, that were woven into worries, needle pointed into narratives. The story of dread felt real.

That reunion enabled me to melt the madness away. This experience reveals, too, how our sense of self can be so profoundly shaped by the stories our own minds create that focus on the concern about what others think. We'd discover, years later, that an active midline "default mode" circuitry in the brain, one we'll discuss in depth later on, appears to act as the neural nexus of that incessant chatter about the self and other. I call this the OATS circuit as it directs our attention to *o*thers *a*nd *t*he *s*elf. My OATS circuitry was having a heyday worrying about how I'd be disappointing Tom and how he'd likely reject me.

After that time of reconnection and adjustment of my OATS madness, Tom and I would attend therapy conferences together, and I felt very close to him and Peg. Our connection was re-established, and for those years of my psychiatry training we all kept in touch with a feeling of ease and comfort.

Now, more than 10 years since those Olympic games and my narrative OATS circuit reframing, I sat staring in space after Tom's call, feeling heavy and depleted, a deep sadness rising inside my body, keeping me down in my chair for what felt like eternity.

Robert Stoller, another mentor who I quoted earlier about clarity asking not answering, had died just a few years earlier from a horrible car accident. Tom was next. I was in my mid-thirties, but felt that these important attachment relationships, these father figures in my life, were still a defining aspect of who I was. Attachment does not end when we leave home; we need important close figures in our life to whom we can turn for guidance and comfort. Losing these attachment figures felt like losing part of myself. As with Bob's sudden death and the grief that ensued, with Tom's diagnosis I began to feel that heavy, sinking sense of despair and helplessness.

At the time I learned about Tom's illness, I had already made a decision to leave academics. I had been on faculty at UCLA, running the training program in child and adolescent psychiatry. After exploring my ideas with academic advisors and realizing, or more like sensing, that something needed to change in my professional course, that something wasn't sitting right in my body, I chose to leave.

I had always thought I'd become a full-time professor, to work and live an academic life, but things had changed. My interests in broad interdisciplinary conceptualizations did not fit well with the understandable drive of a modern research-based institution, which was to have highly focused, narrowly defined empirical investigations be the primary goal for full-time faculty. I was fascinated with ideas, loved scientific discoveries, and passionate about integrating empirical knowledge with practical applications, but did not want to focus only on one area, topic, or research project. So I decided to leave.

From the late 1980s, when I was in my clinical training and drawing from brain research on memory, to my research fellowship studying attachment, narrative and development, I was fascinated with how the mind grew toward health or un-health with the influence of a range of interpersonal experiences. I had proposed that some aspect of the way the brain functioned might be at the root of unresolved trauma, negatively impacting a person's subjective life and relationships with others. In the course of my time in academics after my fellowship, I had agreed to write academic journal articles, chapters, and a textbook based on some of these ideas about memory, trauma, and the brain that had found their way to the teaching I was doing outside the university. But why would I write an academic text if I were leaving academics? Why not just take on a life of private practice when I loved to take care of patients and found clinical work so fascinating and fulfilling? Why bother to write a book? These questions had been looming in the back of my mind before I heard from Tom and set up my trip to go back to the east coast to visit him.

As I boarded the plane to cross the country, I was filled with images and sensations, memories and thoughts, of life and loss. With a new notebook in hand, a green unlined journal I bought for the trip, I began to write. I felt there were so many things to say about this personal trauma of losing Tom as a father figure in my life, and so many things I had been learning as a psychotherapist about grief and healing, things learned in my clinical and research training about attachment, trauma, and the brain, about the mind and relationships.

During the six-day journey back east, the writing just poured out of me. I almost wrote, "pored out of me," and that is what it felt like: Out of every pore seemed to come a story blending the subjective reality I was experiencing with the scientific research I loved so much. I wrote the first draft of an entire book in that green journal, called *Tuesday-Sunday*, in which each day of that trip was a chapter about the mind's struggle with traumatic loss. I wrote about the brain, relationships, and the personal pain of losing Tom—about seeing him in the hospital, memories of our relationship, the narrative of our lives—including the projection of my internally generated fiction that had distorted the actual facts. It was all there, and I shared some of this emerging writing with Tom as he rested in his hospital bed. He seemed pleased with what he heard, and offered to me, with his reassuring, slow Virginia grace, his usual support that I had regained through all our journeys together: "That a way to go, Dan'l!"

Tom died a few months later. I finished translating the flow of experience into words in the green journal, one that wove personal reflections with scientific discussions, and turned it into a completed manuscript. The book was sent in to the publishing house by its original due date. After a few weeks, I received the publisher's response: What in the world had I written? This was not a textbook as my contract required—it was a memoir. I owed them a textbook.

A profound sadness rapidly set in, deepening my sense of despair with the grief of losing Tom, and moving into a depleted, helpless sense of disconnection. Tom was gone. Bob was gone. My membership in an academic community was gone. My longtime drive to connect the inner and outer, subjective and objective, was slamming into a seemingly impassable wall. I felt that I'd never be able to find a way to connect the personal and professional, and the two, the subjective and scientific, would need, in my efforts at least, to remain split apart. That despair removed the air from any space of mind to be able to express my grief, or feel any hope, to see any way out. I felt suffocated and paralyzed.

I was in a dark cloud and went for long walks wondering what to do. That beach that had been so inspiring during my childhood and in reflecting on the nature of mind did little in that state but be a place to wander without hope. I felt lost. I decided that what I needed to do was to separate out my personal life from my professional pursuits. Somehow I'd divide myself up, keep the world a fractured whole, mind distinct from matter, subjective experience separate from objective science.

But the ongoing professional work with patients during that time kept bringing me back to the reality that the subjective was real. I could not treat them well unless I worked directly with their subjective realities. My scientific training kept me convinced that there must be a way to connect empirical insights with emotional knowledge. The scientific and subjective had to find some common ground. The professional and personal did not need to be perpetually divided. Yet I could see no way to proceed, no path to move the two worlds into some kind of collaboration.

At the same time, a number of books were being published that emphatically stated with selective scientific reasoning that parents had little influence on how we develop, short of the genes we inherit from them. As a person in a relationship to Tom and Bob, a son, attachment-trained scientist, clinician, parent in the trenches, and concerned citizen, I felt incensed by the inaccuracy of these propositions, even in the face of their proponents' conviction embedded within these scientists' unambiguous statements.

My own training in science, my knowledge of the scientific literature, strongly suggested that those authors were wrong when they stated that parenting does not matter. At the same time, funds were being removed from community-based programs supporting high-risk families with the rationale that they shouldn't waste the public's money as nothing can be done to help children by supporting their parents—it's all in their genes. That frustration fueled my healing, and those parenting-doesn't-matter statements became the most potent motivators I could have wished for. My fog of confusion lifted and I resolved to attempt to offer a counterpoint to that conviction.

I took from the closet that old green notebook, turned on the computer, and pulled up the original *Tuesday-Sunday* manuscript. Over the next several years I began to extract what I had learned from personal reflection from what I had learned from hard empirical data and careful scientific reasoning. I knitted these strands together into a new manuscript; this one with the purpose of offering what I hoped would be an ironclad document scientifically illuminating how our relationships matter—including ones we have with our parents. That book came to be called *The Developing Mind*. I am happy to say that it became a useful defense against the removal of funding from community programs, including those focused on helping high-risk children and their families.

I share this story because it was a crucial part of my journey to understand the mind. The world of academic publishing usually considered personal, inner reflection "inappropriate" in a professional or science-based book. In peer-reviewed articles, too, offering an inner view of mental life seemed out of the question. One publisher once said it was "unprofessional" to reveal my inner mental life as a therapist—that I should save that for my own personal journals.

Their position is understandable if we consider that putting knowledge into the world should be based not on personal opinion, but on careful observation of phenomena. In fact, the field of academic psychology, as we've seen, took a similar stance about a hundred years ago. Subjectivity apparently had left the realm of "legitimate" data to study the mind in the world of academia.

But what if the mind is the phenomena being studied scientifically? What if subjective reality is one true facet of mind? How do we then explore the mind without diving into subjective experience and trying to articulate the phenomena we experience? Wouldn't a subjectivity-devoid approach miss something central to mind? Wouldn't a stance of leaving out subjectivity so central to mind take the mind "out of context?" This we'll explore in greater detail in our next entry.

After *The Developing Mind* was published, I was asked to teach workshops for parents on how to make sense of all the attachment research and apply these ideas in raising children. That book's practical guide, *Parenting from the Inside Out*, which I wrote with my daughter's preschool director, Mary Hartzell, after we began teaching parenting workshops together, received several rejections from publishers. When we inquired about the common reason for being turned away, we were repeatedly told that the public does not want you to tell them how they need to 'look inward' for a deeper self-understanding to be better parents; they want you to tell them what is wrong with their kids and how to act to change their children's behavior.

Mary and I knew that the research, instead, supported an inside-out approach, so you can imagine how frustrating this conflict and these rejections were. The science is clear: The best predictor of a child's attachment—which is not the only factor, but is a documented robust predictor of the healthy development of the child's resilience and well-being—is the parent's inner self-understanding of how their childhood experiences have influenced their own development. We stuck to the science, and happily we found a publisher.

The science of mind requires that we include internal reflection on subjective, personally experienced reality. Making sense, creating meaning, reflecting on memory, self-awareness, regulating emotions, having an open presence of mind, these are all subjective mental experiences of the parent that support the development of resilience in the child. From a practical perspective, these are all teachable capacities that can shape how a parent's children develop. The science of mind, as we are seeing, shows how our relationships, as much as our body and its brain, shape who we are and who we can become. We can view this as the self-organizing aspect of mind that is fully embodied and relational. Our relationships with one another shape the direction and nature of energy and information flow—between and within us. These relationships shape us throughout our lives.

The repeated rejection of the parenting manuscript at that time was a direct and familiar message to me: "Just the facts, no feelings." But how could we really speak to each other about the mind without the subjective feelings of mental life included in our discussion? I had the task to write about the mind in trauma and in healing for the original textbook assignment, and the effort to include both facts and feelings and still create a book someone was willing to publish had failed. That failed attempt seemed to deepen the sense of grief of losing not only Tom, but also what our relationship had meant for all those years. Later, the repeated rejection of the book's sequel on the inside-out parenting approach reminded me of that sense of despair. I felt that stuck sense of helplessness.

Losing Tom was a painful part of my life as it was unfolding, moment-by-moment. But in our relationship, I had gained so much not only in our connection, but also in my own inner strength. Life is full of loss and gain, and now I was filled with pain. Attachment figures shape who we are. But what do we do when we lose such an important cornerstone of our life, of our mind?

Life flows as energy and information transform in each moment of now across space and the wide array of emergent possibilities. We unfold as we—energy and information flow—continually emerge, from potential to actual and back into that sea of possibility. Is that an intellectualization, keeping me distant, even now as I write this, from the pain of the reality of the loss at the time? Losing someone you love, and someone who has shaped you, feels horrible.

Yet it is true, we can never halt the flow of these moments, one after another. We can live life a rock, feeling nothing, connecting to no one. These bodies we inhabit do not have infinite moments, it is true—

we can lose our own lives, even as rocks. Yet this tension of longing for permanence, for holding on to those we love and to our own lives in the face of life's inevitable transience, seems a deep human struggle. The nature of this was beautifully articulated by a dear family friend, the late poet and philosopher John O'Donohue. Just prior to his death, when asked during a radio interview if anything still bothered him, he said that "time was like fine sand in his hand," and no matter what he did he could not hold on to it.

As I write to you now about Tom, I can recall that same sinking feeling I had then. The recollection of Tom brings despair, and it expands as I talk of Bob and of John, for they are all grouped as one set of losses, of people who meant, people who mean, so much to me. Grief takes time to resolve, they say. But as time may not be real, what is it, really, that grief takes? Grief made me feel stuck, and at times I became filled with sudden intrusions of sadness and images from the past. There were no more open moments of planning a meeting with Tom, no more options for connections. There was some way in which active grief was filled with a lack of ease that simply felt bad. I wouldn't call it a disease, but it was filled with "dis-ease." I was not in order, but in "dis-order." I was at times not able to function, filled with "dis-function." I was out of kilter, and not in wellness.

What did this mean to have a mind that was not well?

What is wellness made of, anyway, we can ask now. What does it mean to be "not well?"

You might say this was natural, given Tom's death, and I would agree with you. "Dan, don't be so hard on yourself, you lost a person who was like a father to you." Yes, you are right, and thank you. But what can be learned from that experience, I wondered. Could grief be a window into a temporary form of challenge to well-being? Could we learn something from the subjective experience of grief that might inform an objective view of what a healthy mind might be?

All those things happening at once shook my sense of balance. As the months unfolded, some relief crept in as I came to accept that the reality of loss was accepting that these moments of the past, my connection with Tom in real time, were now fixed and could not be changed, no matter what anyone did. That was a kind of truth that was filling me. Grief, and forgiveness, is coming to accept that you cannot change the past. Moving through that process required that I let go of some kind of personal self that was shaped by Tom, to enable a new emergence of a "me" that could cherish Tom and our connection, even if I could not call him up or plan a trip to see him.

The work on the first book over those years was also about some kind of truth, it was a project bigger than this personal self. Out of the green journal was emerging a new set of writings, things I had learned from reflecting on losing Tom. There was something about making sense of my relationship with Tom and our scientific and societal view of who we are and what shapes us that all seemed to be woven of the same stuff.

But what was happening? What could this grief tell us about the mind?

Making sense of our lives, relationships, and inner experiences seemed to pull on many aspects of reality. It seemed to take the past, weave it with the present, and help shape the future. These mentally constructed representations of time simply shape how I am expressing this idea, from my mind to yours, with our common conception of time as something that flows. But if there is only now, then being fully present in the moment includes not only feeling the sensory fullness of now, but also being open to whatever reflections arise on past nows, the fixed moments we've experienced and call "past," and the open moments we can anticipate that await us when next becomes now, what we call "future."

The sense of despair and loss, of helplessness and hopelessness, shut down my sense of the future, my sense of openness. I seemed to get either stuck in feeling there was no way out, or filled with intrusive memories of the past. I was in this strange oscillation between rigidity and chaos that filled the grief with a sense of dis-ease and dis-stress.

As this fundamental question of what makes for a healthy mind haunted me, I reflected on all the patients I was seeing, and what I was experiencing. What was the distress of losing Tom all about? What was the distress of not being able to combine the personal and professional, subjective and objective, inner and relational? What could it be that was going on within and between, in health and in un-health, in ease and in dis-ease? I turned to the view of mind from our study group of 40 academics to try to make deeper sense of even this making-sense process.

Profoundly useful insights came from the mathematics of complexity, the discipline that gave rise to a deeper exploration of emergence and self-organization, as we've introduced in the prior entry and explore more below. One of the ways we experience self-organization is in the stories that arise within and between, narratives that help us make sense of our lives. Grief is a deep sensory immersion in making sense of being, and making sense is the drive of our narrating minds.

My old narrative professor Bruner would say that our life stories arise when there is a "canonical violation"—meaning when we have a violation to our expectations (Bruner, 2003). Loss is such a violation from what we expect. Grief is the wrestling of the mind to attempt to deal with those shifts from what we expected. Bruner would also describe how narratives have a landscape of action and consciousness. In our seminar, we'd discuss how this meant our stories focus on the physical aspect of events and inner mental life of the characters in those events. All of this narrating, this story-telling, is an effort to make sense of our lives from the inside out.

As the subjective experience of losing Tom filled my awareness, I felt a narrative drive take over. It was beyond my control, to express externally what was going on internally. At times it felt as if the story was being written through me, not really by me. The writing in that green journal became a repository of this narrative drive to make sense of what was happening in my mind, within and between.

After Tom died and after the *Tuesday-Sunday* manuscript was rejected and the project turned from personal to more universal within the writings, as the need arose to contribute something professionally that could express the scientific findings that parents matter, our relationships influence not only how we develop in childhood, but also our health throughout the lifespan, it continued to feel as if the story was using these hands to write itself. I had never before experienced such a sense of being taken over by a project, consumed by a passion, as with the emergence of the narrative. The narrative felt like it was using me to express itself, using this inner personal experience to tell a bigger story, something more than simply my own private journey, something that belonged to a wider reality.

That, I imagine now, is what Bruner meant by narrative being a social process. I don't really know how to articulate this, but it felt like my own mental life was a part of some larger process than what was happening in this body. Friends would ask, with all of those rejections, why I would continue. All I could say was I was no longer really writing the book, the book seemed to be writing me. Yes, the narrating and the story itself were a function of my inner, personal experience. But I felt this compelling drive to make sense of a violation to expectation, perhaps a violation to a vital truth with all these public statements that parents don't matter resulting in the removal of funding for children in need, a violation to a sense of social responsibility. I couldn't stop the writing. I don't know how else to describe the sense I felt of the story being written through me, not really by me. It was like being a servant for something beyond a personal, private self. The biological scientist in me knows that sounds bizarre, but that's what the subjective sensation was like. It feels, now upon reflection, that as a professional immersed in psychotherapy, as a science-trained psychiatrist, defining the process of what happens when there is a violation of a reasonable expectation was deeply important. There was a need for the scientific and professional community, for parents and the public as a whole, to recognize the importance of relationships. These truths may have created the feeling in this body, in me, that the field of mental health needed to make sense of the mind as much as I needed to make sense of losing Tom.

I was not separated into a professional and personal version of self, even if the publishing world was pushing me to be divided.

Still, I elected to divide those two worlds of the personal and professional. One narrative drive would be to scientifically depict the importance of subjective experience in attachment relationships; the other would be personal reflections that I'd keep to myself. Divide and conquer the conflict.

This epoch of loss and gain was wild with pain and passion. I felt deeply as a person, and could envision vividly the core concepts as a scientist and clinician. Taking a deep breath, I removed all references to my own personal journey and wrote that professional textbook, *The Developing Mind*, in a scientific manner consistent with the expectations of the publishers, and perhaps the academic audience who were the intended readers.

But how could science be blended with the equally if not more real aspect of life we call subjective reality? What would such integration be like? I imagine that this is the narrative drive that emerged within me following the initial book.

It was fine, it seemed to me, to feel sad when sadness is what is present. Why not be able to simply express that, to be open and communicate facets of our human reality that we all share—as person and professional? Couldn't we be fully human, fully present, for life, for communicating, even in writing, to one another? And wouldn't it be fine, too, to tap into the energy and excitement when clarity called, even in the form of questions, jokes, or paradoxes? It seemed fine to take a chance and be connected honestly to what is, and reach out and connect with others in that same way. Couldn't acceptable science writing, especially when dealing with the mind, include subjectivity?

In many ways, I can see now that the journey during those years was a way of living the struggle of self-organization. Making sense of the mind—what it is, how it develops, how we can support one another to find mental health and thrive—was the fuel of a fire that propelled me forward.

Differentiation and Linkage: The Integration of Healthy Minds

While the mind is multi-faceted, with processes including consciousness and its subjective textures along with information processing, defining an additional aspect of the mind as a self-organizing process might not only offer a working place for us to deepen our discussions, it might also enable us to state what makes for a healthy mind.

If we only stayed at the usual description level of mind, delineating its activities such as feelings and thoughts, could we really say what a healthy mind is? What would healthy consciousness be? What, in general, could we say about a healthy subjective life? Without knowing what thinking and feeling truly are, how could we even say what healthy forms of thinking and feeling might be?

As we've suggested earlier in our journey, perhaps self-organization relates to consciousness, thinking, feeling, and behaving. Perhaps it will even be found to illuminate the nature of our subjective sensations. But perhaps we'll discover these things are not related at all. Whether ultimately these elements of mind so often found in descriptions of mental life are, or are not, understood to ultimately be aspects of self-organization, we can at least explore this regulatory facet of the mind now that it is defined. The mind is not unitary, about that there seems to be a universal consensus. Self-organization, we are proposing, is but one aspect of the multifaceted mind, though we've seen it seamlessly links inner and outer as the location of mind.

Defining at least this one aspect of mind as *the emergent, self-organizing process of the complex system of embodied and relational energy and information flow* might help us to not only offer a working definition of the mind for us to discuss and debate, but also help us go one step further and define what a healthy mind might be. If the mind is a self-organizational process, what enables optimal self-organization? Asking this question and exploring this view of mind have enabled new insights to be gleaned about mental life, and mental health.

A pattern had become clear that puzzled me for years as a clinician during my training and beyond, before I had even heard of complexity theory, or had the experience of losing Tom. Here is what I had been observing: Patients seemed to come in for help with either an experience of rigidity in their lives—many things being predictable, boring, without vitality—or *chaos*—life being explosive, unpredictable and filled with distressing intrusions of emotion, memory, or thought. Whether a part of traumatic aftermath, or inherent disturbance, rigidity and

chaos were the common patterns of distress that seemed to "organize" my patients' lives and fill them with suffering.

What could this chaos or rigidity be arising from, I wondered.

As the clinical training years merged into a research fellowship, these questions possessed my mind. What was the mind? What could a healthy mind be? Why did people in dis-ease, dis-comfort, dis-stress, and dis-function all seem to share the pattern of chaos, rigidity, or both?

When I invited those 40 scientists to gather at the beginning of the Decade of the Brain, I didn't know that trying to facilitate that group discussion of what the mind might be would be pivotal in helping to define what a healthy mind might be. With the proposal accepted in that group of the mind as an emergent self-organizing embodied and relational process that regulates energy and information flow, the question for me as a clinician, and scientist, was this: What is optimal self-organization, and could chaos and rigidity be a part of self-organization?

Reading more deeply and broadly in the scientific literature, and discussing complexity theory and systems science with mathematicians and physicists, I came to learn that when a complex system is not optimally self-organizing, it tends to move toward one of two states: chaos or rigidity.

Amazing.

I went to the bible of psychiatric disorders, the *DSM, Diagnostic and Statistical Manual of Mental Disorders* (American Psychiatric Association, *DSM III*, 1980; *DSM III-R*, 1987; and *DSM-IV*, 1994 were the editions I reviewed around that time; now there are later editions, *DSM-IV-TR* in 2000 and *DSM-5* in 2013, in which the same patterns I'm about to describe are still observed). There I found that every symptom of every syndrome could be re-envisioned as an example of chaos or rigidity. Stunning. Patients with bipolar disorder, for example, had the chaos of mania and the rigidity of depression. Those with schizophrenia had the chaotic intrusion of hallucinations and the paralyzing rigidity of fixed false beliefs called delusions, and the emotional numbing of what were considered the negative symptoms of social isolation and withdrawal. Individuals who'd experienced trauma that remained unresolved, those with posttraumatic stress disorder (PTSD), could be seen as filled with both chaos (intrusive bodily sensations, images, emotions, memories) and rigidity (avoidance behaviors, numbing, amnesia).

Whether it was innate in its origin, or caused by an overwhelming experience called trauma, having psychiatric suffering seemed to be manifested as chaos, rigidity, or both.

What would happen, I wondered, if we were able to peer into the brain of an individual and look for signs of impaired self-organization? Could we see that functionally? By reading the science, talking with various scientists, and reflecting on all these experiences, it seemed to me that complexity theory could be translated into common language by revealing a core truth: Optimal self-organization arose when the system was having two interactive processes occur. One was *differentiating* elements of the system, allowing them to become unique and have their own integrity; the other was *linking* these differentiated elements of the system. The common term we could use for this linkage of differentiated parts was *integration*.

Integration enabled optimal self-organization so that a system functioned with flexibility, adaptability, coherence (a term meaning holding together as a whole, having resilience), energy, and stability. Expressing this integrated flow could be embedded in a new acronym, *FACES*. A FACES flow of integration creates harmony, like a choir singing together by differentiating their voices and linking together with harmonic intervals. You may know the feeling of hearing or singing a song in harmony—it's exhilarating and full of life.

A metaphor came to mind of a river of integration that would have harmony as the central flow bounded on either side by one bank of chaos, the other a bank of rigidity.

Could it be that well-being emerged from integration?

Might individuals with mental suffering be found to have impaired integration in their brains and bodies, the within, and in their relationships, the between? Could we help people, no matter if the origin of their suffering is innate or experiential, alter their brain's integration so they could learn to optimize self-organization toward a FACES flow? Nothing was available at that time to answer these queries, but they became burning questions in my mind.

Through a long line of reasoning and scientific reflection that I wrestled with in writing *The Developing Mind* and explored in clinical practice during those years, it became clear that self-organization pushes toward something called maximal complexity by way of these two interconnected mechanisms of differentiation and linkage. It is this *linkage of differentiated parts* that maximizes the complexity of the system, the state of that FACES flow. Self-organization is the natural drive of a complex system to maximize complexity.

Maximal complexity is an unfamiliar term for most, and can even seem foreboding. I want my life to be simpler, not more complex, you might be thinking. But the reality is that often we actually want our lives to be less complicated—which is not the same as being less complex. As mentioned earlier, the reality is that complexity is actually quite simple in its elegance. Optimizing complexity has the feeling and physical reality of harmony. It moves across time, space, or potentialities—the flow of moments—with a FACES flow that is literally what a choir singing in harmony feels like as the individual members differentiate their voices in harmonic intervals and then link them as they sing together. Integration feels connected, open, harmonious, emergent, receptive, engaged, noetic (a sense of knowing), compassionate, and empathic. I clearly have an acronym addiction. This one is where the

River of Integration

term COHERENCE stands for the very features of a coherent flow of FACES. (Apologies to those who don't like these word-memory tools).

Maximal complexity is achieved through the linkage of differentiated parts. Though mathematicians never use the common language term in this way, we can, and that is the word: *integration*. Integration is the way self-organization creates a FACES flow of harmony across time.

(By the way, the reason this term is not used in math is that for mathematicians, *integration* means addition—the summation of the parts into an added-together whole. But here, *integration* as we are using the term implies that the parts maintain their differentiated nature and become functionally linked to one another. In this very specific manner, *integration*, as we are defining the term for this usage, means that the whole is greater than the sum of its parts. In math, the word, *integration* simply means the sum. A related but less specific process is *synergy*, how these elements connect with one another and bolster their individual contributions by something that is larger in the joining. As integration naturally emerges with self-organization unrestricted, we could call this an "emergent synergy" as the mind unfolds and self-organizes energy and information flow.)

Integration may just be the basis of health.

If integration was the way self-organization created a FACES (flexible, adaptive, coherent, energized, and stable) flow, could that be a working definition of a healthy mind? Could integration be seen in our relational lives? Could we see integration in the brain and body as a whole—a whole with differentiated parts working well together?

Later research would emerge, with advances in technology in the new millennium, which supported this hypothesis of integration as the basis of health. For example, Marcus Raichle and colleagues at Washington University in Saint Louis (Zhang & Raichle, 2010) would reveal years later that in patients with psychiatric conditions not caused by experience but innate in their development, like schizophrenia, bipolar disorder, or autism, important and distinct interconnected regions of the brain were, in fact, not integrated well. This could mean that differentiation or linkage, or both, were not anatomically or functionally established. Even in conditions not inherent to the individual's make-up, but something acquired through experience such as childhood abuse and neglect, studies by Martin Teicher at Harvard University (Teicher, 2002) would also later reveal that the major areas affected by this developmental trauma are the regions that link differentiated areas to each other: the hippocampus which connects widely separated memory areas to each other; the corpus callosum linking the left and right differentiated sides of the brain; and the prefrontal regions, which link upper and lower areas with the body proper and social world.

These are all findings that later would support the basic hypothesis that integration was the basis of health. The basic question back then, before the technology could allow us to see into the brain as we now can, was this: Could it be that each of these conditions, whether acquired through painful experiences or some innate constitution, was a compromise to self-organization? If optimal self-organization that allows COHERENCE and a FACES flow to emerge depends upon integration, could integration then be the basis of mental health?

Here is the idea: Integration is well-being.

A healthy mind creates integration within and between.

So far, research that has emerged in the two-dozen years since this idea was first proposed has supported this proposal. Support is the term, not *proven*, as all we can say at this point is that the hypothesis continues to be supported by emerging findings from a wide array of studies, ranging from those that examine the brain to those studying relationships.

Here are some fun factors about integration. Integration is both a process and structural feature: We can see the flow of energy and information from differentiated regions becoming linked (a verb-like process), and see the structural linkages of differentiated parts (a noun-like structure). In the brain, for example, we can now see the functional linkage of differentiated parts within computerized electroencephalogram studies or within functional blood flow scans, such as fMRI (functional magnetic resonance imaging). We can also see the structural integration in the brain, now more than ever before, with a range of new techniques that reveal what is called the connectome. A study just emerging from the human connectome project reveals, in fact, that a more integrated connectome is associated with positive traits of life, while a less interconnected connectome is associated with negative life traits (Smith et al., 2015). But, back then, some of these more advanced techniques were only ideas in the imagination of creative neuroscientists, and we did not have this exciting empirical support for the proposal that integration is the basis of health.

Integration in a relationship can also be studied. The way we respect differences and then cultivate compassionate communication as a linkage is one way of revealing an integrated relationship. Secure attachment, for example, can be conceptualized as an integrated relationship. Vibrant romantic relationships can be seen as integrated.

Amazingly, as I review in the second edition of *The Developing Mind*, the studies that exist generally point to the following possible, though wildly simple notion: Integrated relationships stimulate the growth of integration in the brain. Integration in the brain is the basis, so far as we can tell from all the existing research, of self-regulation at the heart of well-being and resilience. How we regulate our emotions, thoughts, attention, behavior, and relationships is dependent on integrative fibers of the brain. Integration in the brain enables the various regions to be coordinated and balanced—to become regulated. In this way, neural integration appears to be the basis of self-regulation, and integration in the brain can be shaped by integration in our relationships.

Why would that be true? We'll see more as we move along, but energy and information flow within and between, and in this way integration happens wherever mind happens, within us—in our bodies—and between us—in our relationships.

Integration is how to optimize self-organization, within and between.

Self-organization has a natural push within complex systems to link differentiated aspects of the system. This means that we have a natural drive toward health. This drive toward integration happens internally and relationally. Recall that the system is both within and between us. The system is not in one place or the other, and it is not really in two different places at once. The system is both inside and between, one system, one flow. Energy and information flow is not limited by skull or skin. We live in a sea of energy and information flow that happens within the body and between our bodies and the larger world of other people and our environment.

While mathematicians don't have a name for the process of the linkage of differentiated parts that they've identified as what makes self-organization maximize complexity, to optimize regulation of the complex system, in plain language we are calling this process *integration*. Integration is a natural drive of the complex system of the mind. That is what self-organization "does" as it emerges from the complex system and regulates its own becoming. Maybe that is why I felt that the book was writing me—self-organization was pushing for integration by way of the making-sense process of narration. My grief initiated a state of chaos and rigidity, the signs I was not integrated. The narrative drive was an effort to move my life, the system in which "I" emerges, back toward harmony by integrating within the making-sense of my experience. For me, as person and professional, integration was happening on many, many levels all at once. It was happening *within* me, and it was happening in my connections to others, *between* the world and me.

On that journey to see Tom one more time, my mind, on its own without intentional effort or decisions on my part, began a process to make sense of my life, my relationship with Tom, and his illness and impending death. Making sense arises from the fundamental push of the mind to integrate within and between, and to integrate our sense of past, present, and future. Now we can see this as the natural push of self-organization to pull me out of the states of rigidity and chaos, the two states that complexity theory tells us systems enter when they are not optimizing self-organization. When I was in acute grief, my states of rigid depletion or chaotic intrusions of memory and emotion revealed these impairments to integration.

Healing is integration.

The healing of grief involves the transformation of a state of loss through the self-organizing process of integration. Being flooded or stuck, we are not linking the range of distributions from openness to certainty and back again as we change, as we emerge, as we flow. We keep returning to our original state of connection to the lost one, but he or she is no longer there. Our emergence moves us toward rigidity and chaos. As the healing process unfolds, we move toward the wholeness at the root of health and healing. Becoming whole means linking different parts to one another. Healing comes from integration.

On a broad scale, the feeling of this integration had a sense of wholeness. The writings of the physicist, David Bohm, in his classic work, *Wholeness and the Implicate Order,* states: "Now the word 'implicit' is based on the verb 'to implicate'. This means 'to fold inward' (as multiplication means 'folding many times'). So we may be led to explore the notion that in some sense each region contains a total structure 'enfolded' within it" (1980/1995, p. 149). He calls this enfolded or "implicate order." Bohm goes on to postulate that "the *whole implicate order* is present at any moment, in such a way that the entire structure growing out of this implicate order can be described without given any primary role to time. The law of structure will then just be a law relating aspects with various degrees of implication" (p. 155). "*What is* is always a totality of ensembles, all present together, in an orderly series of stages of enfoldment and unfoldment, which intermingle and inter-penetrate each other in principles throughout the whole of space...if the total context of the process is changed, entirely new modes of manifestation may arise" (p. 184). As we'll see, this perspective offers us a deep dive into the layers of reality not seen by classical physics, but illuminated over the last century by quantum views. Bohm asserts, "the implicate order gives generally a much more coherent account of the quantum properties of matter than does the traditional mechanistic order. What we are proposing is that the implicate order therefore be taken as fundamental" (p. 184).

In many ways, the notion of wholeness and an implicate order in the world invites us to think in systems terms—the ways basic elements interact to create emergent phenomena—rather than simply one part interacting with another part in isolation. This systems view is not always easy to grasp at first. But from a mathematical perspective, this notion of wholeness helps us see how the emergent property of self-organization could only be understood by sensing its enfolding or implicate nature, a fundamental aspect of the whole of the complex system's emergence.

Back in that Decade of the Brain, many questions arose about self-organization: Would we eventually be able see impaired verb and noun forms of integration in the brains of those with challenges to their mental health? Could we cultivate integration in the brain and look for this in pre- and post-intervention evaluations? Might we use our relationships to inspire people to rewire their brain toward integration? Could we use the mind to promote integration in our relationships and brain?

Preliminary findings were beginning to emerge to suggest that this basic notion, derived from scientific reasoning, clinical experience, and personal reflection—that integration was the basis of health and impaired integration the basis of dis-ease and dis-order—might indeed have merit in empirical research findings. But those early views from mathematics and the scientific findings of memory and emotion we'll discuss in later entries that supported the proposal that impaired integration is associated with challenges to well-being were only arrows pointing us in a general direction. We'd have to wait for more studies, and especially more advanced technology, to see if these ideas were valid.

With this specific definition of mind as a self-organizing process that was embodied and relational, and with the *definition of health as integration*, it seemed we were in a position to set hypotheses out into the world that could be explored with reflective personal experience, open clinical interventions, and carefully conducted empirical studies. Those in the fields of mental health, neuroscience, and other disciplines could explore this notion of a healthy mind emerging from integration within us—in our bodies including our brains—and between us—in our relationships with other people and the planet. In these many ways we could learn more about the nature of human reality. These definitions of the mind and mental health created a broad platform from which to continue to explore and experience how we might cultivate well-being in our lives.

Reflections and Invitations: Integration and Well-Being

I invite you to consider times in your life when rigidity or chaos arose in your day-to-day experience. Perhaps you had an altercation with a friend in which you felt you were not heard and then intense emotions arose that surprised you. Perhaps it was when something you expected to happen didn't, and you felt out of sorts and couldn't "get back in gear." Or there may have been times your mood simply became flattened and you couldn't think clearly, a change of mind that had no apparent cause. These states away from the FACES flow of harmony may be transient, lasting a few seconds, minutes, or even hours. You no longer feel flexible, adaptive, coherent (holding well together fluidly over experiences), energized or stable. In our daily lives, something may happen that creates such transient movements toward the banks of our metaphoric river of integration, but

we are not stuck there for long. We experience moments of chaos or rigidity, but are not stranded outside the flow of the river for prolonged periods of time, unfolding moments without end. They are transient, short-lived excursions out of the river of harmony's flow. They are simply a part of everyday living.

But at other times in your life, you may find that for more extended periods you have found yourself stuck in chaotic floods of emotions, memories, or behavioral outbursts; and at other times, rigidity has set in and you feel like you cannot escape from a sense of repetition of thought or behavior, as in addiction, or feel a malaise and loss of excitement for life, as in demoralization, despair, or depression. If those prolonged states continue repeatedly and for extended periods of time, something in your life may not be integrated.

Integration creates a FACES flow; impaired integration leads to chaos and/or rigidity.

Integration is the linkage of differentiated elements into a coherent whole. Integration, as mentioned earlier, is the source of the notion that "the whole is greater than the sum of its parts." This is an emergent synergy of the function of many aspects of you, inside your body, including the many regions of the brain, and within your connections to others and the larger world in which you live. When you are moving in an integrated flow, there is a sense of being whole, full, at ease, and receptive. Integration is the source of the experience of harmony.

When integration is present, the unique nature of the parts themselves does not disappear with linkage. In this important way, linkage is not the same as addition or fusing or blending, and integration is not the same as homogenization. Integration is more like a fruit salad than a smoothie.

If either differentiation or linkage is not present, integration is impaired and you are likely to experience states of rigidity or chaos. For example, if aspects of your life are not respected for their differences, then differentiation is impaired. If different aspects of your life are not freely connected, then linkage is impaired. These impediments to integration may be transient, and we move toward or even into chaotic or rigid states, temporarily. But these impediments may also be long lasting, and chaos and rigidity become a regular part of our lives. They come often, and stay for long visits. Detecting these compromises to integration is the first step toward the movement into health. The next step is directing one's attention to enhance differentiation if it is impaired, then promote linkage if that is compromised.

In my own life when these states of impaired integration occur, I find that asking the question, "What in my life right now is a source of this rigidity or chaos?" is a good starting place. With at least nine domains of integration to explore (which I've discussed in depth in *Mindsight,* and will describe briefly here), I can have a kind of checklist for seeing what might be going on in my life and then how to move from states of impaired integration toward integrative well-being.

Here is a quick example of one simple intervention. After my dear father-in-law, Neil Welch, passed away, I was filled with grief. As the making-sense process emerged, I simply let the sadness and loss fill my awareness. Over many months, that feeling of depletion and heaviness would come and go, and soon, a year after his passing, I felt lighter and more filled with vitality. But one day I awoke thinking of him, and felt quite heavy. What was going on? I decided to try a simple integration technique. I thought of every emotional state I might be feeling, even out of awareness, and "named it to tame it" by stating the word and then tapping on my left and right shoulders with my crossed arms alternately. I'd cover as many emotions as I could think might be present, and, to be complete, did so alphabetically. Anger, apathy, anticipatory anxiety started the negative 'A's'. Awe, appreciation, and attachment were some of the positives. On and on it went, and though I couldn't think of a 'Z' emotion, I looked for them anyway.

By the end of this differentiation of emotional states and linking them in consciousness, coupled with the embodied bilateral tapping, I felt actually quite exhilarated. I had a great day.

You can try this too. The amazing thing about integration as a concept is it is so simple and direct, as an idea and practical application. Chaos or rigidity reveal challenges to integration. Differentiation and linkage are likely impaired and what are needed to create integration. You'll feel that FACES flow of COHERENCE emerge as integration returns to your life. Try it out and see what you experience.

The following domains of integration began to emerge as patterns in my own clinical practice back in the Decade of the Brain. Personally, I was grappling with the loss of Tom; scientifically, I was grappling with the group of 40 in our study group of the mind; clinically, I was grappling with how to combine these emergent ideas of the mind and mental health with my patients; educationally, as the director of a training program at UCLA, I was grappling with how I might teach the trainees in the child and adolescent psychiatry program a new way to consider evaluating and treating their patients. Lots of grapple, I see now. Chaos or rigidity seemed to be the universal pattern that people came to me with in their suffering. With the new view of mind as an embodied and relational self-organizing process, I could then conceive of a healthy mind as one that optimized self-organization. How? By promoting integration. Where? Within and between.

As an educator or clinician, could I inspire people to rewire their brains, bodies, and relationships toward integration? If a relationship of attachment between a parent and child could stimulate the growth of integrative fibers of the brain, could a therapeutic relationship between a psychotherapist and client or patient nurture that integrative growth we could now propose was the heart of healing?

I began to work with people coming for psychotherapy with these ideas in mind, beginning with assessments of chaos and rigidity rather than lumping a person into a possibly restrictive diagnostic category alone. Then I would try to assess in which aspects or domains of their life might this impaired integration be? Once this evaluation began to illuminate the nature of compromised integration, we could then focus treatment interventions very specifically on those domains in need of differentiation and linkage. What amazed me in that Decade of the Brain was that it seemed people who came for therapy who had not been changing, either with me or others they had seen prior, began to improve and move their lives toward well-being as they became more flexible, adaptive, coherent, energized, and stable. This FACES transformation was a focus on vitality and well-being. Mental health now could be seen as some COHERENCE emerging from integration: connected, open, harmonious, engaged, receptive, emergent, noetic, compassionate, and empathic. Instead of therapy aiming to reduce symptoms alone, without anything we were aiming toward but rather aiming away from, now this integration approach provided a working definition of health we could aim for.

With the emerging domains in mind, developing specific approaches for each person based on how each of these areas of their lives were differentiated and linked, or not, became possible. With diminished differentiation, we'd need to work on distinguishing and developing the areas of that domain that were not well formed. With diminished linkage, we'd need to find creative ways to focus our attention on bringing those differentiated areas of a given domain into connection and collaboration. Therapeutic intervention became an intentional and strategic cultivation of the emergent synergy of the integration that promotes well-being.

With this integrative approach as a framework, I could work collaboratively with an individual, couple, or family to find ways to assess their present state of chaos or rigidity, identify the domain, or more often domains in need of work, and dive into the process of cultivating integration. One powerful aspect of this approach was that it was health-based. We are all on a journey toward integration; we never arrive and are never done. In this way, we join a common humanity in the lifelong path of discovery and unfolding. Integration is a direction, not a destination. This health-based view also empowered people to find their own inner direction, and I sought to teach them techniques that would last a lifetime as a life-affirming dance of well-being.

Integration is empowering.

As we explore these domains, I invite you to consider how each of these may relate to your own life experience. You may reflect on times when chaos or rigidity became dominant in your life. What aspect of your inner or inter mental life was not being differentiated and then linked? Examining these times now, can you sense how back then there may have been something not integrated in your life? Simply gaining a sense of the origins of chaos and rigidity is a great starting place to explore how integration plays a role in your day-to-day living. Sometimes the intervention can be quite simple, like the example of integrating emotional states in awareness I described above. Other interventions may be more elaborate, and take more steps and time.

There are many ways to conceptualize reality and divide a whole into its many parts. For me, these nine domains capture the wide breadth of issues that my patients have faced, or colleagues have described, or I have experienced personally. You may find the list too long and may simply prefer to focus on integration as a whole with a broad scope; or you may prefer 28 domains. It's your choice. This is simply a way of dividing things up that over the last twenty years I have found useful and comprehensive.

A general comment: Whenever we ask about a domain of integration, it may be helpful to first imagine what could be differentiated within that aspect of life. Then, once differentiation is made conceptually clear, we can ask the question, how can these different aspects of this part of life be connected to one another? That's differentiation and linkage—that is what integration is. It may not always be easy to achieve, but it's that simple. Link differentiated parts and harmony unfolds; block that integration and chaos and/or rigidity ensue.

So let's see how these domains feel for your own life as we go though them:

Integration of consciousness is how we differentiate the knowing from the known of consciousness, then systematically differentiate and link these as we move the focus of our attention on to the various elements of the known (the first five senses; our internal bodily sixth sense of interoception; our "seventh sense" of our mental activities of thoughts, feelings, images and the like; and our relational "eighth" sense of connectedness to other people and the planet). Consciousness is difficult to describe, let alone define, but it may be most effective to think of it as how we are aware. There is the subjective experience of you knowing, and that which you are aware of. In this way, consciousness has a knowing, a known, and even a "knower" sense, ones we'll explore in great detail later on as we'll discuss a Wheel of Awareness that helps integrate consciousness as a direct practice. (The Wheel was actually a table in my office, one I designed during the Decade of the Brain with a central clear glass hub and broad wooden rim with table legs that appear like spokes. Patients would gather around the table that served as a physical metaphor for the mind. See drdansiegel.com to access the practice itself).

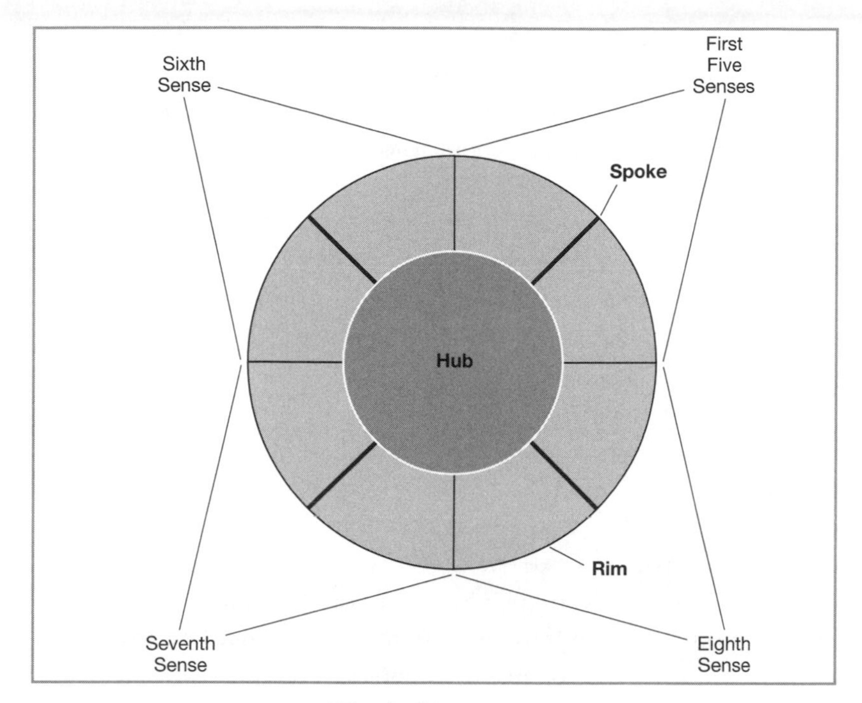

Wheel of Awareness

In your life, the integration of consciousness may seem off-balance if you find yourself "lost on the rim" in which a feeling, thought, or memory takes over your sense of who you are and you lose perspective on a wider vision that comes from the hub. When integration is challenged, chaotic or rigid feelings, thoughts, or memories may dominate your experience. If this is the case, you may find that an integration of consciousness practice will be quite helpful to create more well-being and ease in your life.

Bilateral integration is how we differentiate the left from the right hemisphere of the brain. These differences lead to different "modes" of processing, even if they ultimately share some neural activations in similar processes. We'll explore these differences later on in our journey, but here let me point out that the left mode's logical, linguistic, linear, and literal processing is quite distinct from the more contextual, non-verbal, body-influenced, more emotionally direct right mode processing. Finding a way to honor both and link them to each other leads to bilateral integration. The corpus callosum is one of the major structures linking the two hemispheres. Research reveals that developmental trauma impairs this integrative region's growth (Teicher et al., 2004). Interestingly, mindfulness meditation has been shown to increase its growth.

There are many ways of considering our two-sided brain and differentiated modes of information processing, but one simple way is this: The right mode is filled with an energy and information flow that feels a certain way, filled with context, rooted in bodily sensation, that sends and makes sense of the non-verbal signals of eye contact, facial expression, tone of voice, posture, gestures, and the timing and intensity of responses. In contrast, the left mode's logical search for cause-effect relationships and use of linguistic language will color your experience quite differently. Both hemispheres are important and share neural functions, yet each is also unique. You may find that one "mode of processing" may be dominant over the other in your life, or at certain times in your life. Some people, for example, deal with stress by moving toward a left-dominant mode to distance themselves from their inner emotional/bodily states that feel overwhelming,. This shift to the left during stress may be experienced as rigidity. Others may do just the opposite, reacting to challenges by becoming flooded with the chaos of unregulated, right-mode dominant sensations. Integrating the two modes involves honoring differences and promoting linkages.

Vertical integration is how we connect with our bodies, allowing the internal flow of sensations to rise from "below" into our cortically mediated awareness "above." In science we use the term *sixth sense* for this interoceptive ability—perceiving the interior. Body based therapies and mindfulness practices utilize interoception, an important form of vertical integration.

You may find that if you're like many people, experiences in school reinforce us "living above our shoulders" and we rarely take time to soak in the body's sensations. Yet vertical integration would invite you to enable the body's signals, from heart and intestines, for example, to rise up in your mental life, in your consciousness, and be respected for the important source of bodily wisdom they are. You can sense, perhaps, how chaotic states may be where you cannot regulate the body's signals; rigid states would be where you cut yourself off from the body's signals. In contrast, an integrated vertical state would be opening to the "wisdom of the body" and taking in those important interoceptive signals but not being overwhelmed by them.

Memory integration is how we take the differentiated elements of implicit memory—perceptions, emotions, bodily sensations, and behavioral plans, as well as mental models and priming—and link them together into explicit forms of factual and autobiographical memory. The encoding of implicit memory starts early in life and continues throughout the lifespan; explicit encoding often begins after our first birthday and enables us to integrate implicit elements into larger puzzle pieces of memory so we can flexibly see a larger picture of our lived experiences. The hippocampus is an important region for memory integration, one that is impaired in its growth with developmental trauma. Trauma, in general, may impair this integrative function and leave implicit encoding of bodily sensations and emotions intact but not integrated. The result can be intrusion of these memories without a sense of something being recalled from the past, a free-floating form of chaos that can be quite disturbing and feel like something is happening right now, even though it is an implicit memory. Integration into explicit memory can make those recollections identifiable as coming from something we've experienced in the past and may be an important part of integration at the heart of resolving trauma.

Have you ever been filled with emotions or behavioral responses that you couldn't easily understand? Sometimes, these activations of un-integrated implicit memory can be confusing and even distressing. At other times, they can simply shut us off from being open to new experiences. Identifying these states as potential un-integrated implicit memories can free you to focus attention on these memory activations so you do not have to be disabled by them again. Writing in a journal can be a useful place to start, reflecting on your internal experience with an open mind as to whatever might be its origins.

Narrative integration is how we make sense of our lives by weaving the distinct elements of memory of lived life together and then extracting meaning from those reflections. As we've seen, narrative may be an innately integrative process, and it draws upon other domains of integration—such as those of memory, consciousness, vertical and even bilateral integration, linking the left's drive to tell a logical linear sequence that looks for cause-effect relationships of things in the world with the right hemisphere's dominance for autobiographical memory. Narrative harnesses many aspects of our selves to integrate a making-sense process.

You may find that you have a fixed story about who you are that feels familiar and comforting while at the same time may be restricting who you can become. If you reflect on where you've come since childhood, you may find that your life story has not changed all that much. Identifying the structure of your narrative sense of identity can be explored well, too, with journal writing. You may be surprised by what emerges. Opening your narrative to new input, even from reflecting on your own reflections, can be a powerful way to free your mind. In many ways, we can live our way to new stories if we have the courage to break away from the familiar and the predictable.

State integration involves the many states of mind that we each have, those differentiated ways of being that can then be linked to one another to make a continuous yet not homogenous sense of self across time, creating mental coherence within a given state and across many states of mind. A state is a way in which many functions coalesce into a way of being. States can involve our narratives, memories, emotions, and behavioral patterns. If they are repeating and shape our identity, we can call them a self-state.

For example, you may have a repeating aspect of yourself that enjoys being social while another part of you thrives with solitude. Both states define you. How do you resolve the conflicts that arise when choosing between time with people and time alone? This aspect of state integration reveals that we can honor these differences and promote linkages by giving respect to each aspect of our lives, to the many parts of a heterogeneous self that defines who we are. We can also learn to integrate within a given state, such as keeping a playful part of us alive and cultivated well as adults. State integration empowers you to detect what is going on at many levels of your experience, honor those distinct ways of being, and then cultivate internal communication as well as an external calendar that promotes compassionate ways of respecting distinct, differentiated needs of each part of who you are.

Interpersonal integration is the way we honor and support each other's differences within relationships and then promote linkage through respectful, kind, and compassionate communication. We have many kinds of relationships, from close one-to-one connections, to belonging to larger groups in families, schools, communities, and cultures. Whatever the dimensions of our connections, the principle that integration creates well-being seems to hold well and be a useful way of not only conceiving healthy relationships, but cultivating them effectively.

When you reflect on your current relationships, how do you sense this fundamental process of integration, of respecting each other's differences while promoting compassionate connections? How have your relationships in the past been, going back to those you had in your childhood, in supporting an integrated way of being with others? Tuning in to the internal world of another and honoring that subjective experience enables us to integrate in close personal relationships; honoring differences in religion, ethnicity, gender, sexual orientation, learning styles, economic background, and educational experience, we can promote integration and therefore well-being in our communities and larger culture.

Temporal integration is how we address the existential issues of life that our mentally created and cortically mediated sense of time creates: The longing for certainty in the face of uncertainty; the longing for permanence in the face of the reality of transience; and, the longing for immortality in the face of mortality. Our experience of time may emerge from an awareness of change, and when we build upon our capacity to represent that change conceptually as past, present, and future, we can envision how we humans are faced with a fundamental challenge: How do we find peace and purpose in the face of an awareness of life's transience?

How do these existential issues influence you now, and how did they influence your life in adolescence? How we wrestle with these fundamental issues of being an awake, aware human being on this fragile, fast-paced planet is a part of how you may find challenges to temporal integration surfacing beginning in your teenage years. Being a human is not easy; finding a way to embrace the tension of opposites—longing for certainty but accepting uncertainty; longing for permanence but accepting transience; longing to live forever, but accepting mortality—is the essence of integration. Reflecting on these challenges can help bring integration by learning to embrace the power and importance of holding and honoring these paradoxes of our human existence.

Finally, there is something that can be called transpirational integration, or, *identity integration*, that emerges as we "breathe across" the other eight domains of integration to emerge with a more expanded sense of who we are in our lives and the world. Identity integration is all about the notion that we have a personal interior as well as an interpersonal exterior. This is the within and between. Honoring the distinctions of a self that is both within and between, personal and connected, enables identity integration to unfold.

Can you sense in your own life a private me as well as an interconnected we? If we only have one or the other developed in our lives, we may find chaos or rigidity filling our days. When you reflect on how your identity has unfolded in recent times, as a *me* and *we*, does there feel to be a space in the mind for envisioning a new integrated identity, playing with the linking of a differentiated *me* with a distinct *we* called a "MWe"?

As MWe move along in our discussion, these nine domains of integration may serve as a helpful frame of reference for considering the many ways in which our embodied and relational minds contribute to our well-being across the lifespan. In many ways, too, these practical divisions illuminate how energy and information flow can move in an integrated way with a state of well-being, or in a non-integrated way in states of un-well-being.

Reflecting on that time so many moments ago now, I realize that each of these domains of integration was challenged in my state of grief. With loss, we sometimes become lost on the rim of the wheel of awareness, filled with the images, emotions, and memories of the rim's known, not being able to access the more flexible and spacious hub of knowing. When our right mode becomes filled with autobiographical memories and the externally focused left mode attempts to reach out to the lost one in vain, the inability to reconcile this conflict fills the two hemispheres with a lack of connection and coordination. In vertical integration, the flooding of the feelings of the body, of gut feelings and signals from the intrinsic nervous system of the heart, fills us with those pit-in-the-stomach and aching-heart sensations. The attachment figure gone, our limbic area that mediates our close relationships can also find no rest in the cortical reality of the loss. Each of the other domains, memory and narrative, state and relational, are all no longer able to embrace a differentiated state that then becomes linked. We have intrusions of memory, lapses in the integration of past-present-future in the telling of our stories of the one we're grieving; our internal states that found a grounding in an attachment figure, or what some would call a "self-object" who defines who we are, is no longer well-integrated, and even our relational connections are, literally, severed. For our temporal issues of life and death, impermanence and uncertainty, facing a loss challenges these deepest existential issues of our life. In identity integration, I found my sense of who I had been when Tom was alive had a fullness to it that somehow felt shattered with his illness, impending death, and his dying. Even though we spoke just a few times a year, somehow his dying changed my sense of self. Who I was was now different.

Acute grief is filled with chaos and rigidity.

Finding a way to identify those areas in need of integration and cultivating differentiation and linkage may be fundamental to resolving the grieving process. As the millennium was drawing to a close, I tried to work through the grief, embrace the realities of academic publishing, face what was happening in our communities and clinics, and integrate all these ideas about self-organization and integration as health into something that might help in some way. I was trying, as best I could, to find a way to create a pathway of integration in my life.

That struggle, moving beyond the sense of loss and disconnection, rejection and hopelessness, and trying to find a way to express a sense of some truth about the interconnections of all these things, the scientific and the subjective, the brain and our relationships, was what filled me during those final years of the Decade of the Brain. Meeting with colleagues, working with patients, reflecting on my own life experience, and writing were each an immersion, now being woven together, into a sense of what felt grounded, what felt real. It felt that the inner subjective life we have cannot be divorced from the scientific studies of the mind nor from the care of those experiencing the chaos and rigidity of psychological suffering.

We don't have to try to create integration unless something is blocking it, impairing differentiation or linkage. The key is to let the innate drive toward integration not be impaired by blockages to differentiation or linkage. Integration happens internally and relationally. Sometimes this simply means, "getting out of our own way" in order to let integration unfold, rather than "making it happen." But sometimes a blockage needs to be identified and then intentionally undone, released, removed. Sometimes intentional differentiation needs to be initiated. Then the letting of integration simply means enabling a natural push of self-organizing to, literally, emerge. Integration is the emergent self-organizing way a complex system innately links differentiated parts. In this way, integration—harmony, health, resilience—can be seen as natural drives of our lives. These domains of integration, and the self-organization they depend upon, reveal how mental life and well-being can arise from within and between.

Wild as all this is, as presumptuous as this may sound to you as I imagine it might to me if I were hearing it for the first time, if we consider where we are now in our journey, it may be worth just resting for the moment with this strange proposal that integration is the basis of health. If integration is a fundamental process of health, resilience, and well-being, and even of creativity and interpersonal connection, as we'll soon explore, then integration can manifest itself in a myriad of ways. And so this is not a suggestion of one specific way to live life to be happy. It is not prescribing exactly the steps of how to love, nor how to interact with others. It is, however, offering a foundational stance of self-organization and integration that we can at least put in the front, back, and side of our minds as we move ahead together on this exploration.

Self-organization does this without a conductor, without a program, It is an innate feature of complex systems. If we see the proposal that one aspect of the mind is the self-organizing process of energy and information flow both within and between, then the natural purpose of our mind is integration. We can support one another to liberate this innate drive toward integrative self-organization. Might this be a purpose of our lives, a preliminary, cautious suggestion, a possibility, we can reflect upon with an open mind as we move forward? This is a proposal, not a final conclusion. Beneath all the unique patterns and interactions that unfold, cultivating more integration in the world may prove to be an important reason for simply being here in this life.

When we add to this personally empowering experience a focus specifically on our relationships with the planet itself, with our common home, our mother ship Earth, then we can envision how finding some way to view health as core to our reason to be here can actually give rise to not only a sense of meaning and purpose in our individual lives, but also to enhancements of well-being for the larger world in which we live. Integration begets integration. Integration imbues our life with a warmth and gentleness. Kindness and compassion, for self, others, and for the planet, are integration writ large.

CHAPTER 4

Is the Mind's Subjective Reality Real?

THE JOURNEY YOU AND I ARE ON HAS BROUGHT US TO THE TURN of the century, the new millennium. We've explored two fundamental notions of mental life: 1) embodied and relational self-organization; and 2) integration as the natural drive of self-organization that optimizes energy and information flow, a fundamental basis of health. Before we move further into that new millennium and what emerges during that time, let's take a step back, back to the early 1980s and late 1970s, and see what aspects of this journey into mind arose in those years. First we begin with a celebration as the decade turned.

Adapting to a Medical World that had Lost its Mind (1980-1985)

It's New Year's Day, 1980. I am halfway toward my twenty-third birthday, halfway through my second year of medical education, and halfway about to drop out of school. It's been a numbing year, and I am feeling lost and disconnected from what I thought was a career path that had meaning for me. But now on winter break, a high school friend invites me to a party back in Los Angeles, and a bunch of us in our late-adolescence are feasting on homemade food, conversation, and music. I feel at home in the town where I grew up, where I moved from child to teen and spent my years in college. At the gathering I meet a young woman, Victoria, and she and I explore each other's stories, sharing what we're doing, trying to articulate who we are. We talk about what has meaning in our lives, the values that drive us forward. The discussion and connection spark a feeling of being alive that seemed to have died back in the cold, wintery months in New England that year. Right now I feel present, back inside myself, my mind wide-awake. My body is full of energy, and Victoria and I are talking and walking through the early morning streets of this sleepy city. She teaches ballet as a college student, and I am intrigued with her studies in dance at UCLA. I remember loving to dance in high school, being a member of the ballroom dance team in college, and even doing the choreography for the annual medical student show. All of that felt so alive, so real, so connecting in college. But something was shifting in me in my medical training; it felt like I was losing touch with my self—more than simply frozen in the winter chill, something was dying.

When I went back to Boston after the winter break of 1980, I continued with my academic program that now involved more clinical work during our second year course, "Introduction to Clinical Medicine," or ICM. During my first year in 1978, the clinical experiences I had longed for when I first arrived didn't go so well. One of my first patients was a young woman with a severe lung illness. I sat with her and could sense how sad she felt, could feel the despair that filled her as she faced a life of medical ordeals that lay ahead. When I presented her history to my attending, I described the patient's emotional states and got choked up as I spoke. My clinical advisor, a pediatric surgeon, told me that I was "too emotional" and should focus on the patient's symptoms, not their story. I told her I'd try. The attending looked at me with doubt and disdain. A phone call interrupted the conversation, and I wondered what I should say next. When she was off the call and focused her attending's attention on me again, I started to say that my grandmother had just died and I imagined that I was very upset about that loss and that it was . . . and then the phone rang again, and I sat, thinking, what should I do? After she came back she insisted I needed to be more professional, and off I went.

The next week during that first year clerkship, I came back for my afternoon clinical rotation and saw a patient about my age whose body was turning into a sponge, he said, as his bones were dissolving with a rare orthopedic disease. He was hoping to become a physician, but couldn't imagine how going to medical school could ever be in his future given what was happening to his body. He felt helpless and terrified. I tried not to think about being the medical student he might never become. I took notes, checked on his lab data, put together a summary of his history, read about his disease, and trundled off to my attending's office, again. I presented this patient's case with utter dispassion—a focus on facts, full of clinical detail, methodical. I felt totally dead inside, disconnected from the sensations in my gut, numb to the feelings in my heart, and far, far away from that young man. My advisor smiled and told me I had done an "excellent job." I recall looking at her in disbelief. She had taught me how to fit in, how to disconnect from my mind and the mind of my patients, to lose touch with my humanity. Maybe my becoming like her really meant that I had failed. It felt like a bizarre, unsolvable bind. Fit in and fail; find another way and be failed. I felt horrible and never went back to see her again.

At the end of my first year of medical school, I signed up to work with Tom Whitfield, the pediatrician who was to become my mentor, exhausted by the emotional confusion, and now inspired by the summary of his summer preceptorship in the Berkshires, which invited students to be a part of a team helping impoverished families find medical and social support. I was excited to spend the summer with Tom and see if I could revive my original desire to be in medicine. Tom may have been excited too, as I found out later, so much so that when he and his wife drove out to pick me up at the local bus stop, Tom locked the keys in the car. We worked closely that summer. I observed Tom at work in his office and visiting families living far out along the country roads that in the winter left them isolated and alone. We became more than simply close, as I described in the last entry, developing more a parent-child connection. Tom became a father to me, and I a son to him. One of Tom's most powerful lessons was that old adage, "the way to care for patients was to care about them." Tom was full of heart, could recite poetry he had memorized in his youth at a moment's notice, and had a mischievous glint in his eye and a quick, infectious sense of humor. Tom was always on the go with some project or another either at work or home, baking pies or hauling wood. We could talk for hours, or just be silent together working in his garden. Even when I accidently "pruned" the bougainvillea, cutting it off at its root instead of removing the weeds, he gave out a big sigh and just let me know that "accidents happen, Dan'l—just watch out next time." He was kindness in motion. When I left the Berkshires and was Boston-bound that autumn, I hoped I'd be able to hold on to all Tom had taught me.

Back in Boston in my second year, in the autumn of 1979, we were moving out of the classroom and into the hospital. As that year went on and our introductory clinical ICM course began, I was happy to finally dive into more patient-centered work. I enjoyed the academic classes too, having been a lover of science, biology in particular, appreciating learning more and more about the body. But at that young age—in my early twenties and still in what we know now to be the intense period of adolescence with its then unknown adolescent brain remodeling—I began to feel I hadn't really begun to learn how to live. Most of what I was being taught was how we die in so many ways, inside and out—how we die from medical illness, and how we die by disconnecting ourselves from others, even from our own inner life.

Disease and disconnection were everywhere. One patient I saw in the spring of 1980, at the end of my second year, was a young African American man whose brother had died from sickle cell anemia a few years earlier. We discovered that this patient's depression, his despair and sense of hopelessness, came in part from his belief that his sickle cell trait would lead to his own demise. He and I spent over two hours exploring the meaning of his illness as I listened to the content of his stories, the meaning emerging from his facial expressions, tone of voice, and gestures. Words are important, but they tell only part of a much bigger story, the story of who a person is. There was as much meaning in the context of his life revealed in his non-verbal messages as in the text of his words. I tried to help him see that there was much hope given new medical discoveries, and also recognize the important but previously not explained difference between his sickle cell trait status—which did not have death as part of its course— and his brother's full manifestation of the disease that ended his life. Apparently no one had taken the time to be with him, sense what his fear was about, and blend medical knowledge with human communication to bring clarity to his confusion. He said he felt much better after our time together; I felt that I had made the right decision to go to medical school if this is what being a physician could mean.

Now at the end of my second year in 1980, with all the hope and light of the blooming springtime of New England, it was my time to meet up with my attending physician and the other second year students in our group. The others presented their patients' histories, symptoms, physical findings, reviews of the laboratory data, and overviews of the diseases, just as we had been taught since our first year. Each in turn focused on the diseased organ; just as most seemed to be doing in the locker room as our socialization into modern medicine continued. Some of my peers would say things like, "I saw a great liver today" or, "What an amazing kidney." And they weren't kidding. Their perceptions seemed to have become organized around diseases and organs, not about people and their lives. The mind was disappearing fast, replaced by a focus on the physical structure and function of the body. While the objective, observable, physical body was certainly real, the reality of the subjective mind, not palpable like thyroids and livers, not able to be scanned like hearts or brains, was slipping from sight and becoming invisible to the measuring medical eye. Even the bursting sunlight of the new season could not brighten the dimness of those darkened days.

Now I can see how the medical socialization process of dehumanizing the patient took hold. It was a lot simpler, and emotionally safer, to measure bodily fluids than be in touch with the pain and suffering held by mental experience—in the patients, and perhaps even the physicians and medical students themselves. Dehumanizing means removing the mind, in this case from the focus of medicine. A human mind can see the physical world and lose sight of the mind. What is really lost? The heart of being human, for both patient and physician, vanish in the glare of the physical world—and we were only in our second year.

When I presented the case of my patient with sickle cell trait as the last presentation of my group that afternoon, I spoke about his fears, feelings of despair, relationship with his brother and the rest of his family, where he was in his development, and the meaning of this illness in his life. The teacher asked me to stay, and I wondered if he would inquire about how I knew to be so focused on getting to the meaning of the patient's illness, how to connect with him. Here was how medicine could be, even if it wasn't being taught this way. Finally, I wishfully thought, bringing meaning and mind into medicine would be acknowledged and respected.

"Daniel," the attending began, "do you want to be a psychiatrist?" No, I said, telling him that I was just a second year student and didn't know what I wanted to specialize in yet. The only thing I really ever thought of doing was pediatrics. "Daniel," he said, cocking his head in the other direction, "is your father a psychiatrist?" No, I said, thinking about my intellectually inclined mechanical engineer father. He certainly was not a psychiatrist, nor was anyone else I knew in that field. In fact, I didn't know any physicians except my old pediatrician. "Well, this asking patients about their feelings, about their relationships, about their life stories, this is not what doctors do. If you want to do that, go become a social worker."

That afternoon I looked up the nearby social work program, and also researched training in psychology, and wondered if transferring out of medical school might not be a good idea. As the final months of our second year went on, even in the face of frigid winter blossoming into the warmth of spring, I felt colder and colder. I stopped dancing across the river in Cambridge, where each Wednesday local students and community members would gather at an event called Dance Free to disappear into the music, moving spontaneously, alone, in pairs, in groups, to a wondrous range of heart pounding, body swirling, feet stomping sounds from around the globe. (I'd learn years later, that Morrie—of Mitch Albom's *Tuesdays with Morrie* fame—had danced there, a photo of him dancing in our hall placed in the front of that book. That my own memoir of losing Tom was called *Tuesday-Sunday* seemed later an odd and wondrous coincidence of people writing about their mentors facing death). Enjoyment vanished from my walks around the Fens parkway. I stopped feeling the water on my body as I took a shower. I was going numb.

The body is a well of wisdom filled with truth. But my well was dry. My dance card was empty, meaning depleted, and a sense of hope for the future zeroed out. But my logical mind thought, "All is well." My grades were fine, and I was moving on with the medical curriculum. But inside I felt empty. My reflective mind was conflicted and confused. Which part of me should I listen to, my logical mind or emotions and bodily sensations, the intuitive well of wisdom?

At the end of that second semester of my second year, at the end of the spring of 1980, I had my "final exam" for the ICM course. An elderly, white-haired gentleman was my patient: "Good morning, Doc," he said as he plodded into the room and gingerly sat down in the chair. When I asked how he was doing this fine spring day, he said that he had tried to kill himself earlier that morning. All my preparation on the suicide prevention hotline in college, that training I mentioned in our second entry, came into gear. We had learned in our small attic space overlooking the main walk on the upbeat campus of USC that the key to keeping hope alive for someone in crisis was to tune in to their internal experience, focus attention on their feelings, thoughts, the story of their lives. As we've seen, we can now call this SIFTing the mind—focusing on the Sensations, Images, Feelings, and Thoughts that are the pillars of our internal experience in any given moment, the core of who we are, what gives us meaning, our internal subjective reality.

So I SIFTed this man's mind, exploring with him what had brought him to such a place of crisis that he tried to end his life. Soon a poke on my shoulder from the attending drew my attention to his head looming by my ear—"Just do the physical exam!" he demanded in an irritated whisper, akin to a whimper. So I did do the physical exam, and on the examining table I witnessed, for the first time, someone having a grand mal seizure. The attending seemed unfazed, simply keeping the man from falling to the floor until the seizure subsided, and then he said to me, "Just finish the exam." When I was done I asked if we could at least take him—the patient (but perhaps it would have been just as needed and appropriate for the attending)—to the psychiatry clinic, and the attending agreed.

"These drunks just do that—have seizures, try to kill themselves. You spent too much time talking to him about his life, about his feelings. But you did a fine job on the physical, so you passed your exam."

As I would later explore in *Mindsight*, these experiences were like nails in the coffin of my young mind seeking some kind of meaning in a world of medicine that seemed to be mind-blind and make no sense. I felt horrible. I felt confused. My body was numb. Was I supposed to become like these attending physicians, attending to what? Wasn't there a place for focusing attention on the inner world of our lives in healing? Did medicine need to be mindless? I was disillusioned and felt utterly lost.

My logical mind continued to tell me that I was in medical school, I was at Harvard University, and that I could figure out how to simply learn from these revered professors, learn to fit in, become a part of what logically must be the right way of becoming a physician. But logic could make no sense of the numbness I experienced, and logical reasoning could not explain the physical sensation of disconnection that kept arising in my awareness, the images of wanting to hop on a train and escape, the emotional experience of despair, or the thoughts that all this was just plain wrong. If I could SIFT my mind with any clarity, I'd realize something deeply disturbing and disorienting was arising.

After some intense and painful reflection, taking in these sensations, images, feelings, and thoughts, and with some gentle urging from an older student to whom I feel eternally grateful, I decided to drop out of school. It wasn't even a thought, really, it was just some kind of deep knowing that arose without confusion.

On the surface, I was so confused and lost; with reflection, I realized that I had to stop the madness. When I met with the Dean of Students, she urged me to take a leave of absence rather than drop out. I told her that I had no intention of ever coming back to a place that made you become so inhuman. She said, "How do you know what you'll want to do in a year?" I paused, looked her in the eyes, and said that I had no intention of returning. "But," she gently repeated, "how do you *know* what you'll want to be doing later on?" She was right. I was lost and had no idea what I would do now, let alone in a year. So I agreed with her, that in fact I didn't know, and that I actually seemed to have no idea about anything anymore. That much I was sure of.

"Okay" she continued, "So now, you need to write an essay on what you'll do for your research." Research? I asked her. "Yes, this is a research institution, and you can only get a leave of absence if you do research." I paused for a moment, baffled. Better to just drop out, I thought. I looked into her warm, supportive eyes, asked her for a blank piece of paper and a pen, and wrote my one sentence research essay. This is what I recall writing: "I am going to take a year's research leave of absence to figure out who I am."

She looked at the note, smiled, and said, "Perfect."

During my time away, I tried out many things. I started classes in ballet, and in modern and jazz dance. I looked into choreography training. I took off for Canada, hopping on a train and heading out across the autumn landscape into the Rockies, and then further west to Vancouver Island. For the first time in my life, I gave myself time to not be organized by others' plans, let meaning emerge from within, not always governed by external expectations from the world around me. As I mentioned earlier, I had studied fish biochemistry in college, and was fascinated with how salmon could go from fresh to saltwater and not die. How did they do that? In my college lab, we discovered an enzyme that might explain the salmon's strategy for survival. It seemed there might be some connection between the enzymes that allowed salmon to live and adapt and the empathic emotional communication we had been taught on the suicide prevention service that could make the difference between life and death for someone in crisis. Something fundamental seemed to be shared by enzymes and empathy. I wondered about how our physiological and mental worlds might coexist, how they might emerge from the same essence of something. But I had no idea how to articulate these questions with any clarity. On this journey, I took off to find the salmon in the Pacific, but was really looking to find myself.

One of the many personal discoveries that emerged during that time began as one of those life coincidences we cannot predict nor plan. I returned to Los Angeles, and my new friend, Victoria, had a picnic in her backyard where I met her next-door neighbor who was beginning to teach a course using a book just released by Betty

Edwards called *Drawing on the Right Side of the Brain*. It was based on interviews she had conducted with soon-to-be-awarded Nobel Prize winning psychologist Roger Sperry on his work on "split-brain patients," research revealing the distinct processes on the right and the left sides of our brains. While technology allows us now to see more overlap than ever before, the research as a whole remains quite robust on the distinctions between the two hemispheres (McGilchrist, 2009). Whatever the controversies are now, my own experiences then were quite clear: With Edwards' exercises, I could immerse myself in a new way of seeing the world. Instead of a dominance of analyzing and categorizing things, naming and grouping them, I began to see textures and contrasts in the world that before were not a part of my perceptual experience. Instead of breaking the world down into small parts, a view of the whole came to mind with a new sense of clarity and vitality. I seemed to perceive the world with new eyes. Time also seemed different while doing those exercises; two hours could pass, immersed in seeing and drawing, and it would feel like no time had transpired at all. Perceiving in this way did more than melt the sense of time; it gave me a revived experience of feeling deeply connected to the world around me.

I became involved in the art performances that Victoria and her colleagues were filming at UCLA, as I took on the role of the guy in charge of the microphones. Holding the sound boom, I began to hear things with more richness. It's hard to describe even now, but there was, in my own direct, personal experience, something quite profound shifting in me. I felt more alive, connected, more a member fully belonging to a world with new depth and detail. I made new friends in many fields, from dance to poetry, and life became full.

With these new ways of experiencing life, I somehow felt an emerging clarity, enough to consider what the next phase of my education might be. Where was I to focus my energy, my time, my life? Loving dance and being a part of filming dance and other performances, I came to understand that I was enthralled with the inner experience and not so interested in the appearance of dance. That realization made it clear that I would likely starve to death if I chose dance as a profession, either as performer or choreographer. In this time I also helped my grandmother care for my grandfather in Los Angeles during his final months. I began to feel restless and ready to start a new chapter of my life.

It seemed that everything I had experienced, from my college journeys with the suicide prevention service to my struggles with medical school, were all about the nature of our internal worlds, our minds. During that year away from school, I made up a word, *mindsight*, for how we see the mind, for how we perceive and respect the mind of ourselves and of others. I needed some kind of continuing clarity, strong idea, linguistic symbol that would serve as a piece of information I could hold on to, something that could protect me on the journey ahead, wherever that would take me. I decided to go back to medical school, and the notion of mindsight might help me survive when I once again would become immersed in the socialization of medical training.

Mindsight embraces three capabilities: the capacity to cultivate insight, empathy, and integration. We all may have these potentials, but we may develop more or less of these abilities. Insight is being aware of one's inner mental life. Empathy is sensing the inner life of another person. And as we've seen, integration means linking differentiated elements into a coherent whole. For the mind, integration means kindness and compassion. We honor one another's vulnerabilities and offer to help others relieve their suffering. How do these three elements of mindsight work together? With insight, we are kind and compassionate to ourselves. With empathy, we see the mind of others with respect and care. That is how mindsight contains all three elements of insight, empathy and integration. The notion of mindsight was a deep and sustaining idea that gave me the courage to return to school and face the medical socialization system, now with a new sense of strength and commitment.

I needed to know who we were, how we came to be this way, and that the mind is real. Through all of these journeys, particularly this time spent without externally shaped structure, without others' plans and expectations, this time that enabled me to get in touch with the presence of my own freely emerging mind, it had become crystal clear that the mind's subjective reality was indeed real. We gain a view of our own mental life with insight; a view into the inner life of others with empathy; and we could respectfully connect to each other with compassion and kindness through integration. Even if the external world of teachers' views or cultural values in modern medicine were acting as if the inner world was not real, valid, or existent, the notion of mindsight would hopefully remind me that we can see the mind with a different perceptual lens than how eyesight sees the physical world. We have different senses, different sights, and mindsight was an idea and capacity that helped us see the mind itself—in others and ourselves. Mindsight might help me preserve my sanity in the face of the socialization ahead that acted as if the mind did not exist. Even if medicine had lost its mind, mindsight might help me protect the reality of mind in the years of training ahead.

Mindsight in Health and Healing

Studies published decades later would reveal that a physician attending to the internal subjective experience of a patient coming in for a brief visit, even for a common cold, could actually mediate a healing interpersonal interaction in which the immune response of that patient was more robust—the duration of that cold would be a day shorter. The simple act of being empathic—of what we can call showing mindsight for the inner subjective experience of another—directly shapes our physiology (Rakel, et al., 2011).

Other studies would reveal that teaching physicians about the mind and how to balance their emotions with mindful awareness training would help them maintain their empathy and reduce risk of burnout (Krasner, et al., 2009). Studies would also reveal that empathy could be taught to medical students and help them prepare their work as healers (Shapiro, Astin, Bishop, & Cordova, 2005).

Yet when I was in school, I heard repeated statements that empathy—tuning in to the feelings and thoughts, memories and meanings of another person—was not appropriate for clinical work in medicine. What I was told repeatedly as a student is not only wrong and misguided for physician training, but is a form of sub-optimal treatment

for patients. Now we know better scientifically, but the socialization system of modern medicine is slow to catch up to these "new" discoveries confirming old wisdom about good clinical practice: The way to care for a patient is to care about them. We experience and express our care through mindsight.

One implication of these findings is to lend support to the assertion that subjective reality is not only real, it is really important.

But why? Why would the focus of attention on the reality of one another's subjective, unmeasureable, not directly externally observable inner reality be so vital for well-being? What we need to explore is why and how the focus of one person upon the internal subjective experience of another is so fundamental to health and health-promoting relationships.

Why would a focus on subjective, inner felt reality matter so much?

One simple answer to this basic question is this: When we sense the inner life of another, we can truly differentiate one person from another and then, in seeking to know that subjective experience, we link together. Here is the proposal: Focusing on subjectivity is the portal for interpersonal integration. Two separate entities, two individuals, become linked as one connected system when subjective experience is attended to, respected, and shared.

Connecting by attuning to subjectivity creates integration.

Integration is the way we optimize self-organization.

And so we feel better, think more clearly, and our bodies function better when subjectivity is honored in the focus of our shared attention.

In math terms, when two people share their internal subjective states, the level of complexity is raised in comparison to two people coldly talking with each other, or one person ignoring the internal state of another. Maximizing complexity is the mathematical way of describing the drive of self-organization in a complex system. In perhaps more accessible terms, this is how attuning to another person establishes the interpersonal integration that generates a state of harmony. That's why insight, empathy, kindness, and compassion are so powerful; they arise with integration and are part of optimal self-organization.

Subjective experience is real and is the gateway for interpersonal connection, for integrating with another person. This is why, I'll suggest to you, empathy within a relationship promotes improved immune function and a deep sense of well-being. When your subjective experience is seen and respected, and you receive communication of that attunement, then you and the other person become connected as two differentiated individuals becoming linked. This interpersonal integration raises each individual's state of integration, which feels good and is good for you. You achieve, in complexity science terms, a higher state of integration than either person alone could achieve. This is the notion of the whole being greater than the sum of its parts. That's why subjectivity is so profoundly important. Attending to one another's subjective lives increases integration and therefore increases harmony and health.

As we move ahead in our journey, we will keep a mindsight lens on the mechanisms of mind at work in our attuned relationships. As we'll explore soon, when mindsight permits the internal state of one person to be aligned with the internal state of another, that joining may have a profound impact on the system of the two individuals, now linked. Empathic attunement enables two differentiated individuals to become linked. These empathic connections are a form of integration.

To understand the positive effect of such joining, it is scientifically helpful to consider the mind, as we've been doing on our journey, as a part of a broad system, one that extends not only beyond the boundaries of the skull, but also the skin. Connecting minds transform bodies—we grow and heal in connection with one another. But clearly interpersonal joining is not only when we hold hands physically. This happens when we connect in aligning our internal subjective experience—something you cannot see with the eyes so much as sense with mindsight. Mindsight is the mechanism beneath social and emotional intelligence. Knowing another's mind is the basis for inner and interpersonal well-being. We join by communication that is attuned and empathic.

But what is it that is actually the substance being linked in this joining? If an enzyme acts on the structure of a molecule to change its shape and function, what does emotional communication actually connect?

One answer to these questions is energy and information. If we consider how their flow arises within us and between us, we can look toward integration to understand well-being that arises from the linkage of differentiated elements within us, and between us. And when we look toward subjective experience, we can see that this may, too, be an emergent property of energy and information flow as we've described earlier.

From a philosophical and scientific perspective, we are seeing this as one system, part of a "monistic" view, rather than one that divides mind from body, called "dualism." Roger Sperry described this back in 1980 in an article called "Mentalism, Yes; Dualism, No.":

> By our current mind-brain theory, monism has to include subjective mental properties as causal realities. This is not the case with physicalism or materialism which are the understood antitheses of mentalism, and have traditionally excluded mental phenomena as causal constructs. In calling myself a 'mentalist', I hold subjective mental phenomena to be primary, causally potent realities as they are experienced subjectively, different from, more than, and not reducible to their physicochemical elements. At the same time, I define this position and the mind-brain theory on which it is based as monistic and see it as a major deterrent to dualism." (p. 196)

Sperry went on to explore how this view of the importance of subjectivity builds well on a solid biological perspective, one that the field of medicine might be able to fully embrace:

> Once generated from neural events, the higher order mental patterns and programs have their own subjective qualities and progress, operate and interact by their own causal laws and principles which are different from and cannot be reduced to those of neurophysiology... The mental entities transcend the physiological just as the physiological transcends the [cellular], the molecular, the atomic and subatomic, etc. (p. 201)

We can link this notion that subjective mental life arises from and can influence neural behavior to our view that subjectivity is primary in both our internal life as well as our interpersonal world. When we then apply the proposal that both subjective experience and self-organization are each emergent properties of energy flow, then we can connect subjectivity—a primary experience of mind—with integration—the primary self-organizational flow of mind. Let's turn toward an existing field of science, a branch of psychology, that may offer some relevant empirical findings.

When we examine the field of positive psychology through the lens of integration, we come to the following notion: positive emotions, like joy, love, awe, and happiness, can be seen as increases in the level of integration. This is why they feel good. Negative emotions, like anger, sadness, fear, disgust, and shame, can be viewed as decreases in integration. They feel bad. When such negative emotions are prolonged and intense, we become prone to states of rigidity or chaos as integration is lowered over longer periods of time.

This view builds on the fundamental stance that emerged in the 1990s about emotion itself that I described in *The Developing Mind*. Emotion can be seen as a shift in integration. When integrative levels change, we feel emotional. If integration is increasing, we feel positive, we feel good. Positive emotions are constructive because they build our state of integration. If integration is decreasing, we feel negative, we feel bad. These negative emotions can be destructive. Often arising with threat, we shut down our connections to others, and to ourselves.

I wonder nowadays about the medical socialization process that can cause those young students, and even their attendings, to not feel the pain of others and perhaps even their own helplessness. Without the proper training and support, we can understand how there would be a survival adaptation of simply shutting down to protect them from experiencing, within awareness, their inner sense of despair. While ultimately not good for anyone, that shutting down is an understandable but desperate and often non-conscious attempt to survive and avoid feeling overwhelming negative feelings. Fortunately, with developing mindsight skills such as compassion and empathy, with education in social communication, mindful awareness and self-understanding, young clinicians can be prepared to cultivate the kind and compassionate connections that benefit everyone, patient and doctor alike.

What we'll explore as we move ahead is how communication, the sharing of energy and information, enables two differentiated beings to become linked as an integrated whole. This may be how and why tuning in to the subjective experience of another person, using mindsight to sense the inner life of another, enables health to unfold. If integration is the mechanism of well-being, then honoring one another's subjective experience creates interpersonal integration and cultivates health. Mindsight facilitates integration. Mindsight in medicine, and perhaps our everyday lives, may be seen as an essential tool to facilitate health and healing.

Reflections and Invitations: The Centrality of Subjectivity

Remember the salmon that can switch from fresh to salt water? In your own life, how does the world in which you live surround, shape, perhaps even create you, and send you in a direction, down a particular path of the infinite possibilities that lay before you? This immersion in what is sometimes called "a social field" (Scharmer, 2009) can influence how your mind functions even without your awareness. Some would call this part of the context in which the mind emerges. Others might say it is the surround that creates who you are. We can say that it is the sea outside that shapes the sea inside. Mindsight lets us see this mental sea, to see the sea inside and the sea that surrounds, the seas that shape who we are.

Awakening to the reality that we are deeply social creatures can be a shock if we've never realized the impact our external environment has been having on our internal experiences from our earliest days. The surrounding mental sea shapes the mental sea inside. This is how the social field can be seen to not just mold our mental life; it is a fundamental source of what the mind is.

Awakening to this reality of our being created by both inner and outer factors can sometimes feel startling if we've never been aware of such sources of our mental life. We may often naturally feel that we own our minds, or at least want to own them, that we are the ones in control, the captains of our own ships. If we place the mind only within the body's brain, or even if we extend the mind to our bodily-defined self, we can capture that longing for control and ownership. In this view, the mind emanates from somewhere in my body, from my brain, from my skin-encased body I call "me." But it may just be the fuller reality, a frustrating or frightening fact for some, at least at first, that our mind, and its sense of self, actually emerges from not only our inner life, but also from our inter-life as well.

If the external factors of our mental lives, the flow of energy and information between our bodily selves and larger social and physical world surrounding us, create conditions that shut down integration,

Photo by Kenji Suzaki

then finding a path that liberates us from such external constrictions may be essential to preserve our sense of self, even our sanity. For me, these were the conditions I experienced in the world of medical training. Without perspective, I simply tried to fit in within that world as best I could. But with some distance, with the mental space and linguistic symbol of the term *mindsight* to bolster my experience that the mind was real, I could re-enter that world and hold on to inner values that were not shared by that larger social field.

What is it that can help disentangle ourselves from the between factors that create us, and now may be suffocating us? The ability to adapt well to a troubled world is not a sign of mental well-being, a poster on a wall in Romania where I was teaching recently declared with a sentiment similar to one often attributed to Jiddu Krishnamurti, "It is no measure of health to be well adjusted to a profoundly sick society." Yes, I thought, how true. So how do we know to pull away from these non-healthful external forces when they are shaping who we are at that moment?

When chaos or rigidity begin to dominate our experience, from the framework we've now constructed we can see that this is a sign that integration is blocked. With the notion that integration is health, and that each living being has a right to well-being, then integration is a true north to guide us even when our external compass governed by our social world is pointing in a different direction. Have you ever been engaged in a world of social interactions and group actions that began to feel unhealthful? Was chaos or rigidity a part of what you may now see as a non-integrated surrounding sea? How did you respond to such lack of harmony in your professional or personal life? If efforts to change the system are not effective, and if change perhaps is not possible, having an internal compass is helpful amidst the chaotic storms or rigid deserts. If change is not possible, sometimes we need to leave that world and then return with a new sense of clarity to bring that world toward a more integrated way of being.

Have you ever faced a conflict between your inner sensibility and outer reality? How did you, or how do you, wrestle with such a conflict? How do you place value on your own internal compass, some interior guide that helps you sort through what has meaning and what is meaningless?

As complex systems, we have internal and external constraints that serve as the many factors that shape our emergence. We have proposed that the mind, at least in part, is an emergent property of energy and information flow. Perhaps awareness and its subjective textures are emergent aspects of this flow. Certainly the information processing that arises from energy flow in symbolic form would be a part of that flow, by its very definition. When we build on the proposal that the mind is also the self-organizing aspect of this flow within and between, embodied and relationally embedded, then we can see how optimal self-organization would arise: We link differentiated parts to create the FACES flexibility and harmony of integration.

Mindsight enables us to reflect on those inner and outer features of life that allow us to differentiate and link. Mindsight is the insight, empathy, and integration that helps us make choices on how to move forward in life, even in the face of internal or external factors of the seas inside and surrounding us, so that we intentionally create higher states of integration and well-being. With such mindsight reflections, we can then alter our course with intention, rather than simply being the passive recipient of how the world is shaping us.

In medical school, making time to step away from what others were telling me to do was not something I was able to do for any extended period of time, at least not until I faced the painful experience that things just weren't going right, they didn't feel right, and I went numb. I didn't know then what we see now, that this was a state of rigidity from the lack of respect for the subjective world—of the patients, students, and my inner experience. Medicine had lost its mind, and I had lost my way.

There may be no logical thoughts revealing how the outside world is shaping us, just an inner sense of disconnection or discontent that reveals itself in dreams, images, and desires to be free from all of which we simply try to ignore and suppress, until they go underground without effort and leave us alone. But their disappearance from our conscious mind does not mean these inner stirrings are gone; they've just halted, temporarily perhaps, disturbing our waking hours, entering our waking awareness. In the non-conscious mind they remain, however, waiting for a moment to more directly reveal themselves. Information processing does not require consciousness to impact our lives. The mind is more than what consciousness makes available to us.

Which raises an important issue: While subjective experience is a part of what mind is, this being the felt texture of lived life that we can experience as part of being aware, the mind is also out of awareness—and therefore has no felt texture. Or does it? If we assume that the definition of subjectivity, of subjective experience, is the felt texture of lived life or something like this, don't we need awareness to have a felt texture? Can we have subjectivity outside of consciousness? If the answer is no, this naturally means that information processing outside of awareness, which we know occurs from a wide array of empirical studies, is not the same as subjective experience. This means that an experience like meaning happens even if we are not aware of its happening.

For some, just as it is when realizing the mind arises, in part, from a social field outside one's control, the knowledge of the powerful impact of things outside our conscious experience is terrifying, as it signifies being out of control. And so whether the influences are non-conscious from within or social from without, for someone interested in being in charge, accepting the non-conscious mind and social mind can be threatening. These inner and outer mental realities mean we are not the boss of ourselves, not fully in charge. The mind can have a mind of its own. Sometimes the first time we realize this is during the awakening of our teen years.

In adolescence we often have a chance to begin to reflect on how life is going. We start to see that the world we are handed is not necessarily the one in which we imagine we'd like to be living. Yet for many of us, when we move from the novelty-seeking and creative exploration of the adolescent period between childhood and adulthood into the responsibility of our later years, we leave this restlessness behind, attributing it to an immaturity or adolescent rebellion that has no place in our lives. In the book *Brainstorm*, I offer the adolescent, or any adult who once was an adolescent (which is most of us), a way to explore these crucial challenges and opportunities of this important period of life. What has been so life-affirming in the experiences of teaching those ideas to adolescents is how open they often are to considering these deep issues of what and who we are.

It may be helpful as an adult to realize that this emerging emotional spark, this passion for life, does not have to be squelched by life's responsibilities as we find our niche in the world, as we fit in to what others expect. The ESSENCE of adolescence includes an *e*motional *s*park, *s*ocial *e*ngagement, *n*ovelty-seeking, and *c*reative *e*xploration. This essence of our adolescent period of brain growth and remodeling turns out to also be the essence of how we, as adults, can keep our brains growing well throughout our lives. That remodeling lasts well into the mid- to late-twenties.

Finding space in your life, in the minutes and hours of solitude you can create in a day, an extended time on a weekend, or taking a break from the routine for more lengthy periods, can offer opportunities to reflect on where you are now and how your mind may yearn for a new kind of clarity, a new kind of life. Though I did this as an adolescent in medical school in my early twenties, taking such time to reconsider your life's path can be important at any age. There is a powerful place in our lives to reflect inwardly, to take "time-in" as we take "time-off" from the daily grind of the expectations of others—and those we have given ourselves—and find a new way of being with our life's path.

As we've seen in this brief set of stories from medical school, sometimes the waking up is not based on logical reasoning. Sometimes the call for clarity amidst the confusion emerges from our body's wisdom, the sensations in our heart or gut, the images and feelings that arise, and the thoughts that may seem irrational and are simply ignored. But tuning in to these non-rational signals and exploring what they may mean can be the most rational and important move you've ever made.

This is not about becoming self-absorbed, but about opening to a journey of personal inquiry and self-discovery.

Perhaps such questioning is a fundamental part of the journey you had in adolescence. That period of time can be filled with the tension between wanting to belong yet find your own way. How can you fit in and have membership while somehow being an authentic individual? You want to blend in and stand out. As we find our way into the larger world, beyond family and friends, we are offered the opportunity to clarify who we are beyond others' expectations. To know ourselves, we turn, in awareness, to what we feel, to our subjective inner life. Without access to that inner world, to that sea inside, we would not be able to develop an internal compass with which to find our own way.

Honoring our subjective experience is vital to living an integrated life.

These subjective sensations are real, even if they can't be detected by someone outside ourselves. In this sense, subjective experience is not objectively observable by someone else; this is why we use the word *subjective* to describe these experiences, as they are only truly knowable by the subject. The classic view is this: Even if you see red and I see red, we can never know if the way you experience that color and the way I do are the same—hence, perception (along with the rest of our mental activities) is ultimately subjective.

We can call this whole entry an invitation to awaken to the choice you have to create a mental space in your life that gives respect to subjectivity, cultivates freedom in your growth and expression, and gives respect to the inner and inter worlds that create who you are. We can awaken to the central reality of our subjective mental life.

As the 13th century mystical poet Rumi urged us in his poem, "The Breeze at Dawn": You are awake now, don't go back to sleep.

> *The Breeze at dawn has secrets to tell you / Don't go back to sleep! / You must ask for what you really want. / Don't go back to sleep! / People are going back and forth / across the doorsill where the two worlds touch / The door is round and open / Don't go back to sleep!* (Barks, 1995, p. 36)

Perhaps those two worlds are the objective world perceived with physical sight, and the subjective world perceived with mindsight.

The only way we get to develop this awakening is by paying attention to our subjective reality. Yet a life or a world, like medical school was for me, which focuses only on external, physically observable reality, can only be structured by logic and others' expectations that leave the mind aside. We can use only physical sight to see the objects in front of our eyes. The socialization process of some cultural worlds pulls us to use physical sight to fit in, to learn the external rules that govern visible behaviors. Focusing on this observable external world is quite distinct from focusing on the inner and inter world of the mind's subjective reality.

We can move beyond the important but not sufficient or complete physical sight that sometimes dominates our lives. We can also develop the mindsight habit of focusing on the subjective reality of our lives by at least starting with the process of SIFTing the mind. Invite yourself to become aware of the richness of the *sensations* from within your body. Have these sensations served to help you gain deeper insight into what was happening in your life? When *images* arise, can you hear an inner voice and sense the meanings of its message, see visualizations with vivid detail in your mind's eye? Images can be in many forms, and tuning in to these sometimes non-worded worlds can offer an important window into your own mind. As emotions arise, can you *feel* their range rise and fall within the affective landscape of your inner mental life? If you consider emotion as a shift in integration as we mentioned earlier, can you sense when that shift involves decreases that bring you toward chaos or rigidity, or increases that bring a sense of connection and harmony? And as *thoughts* emerge, how do these words or un-worded meanings fill your awareness? Thoughts are but one part of a rich and complex mind that may help us know ideas and liberate us from the immediacy of now so we can usefully reflect on the past and plan for the future.

In all these ways, we SIFT the mind as a starting place to explore the subjective world of ourselves and of others. When we approach this subjective sense with kindness and compassion, we are bringing integration to others, and to ourselves. This is how and why connection—honoring and attuning to one another's vulnerable subjective reality—feels so good: It creates a more integrated and therefore more harmonious, vibrant, and healthy way to live. This is how we use mindsight to support our self-organizational movement toward integration and well-being.

What are the sensations like now in your body? What images arise as we come to this place in our journey together? How do you feel? What do you think?

Is subjective reality—the felt texture of lived life—real? And does honoring it in oneself and others really matter?

CHAPTER 5

Who Are We?

WE'VE BEEN MOVING ALONG ON OUR JOURNEY TOGETHER, HAVING explored the notion of the mind as a self-organizing process, one that has a natural drive toward integration—the linkage of differentiated parts of a system that promotes optimal functioning. We've seen how one way of enhancing integration is to focus upon and honor the subjective inner reality of self and other. Self-organization, integration, and subjectivity are fundamental to mind. In this entry, we'll continue addressing the notions of the what's and how's of the mind by focusing on who we are. The answer may not be as simple as we think.

Exploring the Layers of Experience Beneath Identity (1975-1980)

From my early years of college in my mid-adolescence, I, like many teens, felt an inner drive to explore the world and try out different ways of experiencing reality. During the day I was a biology student, journeying deep within the molecular mechanisms of life, working in a lab searching for that enzyme I spoke about in the last entry, a molecule that might help explain how salmon transitioned from fresh to salt water. I was fascinated with life, intrigued by this miracle that we live and breathe, interact with others and reproduce. Many nights, I worked on the suicide prevention phone service, focusing on how emotional communication that created a connection between the caller and listener could save people in psychological crisis from ending their lives. One summer while working in that enzyme lab, I also chose to learn about Taoism and the physical expression of its philosophical tradition in the practice of Tai' Chi Chu'an. Being in the flow of the series of movements and balancing the positions of palms up and down, facing right and left, open and closed, seemed a powerful way to create a sense of harmony. That embodied way of becoming awake, of being in the moment, felt akin to the essence of well-being. Later in college, the flow of our ballroom dance team, gliding along the auditorium floor or on the turf of the coliseum at half-time during our school's football games, felt like some kind of relational connecting that was something real, something at the heart of our lives, something filled with a sense of well-being.

It was an enthralling time.

Enzymes, emotions, movement, meaning, and connection—something about all this seemed to hold a common essence, something about life, love, and wholeness I couldn't put into words.

I was fascinated with these layers of reality. As I'd move from lab to crisis center, Tai'Chi studio to the dance floor, it was as if I was shifting my identity, but all the while sharing some common core of not only who I was, but also what life might be about. I had no idea what to make of it, but took note of this strange amalgam.

As college drew to a close, I worked in Mexico for the World Health Organization on a project to study *curanderos*—folk healers—in a region where the press of modernization of a local dam, La Presa Miguel Aleman south of Mexico City, was changing their communities and local medical services. One morning, on a horseback journey to interview a local healer as part of my project, the saddle on my horse loosened and, with feet still strapped into the stirrups, I was dragged, they tell me, a hundred yards over gravel and rock, my head banging against the ground beneath the horse's racing hooves. When the young and frightened horse finally came to a stop, my riding companions thought I must be dead, they said later, or at least that I'd broken my neck. I

did break my teeth and nose, and damaged my arm. The head trauma induced a state of global amnesia that lasted about a day.

What that meant was that I was wide-awake, but had no idea who I was.

During that 24-hour period, the sensations of everyday things had a very different quality. Drinking a glass of water, for example, was a wild experience that began with observing shimmering light bouncing from this shiny vessel, cool to the touch, liquid slipping into the space in the head, over those top-jagged, smooth-sided protrusions we call teeth with a cool, flowing feeling, diving beyond a moveable fullness of flavoring into the center of the body, cold, chilling, expanding. A mango was not a mango, its color not named as mango's yellow. Its round shape was intriguing, the texture of the surface smooth and shimmery, but with some roughness at the top, the tones of light varying across its layers, the scent intoxicating as it wafted into this head, the feeling of the moist inner substance of it woven together with the bursting flavors within.

This was my day without an identity. I wasn't frightened to not know who I was; I was filled, moment-by-moment, with sensory immersion that felt, well, somehow complete. There was nothing missing, nowhere to go, and nothing else to do but let the experience flow. There was nothing between this body and this experience; there was only being there.

When the identity of being "Dan" gradually came back and I could recall my history, it didn't have the same weight or sense of seriousness it held before. It was as if the external structure of a social convention of "Dan" stopped having the same meaning. I had 24 hours of not being "Dan," yet being fully awake, alert, and attending to stimuli without a clutter of memory about my personal identity nor a blanket of past experiences filtering the world through the lens of prior experience. I felt a different sense of what a sense of self, of what the mind, was all about. If knocking your head around could bang out of your brain a sense of *I*, yet you are fully aware and wide awake, what does that *I* really mean?

You can be awake—perhaps even more fully awake—if your personal identity with all its baggage of history, learning, judgments, and filtering of perception is suspended, or at least not taken so concretely as a blueprint of how to be.

It seemed to me that the day of no-identity was an immersion in "now-here." With the return of "Dan" into the picture, I moved the hyphen over and had a distancing experience at times that felt like "no-where." The water felt more distant, the mango became a mango. Instead of experiencing the fullness of the fruit, I was prone to experience the limits of language.

Later I would hear that this kind of altered perception akin to seeing things for the first time is what hallucinogenic drugs induce, and was parallel to this experience I had after falling off a horse in Mexico. Interestingly, some research projects are finding positive therapeutic benefits from controlled hallucinogen use with individuals with Post Traumatic Stress Disorder. As we'll explore later, mindfulness meditation, or other interventions, may liberate a person from anxiety, depression, or posttraumatic states of dysfunction by changing a fixed sense of identity and patterns of perception.

This experience after the horse accident gave me the opportunity to learn that there is a level of knowing beneath personal identity, personal belief, and personal expectation. I had no idea what to call this change in "me" so I never discussed it with anyone, putting it into a category of some existential wakeup call to lighten up, given life's fragility following that near-death accident, to be grateful to be able to move my neck, be alive, be awake and aware. I didn't think of it then as a gift, but I realize now it was one of those unplanned experiences that are turning points, even if we don't realize their impact at the time.

When I came back to college I was filled with an eagerness to understand these experiences. Head trauma and identity, salmon and suicide, I was excited to know more about how all these layers of reality, these domains of life, might find some common home, some common ground of understanding. When I was accepted to medical school, I thought this next phase of my formal education would enable me to explore these ideas. Little did I know how untrue that would be.

Top-Down and Bottom-Up

Two terms that indicate how information is processed in our minds and brains are "top-down" and "bottom-up." While these terms are sometimes used for the anatomical location of processing (the higher cortex at the top, the lower brainstem and limbic areas at the bottom), these same terms are also used for layers of processing not related to the anatomical distribution of up and down. Instead, they are used for

the degree of processing of information. In the view we will be using here, *top-down* refers to ways we have experienced things in the past and created generalized summaries or mental models, also known as schema, of those events. For example, if you've seen many dogs you'll have a general mental model or image of a generic dog. The next time you see a furry canine strolling by, your top-down processing may use that mental model to filter incoming visual input and you won't really see the uniqueness of this dog in front of you. You have overlaid your generalized image of dog on top of the here-and-now perceptual stream of energy that creates the neural representation of "dog." What you actually have in awareness is that amalgam of the top-down

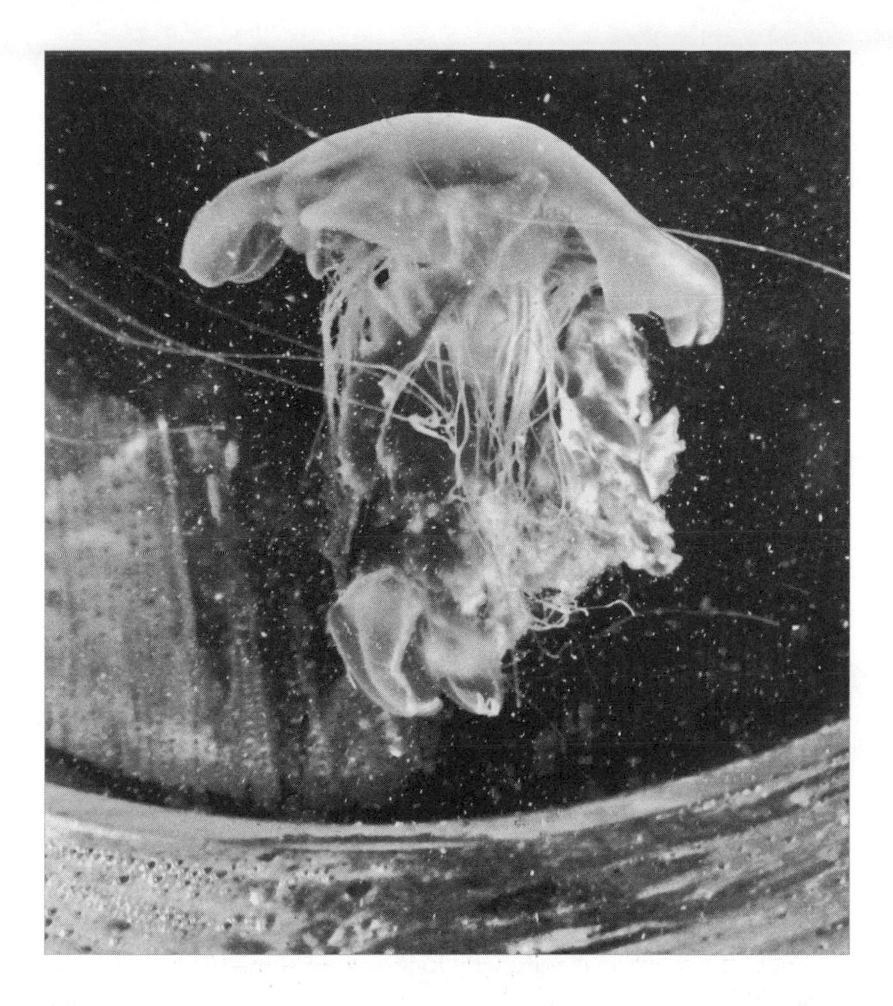

filtering of your experience. So here, "top" means prior experience is activated, making it difficult to notice the unique and vibrant details of what is happening here and now. The top-down generalized notion of dog will shade and limit your perception of the actual animal in front of you. The benefit of top-down is that it makes your life more efficient. That's a dog, I know what it is, I don't need to expend any more energy than needed on insignificant, non-threatening things, so I'll take my limited resources and apply them elsewhere. It saves time and energy, and therefore is cognitively efficient. That's top-down processing.

On the other hand, if you've never seen a spiny anteater before, the first time you come across one on the trail, it will capture all of your attention, engaging your *bottom-up* processing so that you are seeing with beginner's eyes. These are eyes leading to circuitry in the brain not shaping and altering ongoing perception through the top-down filters of prior experience. You'll be taking in as much pure sensation from eyesight as possible without the top-down filter altering and limiting what you see now based on what you've seen before.

When we travel to a foreign country, bottom-up perceiving can fill our journey with a profound sense of being alive. Time seems extended, days full, and we've seen more details in a few hours than we may have seen in a week in our familiar life. What *seen* means for bottom-up perception is that we become more attentive to novelty, seeing the unique aspects of what is, literally, in front of our eyes. Ellen Langer calls this "mindfulness" and has done numerous studies to reveal the health benefits of being open to the freshness of the present moment (Langer, 1989/2014). The novel experience of foreign travel in contrast to the sense of dullness in our lives back home also reminds us of what life is like in familiar terrain: top-down can dominate bottom-up and give us a familiar sense of the same-old, same-old. A street at home with just as much detail seems dull compared to the novelty of a street in a foreign town seen for the first time. This loss of attention to the familiar can be called top-down dominance. Prior learning creates top-down filters through which we screen incoming data and lose the detail of seeing things for the first time. This top-down dominance is one of the side effects, if you will, of experience and knowledge. It's one of the downsides of expertise—we stop seeing clearly because we know so much. We know what a dog is, so let's move on and not lose attentional energy by focusing on something we already know. We save our attentional resources for something more pressing than the familiar. Knowledge from prior experience helps us become selective in what we perceive so we can be more efficient in allocating attentional resources and more effective and rapid in our behavioral responses. But something gets lost with that efficiency. We literally walk next to the roses and pass them by, naming them, knowing them as the flowers they are, but we don't stop to immerse ourselves in their scents or notice their unique rainbow of colors and textures.

One general way of considering the distinction between these perception modes is that with the bottom-up we are experiencing the mind as a conduit of sensory experience whereas in top-down we are additionally a constructor of information. A conduit enables something to flow freely, directing that flow but not changing it much; a constructor is fueled by input and generates its own output, a transformation that changes the fuel into another form: It constructs a new layer of representational information beyond the initial sensory stream.

The mind can be a Bottom-Up Conduit and Top-Down Constructor.

Part of the answer to the question of "Who are we?" is to consider that we are at least a Conduit and Constructor. It may be that if only one or the other is utilized in our lives, we may become blocked in our functioning. Without Constructor, we don't learn; without Conduit, we don't feel. Could this be an extreme Constructor thing to say? My Conduit mind somehow urges me to stay open about this—maybe being only a Conduit is fine. But if I have put these into words, even my Conduit is connecting with my Constructor to stand up for itself—a sign of the importance of both, don't you think, don't you feel? Both are important, each playing an important but distinct role in our experience of being alive. One without the balance of the other and our lives become limited. Differentiate and then link the two, and we become integrated.

The mind, we've been proposing, is both embodied and relational. In our communications with one another, we often send linguistic packets of top-down words with narratives and explanations that are already constructing the reality we are sharing with another. Even when we try our best to use words to describe what we are experiencing, rather than explain what is going on, we are still using the construction of linguistic forms.

And in our brain? Energy and information flow within us as well as between us. The nervous system, including its brain, plays a major role in shaping our embodied energy flow patterns. This is how brain research has so much to offer in illuminating at least the within aspect of mind, even if not the totality of what the mind is and who we are. So we can ask, what neural processing might help us understand these differentiated ways of experiencing mind, conduit and constructor? Two findings may be relevant here.

One is the recent finding that in the brain there are two anatomically distinct circuits mediating each form. A more lateralized (side) process involving sensory input areas includes the anterior insula (which some say is part of the ventrolateral prefrontal cortex) and the consciousness mediating dorsolateral prefrontal cortex (the upper side area in the front of the brain, behind the forehead and above and to the side of the eyes). Notice the term *lateral* in each of these regions. These side circuits seem to be active when we focus primarily on moment-by-moment sensation (Farb et al., 2007). In contrast, we have a more centralized circuit that seems to generate thought. This region overlaps, in part, with what some researchers have called the "default mode" we mentioned briefly in an earlier entry. This is the OATS circuit we discussed that constructs all sorts of top-down chatter about *o*thers *a*nd *t*he *s*elf.

Sensation may be as bottom-up as we get. Since we live in a body, our within-mind experience is shaped by the physical apparatus that lets us take in energy flow from the outside world. We have our first five senses of sight, hearing, smell, taste, and touch, we have our proprioceptive sense of motion, and we have our interoceptive sense of the signals from the interior of the body. These perceptual capacities to sense the outer world and internal bodily world are built upon the physical neural machinery that enables energy to flow. These energy patterns can be symbolic of something other than the flow itself, and information is created with these patterns as ions flow in and out of membranes and chemicals are released as part of the pathways of neural activity.

In some ways, bottom-up, within-mind experiences can be considered as being as close to the present as physically possible, in the literal meaning of it being "pre-sensed." Sensation has already filtered reality through the restricted dimension of receptors and neural pathways. Yet this restriction is a neural architectural reality, not one, at this level, changed much by experience. We wouldn't need to call sensation a top-down process. We can use the term *sensation* to indicate the most bottom-up we can get.

Yet there is a reality to energy and information flow even before our external sense organs, such as our eyes and ears, or internal receptors of our body's muscles and bones and internal organs, send their signals upward to the central nervous system. As these flows of sensory energy stream toward and into the brain, we move from sensation to perception—and so sensation is as close as we get to being fully present in the world. Yes, perhaps we can never fully get to a pre-sensed energy flow, to truly be in the world before our body's sensory receptors' constraints kick in, but we can get close, and we'll simply say that getting as close as you can to that flowing reality is what we'll term *bottom-up*.

In contrast, when we assemble those bottom-up sensations of our conduit circuitry into perceptions, or go even further down the construction process and reflect on the meaning of a sensation or perception, associating it with thought and memory, we are utilizing the activity of a more central circuit that involves distinct areas, including midline areas of the prefrontal cortex and regions such as the precuneus, medial and temporal lobes, lateral and inferior parietal cortex, and cingulate cortex. Please don't worry about all these top-down names if you're not interested in these terms. This "observing" circuitry is a part of something we've identified before and more easily recalled with the term, *default mode network*, an important system of the brain that is a midline front and back set of areas that matures over development as it "integrates into a cohesive, interconnected network" (Fair et al., 2008).

This circuit, as we discussed, is called "default" because at rest in a scanner, the subject has a robust brain firing as the baseline neural activity without being assigned a specific task to perform. What does this circuit involve itself with? Self and others—the OATS system. In fact, some neuroscientists have suggested that elements of this default mode circuitry give rise to our sense of personal identity and may be connected to our mental health (Bluhm et al., 2009; Raichle & Snyder, 2007). Studies of mindfulness meditation have pointed out that this system becomes more integrated with sustained practice. We are reflective and social beings, and it would be natural to focus on others and self as a baseline activity when we're just hanging out with no particular assignment—even in a big blasting brain scanner.

Perhaps it was this OATS system that was temporarily disabled after my horse accident. Without engagement of a more distanced constructor of this top-down circuitry, the direct sensory input of each moment at that time could then more easily fill my awareness. Without the top-down filter of prior experience and personal identity, I was literally seeing things for the first time. The lateral sensory bottom-up conduit circuit and the midline top-down constructor observing circuit have been shown to be reciprocal in their activation: When one is turned up, the other is turned down. With a lateralized sensory conduit system fully intact in the face of a jolted midline construction circuit knocked offline for a day, I could experience a fuller, richer, bottom-up, sensory world through the conduit of my within-mind machinery.

Construction may have many top-down layers. One is at the level of perception, so when we see a familiar dog it is just a dog. We literally sense the visual input of the dog but do not perceive that input in awareness with much detail. We can also have the experience of observing so that instead of being immersed in the sensory conduit flow, we are a bit distanced, having the experience of "Dan is seeing a dog. How interesting for him. Let's move on." Such observation with the presence of an "observer"—Dan— may be just the beginning of the OATS activity. Now there is a personal identity that indicates who is doing the seeing. Once I actively link autobiographical and factual memory together with linguistic forms, top-down has become the active constructor, and the OATS activity is off to the races.

Such a cascade of top-down can easily move from direct sensation to the more sensory-distant observational experience. We are now observing, not sensing. Such observation can then give rise to the presence of a well-defined witness—we witness an event from an even more distant stance. We then move further in a top-down mode to narrating what we are witnessing and observing. Yes, your OATS circuit may have noticed another acronym and you are realizing, Dan has an acronym addiction: This is how we OWN an experience, as we *o*bserve, *w*itness, and *n*arrate an event.

When we observe, witness, and narrate our experience from top-down constructor circuitry connected to prior experience, we become more distant from being simply immersed in the sensory bottom-up flow of our conduit circuitry in the present moment. This is the balance we live day-by-day, moment-by-moment, between top-down and bottom-up, constructor and conduit. Language emerges from this observational flow, and we can see how wording the world can make us distant from the sensory richness that surrounds us. This doesn't make language a bad guy, but it makes the observation, witnessing, and narrating simply distinct from sensory flow—each distinct, important, and inhibiting the other.

A second possible neural mechanism comes from older studies that are less well established but nevertheless may be at work. This suggests a potential pathway that acts not instead of the sensory/observational or lateral/midline distinctions, but in addition to those distinct circuits. This possible mechanism certainly may serve as a useful metaphor, even if its exact neural components are not ultimately verified. But if this turns out to be established as an actual neural mechanism, not simply a useful metaphor, it will likely be difficult to study because it involves micro-architectural features of the cortex, not whole brain regions more readily seen on a scanner as in the lateralized and midline regions.

Vernon Mountcastle and other neuroscientists noted decades ago (Mountcasle, 1979) that the flow of energy in the cortex was bidirectional. The highest part of the brain, the cortex, is composed of vertical columns, most of which are six cell layers deep. The highest layer is labeled number one; the lowest labeled six. Folded over and over itself, the cortex appears thick, but six layers of cells is actually quite thin, like six playing cards laid on top of one another. The cortex serves to make neural "maps" of the world—of what we perceive and conceive, taking in our sensory input of sight and sound and building larger maps of what we perceive, constructing our conceptual thoughts about self and other.

In this possibly true, yet-to-be verified view, bottom-up sensory flow arises from the lower layers—six to five to four.

If we've seen something before, this prior experience shapes the mapping of cortical input by activating a top-down filtering stream initiated from, literally, the top-down—from layers one to two to three, and so on. In this proposed mechanism, when we see a furry animal with a wagging long thing behind it, we take in the sensory pathway from six, to five, and to four. That's bottom-up sensory input. That's our *conduition*, to make up a term for what the conduit does. But if we've seen such an object before, naming it "animal" or perhaps "pet" or even "dog," or maybe even more specifically, "Charlie," then that prior knowledge influences the momentary experience by activating cortical mappings embedded in memory that will be initiated from layer one to two to three. In other words, the higher cortical layers stream prior knowledge along the same cortical column, even as it takes in the sensory stream from bottom-up, six to five to four. Naturally this top-down flow was initiated by some kind of detection of patterns, enabling this flow to be created that is specific to this sensory stream in the moment—and that's the amazing and crucial part of this view.

In other words, some part of this mechanism could detect a familiar pattern and engage the top-down flow. This detector must be directly connected to pattern perception and memory systems. If what is coming in now from sensation matches some pattern represented from prior learning, top-down filtering is initiated.

In this view, our sensory bottom-up conduit serves as the fuel and forms for the perceptual top-down constructor. The "crashing" of input at the meeting of layer three (from top-down) and layer four (from bottom-up) will determine how much detail we actually perceive with or without awareness. If we experience a lot of bottom-up

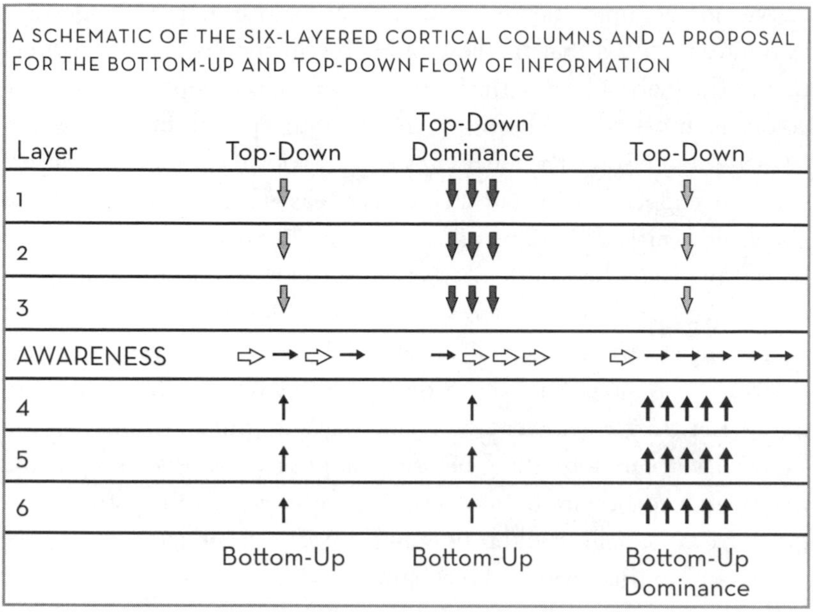

A SCHEMATIC OF THE SIX-LAYERED CORTICAL COLUMNS AND A PROPOSAL FOR THE BOTTOM-UP AND TOP-DOWN FLOW OF INFORMATION

Layer	Top-Down	Top-Down Dominance	Top-Down
1	⇩	↓↓↓	⇩
2	⇩	↓↓↓	⇩
3	⇩	↓↓↓	⇩
AWARENESS	⇨→⇨→	→⇨⇨⇨	⇨→→→→→
4	↑	↑	↑↑↑↑↑
5	↑	↑	↑↑↑↑↑
6	↑	↑	↑↑↑↑↑
	Bottom-Up	Bottom-Up	Bottom-Up Dominance

TOP-DOWN AND BOTTOM UP

conduit input free from top-down filtering, we see a lot of sensory detail. Awareness becomes filled with all the richness of something new. A lot of top-down dominance functions as a constructor to interpret and transform that initial flow from bottom-up into mental models and generalized perceptual maps, and as a result we just know it's a dog, perceive only generalities, and move on. The details experienced in awareness are dull, the richness vapid, and the interest low.

Top-down can repeatedly shut off input from bottom-up as it ascertains what it thinks its sees and no longer "needs" the input from bottom-up. Top-down has done its job and to be efficient, shuts down further attention to the details of the sensory stream. That's just another dog, let's move on.

An interesting finding possibly relevant to this proposed view is from comparative ape neuroanatomy (Semendeferi, Lu, Schenker & Damasio, 2002). The differences between the human and ape brain are found primarily in the cortex. These distinctions include the fact that the two sides of the human brain are more differentiated and their linkages more robust through the corpus callosum. That makes, by the way, a more highly integrated brain with more differentiation and linkage. In addition, in the prefrontal region of the human cortex, the neuropil— cell bodies that perform neural calculations—are much thicker in humans than in our ape cousins at layer three-four. No one knows exactly what this means, but it could mean that we have more capacity to deal with the interface of bottom-up and top-down, a balancing act that may be carried out in just this part of the prefrontal region's neural calculations. Or perhaps it is simply a sign of our top-down heavy constructor function that is bumping in to the bottom up and dominating. Future studies may benefit from examining, for example, if this three-four level cortical column zone is even thicker after mental training that helps highlight the distinctions between bottom-up and top-down, a process which I believe mindfulness meditation can carry out.

To repeat the top-down caution, please understand that this proposed columnar view of these distinctions is just a hypothesis. When I proposed this once to a neuroscientist, the response was, "The brain doesn't work that way." Perhaps he is right. But perhaps we simply haven't found the techniques to examine this level of columnar processing, at least not yet. We need to keep an open sensory mind and not come to a premature closure of possibility. That's the benefit of bottom-up, to keep our top-down categorizing minds on their toes. Just because the constructor has conviction about the truth of something doesn't mean the sensory reality may not prove otherwise when seen from a more bottom-up process of conduition. That is one of the challenges of science in general—as we become expert and make language about things, we may lose touch with how things are. That's a concern we expressed at the beginning of our journey about the caution on using words to reflect on the non-worded world of the mind. Bearing these top-down construction cautions in mind will be a useful mind-set as we continue our journey.

In any event, there is general consensus that top-down and bottom-up are distinct ways we experience the world, whatever neural mechanisms contribute to our mental processing (see Engel, et al., 2001).

Top-down construction helps us deal with whatever we need to deal with in a more efficient manner. It helps us be in the world. When we drink a glass of water because we are thirsty, we don't need to bliss out on all the sensory elements of that experience like I did in Mexico after my accident, or after I meditated for the first time. Sometimes we need to simply drink the water and move on. On the other hand, the older and more experienced we get, the more dominant top-down construction can become. A fully top-down life is so distant from the bottom-up sensory flow of now that it can feel quite disconnected from a sense of vitality. The conduit sensory stream is in the vivid present. We are now-here. But the conceptual constructor, while important, when without the sensory conduit stream can feel quite numb, distant in time and space, and when lived to the extreme, give us a sense of being no-where.

I love thinking and constructing ideas. But living a life only of construction and not linking to a conduit flow is a non-integrated life, making us prone to chaos and rigidity. Harmony involves honoring both conduition and construction.

If we honor bottom-up in our lives, learn to cultivate it well—or revive it from our earlier years of living before our personal identity became dominant in our lives—life can become fuller and richer and more meaningful. We can learn to live more in the present conduit sensory stream than the distanced world of only construction. Bringing back bottom-up may involve each of these possible mechanisms, enhancing our lateral sensory conduit circuit and the bottom-up columnar flow from six to four, and at the same time decreasing the dominance of the flow from our observing, witnessing, and narrating constructor circuits and descending top-down columnar flow from one to three.

We humans have so many varieties of mental anguish because we have an OATS circuit and robust higher layer (1, 2, 3) of our neocortex. Each of these possible mechanisms suggests that the human brain may be a source of impaired integration in the flow of energy and information. What does this mean? If the sensory stream functions as a conduit, we can imagine there could be a natural emergence there, including the self-organizational drive for integration and well-being. On the other hand, the opposing midline top-down constructor circuit and the intricate higher 1-2-3 layers of the constructing cortex may have a set of patterns that constrain the natural emergence of integration, in some people and in certain states. This may help explain why, as mentioned earlier, each symptom of each syndrome of the psychiatric disorders can be viewed as an example of chaos, rigidity, or both. As chaos and rigidity arise from impairments to integration, we can see how impaired integration can be seen as a source of mental suffering. Such anguish manifests in our lives as chaos and rigidity.

So far, every brain imaging study of individuals with major psychiatric challenges reveals impaired integration in the brain.

It might be that the constructor role of who we are is what gets us into mental difficulty. Building up the equally important but often less preserved conduit function as we grow throughout life may be a secret to maintaining well-being. Balancing conduit and constructor, we can live a more integrated life. Whether by acquired adaptation to experience, or some innate, infectious, toxic, or genetic factors that initiate this impediment to integration, it may be the constructor that constructs obstacles to integration. In contrast, since construction and conduition naturally inhibit each other, highlighting the opposing and often under-developed sensory circuitry in the face of overly zealous construction may reveal how cultivating our conduit function permits integration to naturally arise.

In this way part of the art of living fully is to reawaken our beginner's mind, which means strengthening our capacity for bottom-up conduit living and living with uncertainty. As mentioned earlier, Langer's (1989/2014) notion of mindful learning, in which we pay particular attention to novel distinctions, exemplifies this kind of awakening, and her research reveals the powerful health benefits of such an intentional way of harnessing bottom-up experience. When we are in that place of being open to what is happening as it happens, we can embrace the natural uncertainty of life (see Siegel & Siegel, 2014). But when our constructor constructs expectations about how the world should be, we are a set up for disappointment, anguish, and stress. The conduit is in the present moment, accepting what is; the constructor is often far from the sensory present, building models of life that conform to what we've experienced in the past, often making it so we don't really see fully what is happening right now. Such a conflict between what is and what we expect can create great challenges to being fully present.

Having fixed notions of something—such as the idea that the mind is only brain activity—can limit how we are open to new ways of thinking about old issues. Top-down concepts can constrain our freedom to take in new perspectives, even without our knowing this perceptual filtering and imprisonment is occurring. Such top-down processing may directly impact our executive functions, our sense of who we are, and shape the decisions we make. In other words, beyond shaping our perceptions, top-down construction may have deep conviction in its own constructed concepts, its own linguistic output. That's why it is so important to create a top-down way to understand top-down, so we can inform our own constructor and invite it to let the conduit become more active in our lives. We also can learn to have humility—words are simply words.

Top-down is crucial, though, to take stock of where we are and course-correct. Words are fabulous at trying to articulate our experience. We can get a bit of distance from the immediacy of an experience, perceive patterns, and conceive concepts with construction and words that can be deeply empowering. These constructive words are essential at times, in fact, to distance ourselves from direct experience so that we can reflect and make sense of things. Careful observation gives rise to witnessing and narrating, the basis of disciplined study of anything. To "research" is to search-again. Top-down construction is vital in our complex lives, too.

What this discussion brings us to is simply this: Both conduition and construction are important components of the *who* of who we are. We are both conduits and constructors.

Reflections and Invitations: Identity, Self, and Mind

In our journeys through life, unexpected turning points arise, and we often do not understand their long-term impact.. As we move from childhood into adolescence, the ways we've learned about the world become embedded in the top-down mental models that shape how we continue to perceive and make sense of life as it unfolds. If those models remain unchallenged, we continue in a recursive, self-reinforcing way to create in our interactions the kinds of experiences that simply reinforce what we believe we are. Our top-down constructor can filter experience, create executive decisions on actions, and engage with the world in ways that repeatedly shape what we are immersed in and even how we respond. These repeated experiences are often woven into a tale of our identity as we observe, witness, and narrate a story we've told over, and over, and over again about who we are.

At least who we think we are.

Often, few of these top-down constructor processes that shape at least part of the who of who we are come into our awareness.

Photo by Alexander Siegel

Our top-down, constructed personal identity may limit our freer, more detailed, sensory-rich, bottom-up living. Something new—even a traumatic injury—can be an invitation to see things with new eyes. But you don't need to bang your head to shake up your mind. For example, an experience that stirs our sense of who we think we are and opens our minds to new ways of perceiving and experiencing life can be transformed from a helpless trauma to an opportunity for awakening. Sometimes a conversation with another person can initiate more bottom-up; sometimes it is listening to a new poem or song. There are many opportunities to invite the conduit to awaken and constructor to re-evaluate its familiar conclusions about life and identity.

What does awakening the mind really mean? In this case, it means that we have a perspective that we are more than what we think. We are more than what we remember. The notion of a personal self, such as my sense of a "Dan" that lives in this body, is a construction. If repeated knocks on the gravel could knock out my sense of Dan for a day, what was left of the *who* of me? Clearly, I am more than my identity as Dan. Dan is constructed. What was remaining? Conduition. I could experience the profoundly rich vibrancy of conduit living. Conduition is not constructed. It is as bottom-up as we can come in this life.

That experience was a wake-up call for me to experience a shift in perception, to see that the personal identity self is a personal construction.

A turning point in one's life may arise when the top-down filters that shape our feelings, perceptions, thoughts, and actions are suddenly broken down and shaken up, and a new bottom-up experience fills our awareness. In other words, the dominance of top-down from the midline constructor circuits that reciprocally inhibit the lateralized conduit circuits can keep the status quo beliefs about the self, and ways of the self-in-the-world, intact for years, decades, a lifetime. (Wow, those are a lot of words, go constructor!) In short, we can be imprisoned by hidden top-down views of who we think we are. From the cortical column perspective, we can have a top-down set of constraining flows that again and again shape our perceptual experience; we believe that what we keep on seeing is the reality of what is. One way of expressing this is that there is no such thing as immaculate perception. Perception is shaped by top-down learning from prior experience. We need to connect with sensation to begin to liberate us from the potential tyranny of top-down filters often hidden from our awareness. (Go conduit!) Oddly, in writing this paragraph, I can feel some kind of alliance of the two as they engage in what feels like a conduit-constructor duet. Maybe that's a constructive wish about integration; but it feels like a conduit reality, so maybe it's just really happening, who knows?

And that's exactly it. Who are we and who is it that knows?

Can you recall a time when some kind of shift in perspective entered your life? This can be subtle or severe, sudden or gradual, but often this emerging inner sense of clarity may be suffused with a feeling of novelty, as if you haven't seen this object or had this insight quite in this way before.

If you've had such an experience, it may be helpful to reflect on how your mind builds a picture of the world based on what has happened. As we move into our later years, from infancy onward, we live in these bodies whose brains are anticipation machines and pattern detectors, getting ready for next based on what we imagine will happen in the future shaped by what we've learned about the past. What this means is that our raw experience of a beginner's mind, of being in this present sensory now, will be replaced as we age with an expert's mind. Ironically we'll be able to see, literally, less clearly, less fully, with less detail, not more. How sad, but how workable if we understand it.

So who are we? We can say simply we are our minds. But what is this mind exactly? At a minimum we can say that from a bottom-up experience, we are our sensory flow of energy that arises from the outer world and from our inner, bodily world including its brain. This is how we function as a conduit of sensory experience, immersing ourselves in the miracle of being here, in this moment. We are also our top-down experience—ways we filter energy flow into information, symbolic meaning that stands for something beyond the pattern of energy we experience. This is how we are also a constructor of information, not merely a conduit of energy.

Ironically, having a constructed view of this two-part notion of the *who* of who we are can help you cultivate the conduit of your life if that is what is needed. How we reflect on our fixed past experiences, live with openness to the emergence of now, and anticipate and shape the flow of open moments in what we call the future, are ways in which our minds continually create who we are in the only thing that exists, the moment of now.

This is a top-down view of who we are. We can OWN it, observing, witnessing, and narrating it. Constructing our experience of life is an important aspect of who we are. In some ways, this can be seen as how the mind creates information from energy flow, beginning with perception from sensation and then moving in ever-more complex ways into memory, conception, and imagination.

Conduition is equally important, only different in its function. Being a conduit is more like being in an experience, enabling the subjective sense of energy flow before it is transformed into the symbolic form we call information. We can strengthen our conduit self by getting in the habit of SIFTing the mind, checking in with direct sensations, images, and feelings, and even sensing the texture of our constructed thoughts. We can temporarily stop reading, stop writing, and be without words, immersing ourselves in drawing, dancing, singing, and simply being.

Befriending both conduition and construction creates a vibrant life of honoring the importance of these fundamental ways of being. Each alone restricts us, each together in our life frees us to live more fully.

Construction and conduition are a part of being who we are. Can you feel the energy of such a vision in the conduit of your experience? Can you sense the informational insights in the constructor of your experience? Conduit and constructor, we can welcome both of these facets of who we are right now, in the flow of this moment.

CHAPTER 6

Where Is Mind?

WE ARE DEEP INTO OUR JOURNEY, EMBRACING THE FUNDAMENTAL nature of energy and information flow as the source of mind. We've discussed how this flow is part of a complex system, one with the emergent property of self-organization. Integration is how self-organization moves the system in an optimal way, and we've seen that this linkage of differentiated parts, this integration, is proposed as a core mechanism of health. We've also explored how the subjective experience of mind may also be an emergent aspect of energy and information flow. When we tune in to that subjective felt experience in someone else, we promote the joining of two differentiated entities into one coherent, integrated whole. This allows us to join the other—we belong, and there is a vibrant sense of harmony that arises. This is how the whole is greater than the sum of its parts, the outcome of integration. We've also described how energy can flow in the conduit, bottom-up function of the mind; and energy can be transformed into information, into symbols with meaning beyond the energy pattern, by the constructor function of the mind. Conduit and constructor, the mind entails a range of experiences of the moment. In this entry, we'll focus on the *where* of mind by exploring the location of this energy and information flow.

Could Mind Be Distributed Beyond the Individual? (1985-1990)

Before the 1990s, before that Decade of the Brain, I had moved into my post-graduate training, first in pediatrics, then in psychiatry. As "caretaker of the soul," the field of psychiatry seemed to me a good

Photo by Lee Freedman

place to explore what the nature of our heart of being human might be all about. Yet it felt to me as if psychiatry was struggling to find an identity in both medicine and in the broad field of mental health. The journey of becoming a mental health clinician was filled with many surprises, such as learning that the various disciplines focusing on the mind did not have a definition of the mental, or of the health, as we've discussed.

The field of psychiatry that I experienced when I returned from my time away and went through my rotations during my last two clinical years of medical school, back in the early 1980s, was filled with conflict and turf battles. Were you a psychoanalytic therapist or a biological psychiatrist—did you believe in Freud or molecules? Did you believe in therapy or medications? Were you interested in researching mental disease or practicing in the community? These divisions seemed insurmountable, so even though during my time away from school I had learned of my fascination with the mind, given how much I loved children and perhaps because of my relationship with my mentor, Tom, I elected to train in pediatrics.

But I held on to that concept of mindsight—the way we see the mind in others and ourselves—when I returned to medical school, and now in my post-graduate training. Families who focused on the feelings and thoughts of their children, and of the parents themselves, those that showed the empathy and insight of their mindsight skills, seemed to deal more effectively with the challenges of the severe medical illnesses many of the children with whom I worked were struggling with. Mindsight seemed an important and helpful skill that cultivated resilience. I tried my best to help those families cultivate their insight and empathy, their mindsight skills, but medical care and the pressure of so much work made such a focus not practically possible. After my first few months of pediatrics at UCLA, I realized my passion for the mind was better focused in psychiatry, and I transferred programs at the end of my first year.

I loved seeing psychiatric patients. Being able to use the concept of mindsight to focus on their internal worlds, keep track of my own,

and help them clarify their mental lives and find the social connections that could support a healthy mind and reduce their psychic pain felt compelling and meaningful. I had no idea what would happen in psychiatry or the larger field of mental health, but knew from the day-to-day immersion in the work with patients that I had chosen the right direction.

But even in psychiatry, just as in medical school, there was a push for an objective stance toward mental illness, conceptualizing disorders, and clinical intervention that seemed to diminish the importance of an individual's unique subjective experience. The latest edition of the American Psychiatric Association's *Diagnostic and Statistical Manual*, the *DSM-III*, was first prepared when I was in medical school and tested out on us as clerks rotating through psychiatry. One of the stated goals of that document, our professors told us, was to create a vocabulary of objective criteria about what constituted a psychiatric disorder so that someone in Iowa or Indiana could speak with someone in Ireland or India and have the same notion of what the terms meant. Rather than the subjective interpretation of an individual practitioner, the new *DSM* would clearly outline a mental health profession standardization of which set of symptoms were needed to meet the criteria for receiving a psychiatric diagnosis.

This drive for objectivity made sense. It was consistent with the medical model we were taught in school, and it honored psychiatry as being a branch of medicine. It seemed that this step would also allow collaboration among all clinicians of various backgrounds, including those in psychiatry, which ought to be beneficial to patients.

But an interesting tension was inevitably built in to such an approach, in that many important things of the mind were, in fact, subjective. How could we truly offer objective data about something that included subjective processes? Oddly, I'd come to realize years later, a book of mental disorders that was to shape an entire broader field of mental health never addressed the issue of what the *mental* was, of what the mind was, or what the health of the mental might be. How odd, to have a field of the mind that didn't define what the mind is.

As a trainee, I was chosen for an honorary fellowship with the American Psychiatric Association, and as a fellow participated in a debate on this topic: Should psychiatric trainees be taught how to do psychotherapy? At the annual national meeting, my opponent in the debate began with the following argument: Psychiatrists are medical doctors, and as physicians we base our work on medical science. Psychotherapy is an activity that asks patients about their feelings. There is no scientific study that proves feelings are real. Therefore, feelings are not a scientific topic. Thus, psychiatrists, as medical doctors basing their work on science, should not be involved in a non-scientific activity like psychotherapy. The natural conclusion: There is no need for psychiatrists to learn about or practice psychotherapy.

I was dismayed, and sat there frozen as my debate opponent sat down. It was now my turn. What should I say? What would you have said? How could I approach this question of whether our inner subjective experience, the felt texture of lived life, including the feelings of our emotions, was real or not? If these feelings and sensations, thoughts and memories, these mental activities, were not measureable in a quantitative way, or even externally observable, was the mind itself really real? Were feelings and the mind with which we become aware of them scientifically validated entities?

I hesitated, took some deep breaths, and then stood up and walked over to the podium.

I looked out over the packed room of trainees and seasoned psychiatrists in the room. We were the future of psychiatry, as trainees, and we were, they told us, the future of academic psychiatry, we fellows. This debate was a moment in the evolution of our branch of medicine devoted to being a care provider of the soul, the psyche, the mind. What could I say? Here is what I recall happening.

I looked out at the audience, and over to my opponent, and began by simply saying that I felt sad. And then I paused. I took another deep breath and said that I was not aware of any research study that scientifically demonstrated that my opponent was real, and that I had nothing more to say, and made a gesture to sit down.

I stayed at the podium, and after the laughter and applause subsided, we debated some more, looking into scientific methodology that relies on measureable factors, things you can observe and quantify, and I offered some discussion of what we knew at that time, about research into the impact of psychotherapy on mental health. But later I came to realize that an old statement often attributed, apparently falsely, to Albert Einstein, but actually published in 1963 after Einstein's death by the sociologist William Bruce Cameron, would have been so useful to state at that very moment, so relevant in that discussion: "Not everything that can be counted counts, and not everything that counts can be counted." I simply state a similar sentiment this way: Not everything that is measureable is important, and not everything that is important is measureable.

You might think that this division of the objective from the subjective is something new, some part of the emerging of a brain-based science approach to the mind. But I was surprised to learn that this specific issue of psychiatrists being thought of as objective medical doctors who should not focus on inner subjective experience was, in fact, not new at all.

My clinical supervisor, when he heard the story of the debate, smiled and asked who my debate opponent had been. I told him where he was in training, and my supervisor's smile turned into a laugh as he said, "I'll bring you something next week." That week he brought in a mimeographed copy he had kept of an almost identical debate he was in with my opponent's own current supervisor from 25 years earlier. California versus Iowa, round two.

It was as if the culture of psychiatry, and perhaps the larger field of mental health, was embracing some foundational stance that mental life was limited to observable behaviors emanating from one person and perceived by another. Even research psychologists had, in the last century, moved to a focus on objective, externally measureable factors representing mental life, rather than the far less reliable, "difficult and fallible," results of introspection, as William James stated in 1890.

Certainly moving beyond behaviorism, the *DSM's* aim to objectify and make describable our inner thoughts and feelings, though well intended, seemed to encourage, perhaps unavoidably, removing the reality of the individual's uniquely experienced inner mental life from the approach to evaluation in the field of mental health. Yes, the *DSM* discusses signs that are directly observable as well as symptoms that are capable of being articulated by a person, the inner experience a patient reports to be having. But what I mean by this problem is that making lists of symptoms, checking them off during an interview, then using them to label a person with a linguistic name, a diagnosis, is simply not the same as the clinician being taught to tune in to the interior life of another person before and beneath words. In other words, each person is different even if they use the same words to express their symptom profiles. Maybe it was my suicide prevention training, or maybe it was my experience in medical school, but it seemed to me that a clinician needed to connect deeply with the person in their care, not just check off something on a list and come up with a diagnostic label. Each person is unique. But unfortunately our top-down, constructor minds can start to perceive through the bias of the diagnostic category we've been taught, and we stop being present to truly see the person sitting before us. The risk of objective categories is that we can lose sight of the individual reality of our mutually created subjectivity. Recall that no one has immaculate perception. Often, sadly and worrisomely, the constructor mind has great conviction in its constructions. As mental health professionals, we especially need to be mindful of how categorical diagnoses can limit how we sense the mind of the person with whom we are working.

This tension between the objectively observable and subjectively experienced is an old conflict, one that brings cloudiness rather than clarity, competition rather than cooperation. The truth is that one aspect of mental activity, one essential feature of mind, is the subjective reality of our feelings and thoughts, beliefs and attitudes. We can also objectively observe, even sometimes measure, some results of mental life, like behavior. We've gone deeply into this view in our prior entries. We've seen how we can measure the shadows of our subjective experiences, receive responses to checklists of descriptions of symptoms, but we cannot know the subjective world directly in another, even in these ways of collecting reports from an outside perspective. This is not only true for mental health professionals; it is also frustrating for the broad field of science that is built upon close observation and the often-necessary measurements needed to quantify results. Subjectivity is not ultimately measureable, as we've seen. Even if we tick off a line on a page, as in a Likert Scale of *a lot, some, not much*, or *none*, or if we check off a box of colors to indicate what we are seeing, we aren't really revealing what it feels like, say, to be angry or even truly how angry we really are, or what seeing green feels like within our own subjective experience.

When a clinician is not aware of the limits of the constructed reality of a diagnostic category, then it is possible to have their internal, subjective, top-down filters create a perception of the person that is quite distant from what is actually going on for that person. Without awareness of their own mind, clinicians can lose touch with the reality of the person in front of their eyes. The person is seen from a top-down categorical construction, a non-conscious perceptual bias that shapes the shared narrative of the patient-therapist relationship, one that can direct how we live this top-down view of reality, in client/patient settings, as family members, and as psychotherapist. Sadly, this narrative process can reinforce itself, as the story shared by us becomes the story within us which then sculpts the story we live ourselves into, both individually and collectively.

As a trainee, I felt perplexed by these dilemmas. I loved the general idea of the field of psychiatry, but could feel the deep challenges of how to create a careful approach to categorizing mental suffering. Clearly, naming a challenge to well-being could enable us to carefully collect systematic data, to empirically study individuals with that condition, come up with research-proven effective treatment strategies, and then measure successful outcomes. But while these are crucial steps for a field to take, what if the individual was lost in the categorization, or, worse, if the categories were inaccurate and inadequate? What could protect the individual from such self-reinforcements of the potential limitations and inaccuracies of diagnoses? What could we do? One step, I thought, might be to dive deeply into what this mind actually is instead of ignoring the topic all together.

So let's recall that mind has four facets we've been exploring: information processing, subjective experience, consciousness, and self-organization. For some, the information view of mind lands mental life in the head. Mind, often seen as an information processor, is equated with brain activity by a vast majority of academics in the field of mental health and neuroscience, and had been proposed by Hippocrates as long as 2500 years ago, as we've discussed. But as we've also seen, energy flows throughout the whole body; it is not limited to the head. So if our proposal that overall, the mind as a whole may be an emergent property of energy flow is accurate, isn't it true that this flow is not limited by the skull, and flows throughout the body? And so with this line of consideration, we've seen that the mind's *where* may be fully embodied. The mind as an emergent property of energy flow might be located within the whole body, not just in the body's head.

Perhaps the whole body is the location of energy flow, but could it be that the brain in the head enables the energy of neural firing to be transformed into information? If this is true, then the meaning-making mind, the mind as information processor, would understandably be attributed to brain processes, not fully bodily ones. And maybe that's all there is. This view would say that if mind is an emergent process of energy flow, and if information is created only in the neural networks of the head, then the brain in the head is the source of the information aspect of mind. That might be the whole story. But we are now learning that there are neural networks, parallel distributed processing systems, as we've discussed, in the intrinsic nervous system of our heart and intestines. It's possible that this energy flow in these non-head regions may fully create neural firing patterns, neural representations that are re-presenting in their neural network profiles of activation "information" that corresponds with mental representations. In other words, the body-proper may not be just energy flow—it may also involve information flow as well. Information is indeed created by networks of energy patterns. We're seeing that we have a head-brain, heart-brain, and gut-brain, which, at a minimum, would make the mind in both energy and information flow fully embodied, not just enskulled. Let's again keep an open constructor mind about this possibility.

But could the mind be more than embodied? Anthropologists consider the notion of a "distributed mind" and its connection to the evolution of our social brain. Robin Dunbar, Clive Gamble, and John Gowlett (2010) state:

> The social brain hypothesis argues that the complexities of hominin social life were responsible for driving the evolution of the early hominin brain from its essentially apelike beginnings to its modern form. The theory offers a dynamic new perspective for exploring the evolutionary origins of fundamental social capabilities such as the formation of large, cooperating communities and the maintenance of high levels of intimacy and trust. Simultaneously, it informs on the specific cognitive abilities, such as theory of mind, that underpin these capabilities." (pp. 5-6)

These authors further propose "the concept of the *distributed mind* draws from a diverse interdisciplinary base to consider cognition as embodied, embedded, extended, situated, and emergent (Anderson, 2003; Bird-David, 1999; Brooks, 1999; Clark, 1997; Hutchins, 1995; Lakoff & Johnson, 1999; Strathern, 1988; Varela et al., 1991)" (2010, pp. 12). The fundamental implication is that information processing is physically and socially distributed beyond the individual person. And so from the point of view of our question about the where of mind, this view embraces the interactive ways in which both other people and the artifacts they create move the mind beyond the interior of the individual. "The notion of the distributed mind thus affords considerable potential for examining the socio-cognitive relationships structuring homonin and human societies...the extent to which cognition is gradually extended both into the material world and into the social world informs on the nature of what it is to be human" (2010, pp. 12-13).

Philosophers use terms such as extended, embedded, and situated to describe the nature of mind beyond its embodied origins. For example, Robert Rupert (2009) suggests that "embedded models emphasize representations of actions, the self, and relations between the self and environment" (p. 204). An active discussion (see Clark, 2011) about how our minds are extended into information processing systems outside of our bodies builds on this notion of a mind being beyond our interior.

Here we are using the term "relational" to indicate that the mind is embedded within the world around it, extends into information systems outside of the body, and is situated in social contexts. But importantly, we are using the term relational to suggest further that the mind is in constant interaction and exchange with that "external" world, especially with other people and other entities in the environment. This exchange includes the information processing of cognition, but involves much more. This term relational includes more than computational events; it involves entrainment and coupling, as forms of attunement and resonance, for example, that enable the mind of an individual to become linked as a fundamental part of the mind and self of what, traditionally, are called "others" and the external environment.

Though these research fields in anthropology and philosophy often focus primarily on cognition, their inherent propositions are contained within our terminology of an embodied and relational mind that emerges, too, as the self-organizing facet of mind. The "embodied" aspect includes the emergence of internal processes encased by the skin that arise in the whole body as it interacts with the world, not only activity of the brain within the skull. And the "relational" aspect of the self-organizing facet of mind is emerging from—not just shaped by—the engagement with others and the larger world.

If we build into this view the proposal that the subjective-experience facet of mind may also simply be an emergent property of energy flow, then we could say we have the felt sense of our internally experienced bodily lives. Some might argue that the consciousness that permits subjective experience to be felt is a neural function of the brain, especially the cortical brain for "higher consciousness" such as self-reflection and for interoception—an awareness of the perception of the interior of the body. And this would argue for consciousness being a head-brain property (for discussions of this perspective, see Graziano, 2014; Thompson, 2014; and Pinker, 1999. For a contrasting view of consciousness not emanating from the brain alone, see Chopra & Tanzi, 2012; Rosenblum & Kuttner, 2011; and Dossey, 2014.)

Let's imagine for a moment that energy flow streaming through the body can be felt perhaps even more directly in our subjective reality than its representation in the head-brain, in the neocortex—we feel the churning of our intestines, tightening of our muscles, pounding of our heart—could there be a kind of "body consciousness" that permits this fully embodied subjectivity to be felt at its source? And expanding the source of subjectivity even further, beyond the skin-encased body, when you dive into a lagoon and feel your body swim among the fish and sea turtles, is this really a head-based consciousness, or is it something more? This might mean a more direct experience of body energy than its re-presentation as neural firing patterns in the head. In other words, could we have a body-sense, a direct body-based consciousness and subjective experience? Or is there simply a cortically based consciousness that is aware of the body in the sea, and the sea itself, but that is simply head-based? This has been the stance of medicine for millennia, and the common perspective of science, and maybe it is true. As Michael Graziano states, this view "launched two and a half thousand years of neuroscience" (2013, pp. 4). But what of the source of that subjective experience? Even if the experience of consciousness is a cortical construction, might the stuff we are aware of create its own subjective textures? At a simple level, this says that the input from the body and the input from the world are each forms of stimuli that send signals to the head. We create an illusion of our being fully immersed in that lagoon, but really we are in the camera, not the scene.

So far, then, this line of reasoning leads us to an internal source of mind—something that is an inside job, a solo act. Yet information processing outside of awareness is fully embodied, even if consciousness and the origins of our awareness of subjective experience are placed up above the shoulders. These three facets of mind—consciousness with its subjective experience and information processing— have their locations in your head-brain, your somatically distributed neural networks in various organs, and the body as a whole. Your self, as we've seen in the last entry, as a function of your mind, is a product of your whole body at least. From this point of view, the self and the mind from which it arises is bounded by the boundary of the skin.

We now come to the proposal of possibly broadening this notion even further. Could the mind as a whole be more than only something happening inside of us? Even if consciousness turns out to be a product of the head-brain or perhaps even the body as a whole, the mind, as we've defined it, is more than consciousness, more than subjective experience, and more than information processing. The mind also includes self-organization, an emergent property of a complex system comprised of energy flow, which is, like energy itself, not bounded by skull or skin, as we've said. The self-organizing aspect of mind would not be limited to being a solo-act inside the head or the body.

With the self-organizational facet of the mind, as we've suggested from the beginning of our journey, the mind may not only be within us—it may also be between us. This notion may certainly be a stretch for how we usually think of mind in modern science, or what we are taught at home, in school, or in contemporary society. But let's acknowledge those common top-down constructor filters learned with prior experience and see if we can open our minds a bit.

Let's return first to our other facets of mind—information processing, consciousness, and subjective experience, and see if they, too, may have a source beyond the single body. This may be counter to our deeply engrained beliefs that constrain how we view the world. These cognitive filters basically say this: The mind is an inner process. In this case, we own our minds. Some attribute the mind to the brain, and we've extended that source to the whole body. Okay. Either way, mind is inside the individual, and comes from within the body. Sounds reasonable. To point to the self, a product of mind, we point to our body. Fine, we are on familiar ground.

Now let's relax those common constructive filters, and let bottom-up flow begin to emerge more freely. Let us try to be open to this conduit of sensory experience as we let it arise as best we can without the dominance of top-down construction creating what we think we know to be real, true, and acceptable by others who've taught us and by the larger world surrounding us. That's not so easy for anyone to do, but give it a try. Part of this challenge is to simply sense, rather than perceive, conceive, believe, or think. Naturally, even trying to do this can be a construction—taking the constructed concept of conduition and trying to sense rather than think. But letting that go as best as you can, see if you can simply give your mind over to sensing something as directly as possible. This is opening to presence, to pre-sense, as best we can.

What does being present to betweenness feel like? Can you sense the connections you have with others as you let the focus of your attention broaden, like night vision taking in the subtle light that with day vision would be unseen? With the glare of daylight, we see the boldest of images, taking in the obvious scenes in front of our eyes. But in the dimness of nightfall, our visual system adjusts. In the back of our eyes the more sensitive rods of our retinas become more active, and our vision becomes shaped by input different than our centralized cones of the day-vision colors. With this shift we begin to detect the equally real yet more subtle sources of light that before were undetectable to the perception of our conscious mind. The subtle stars of the sky were there all along, but we could not perceive them. If you had asked us during the day if there were stars up there besides the sun, we'd likely say no, that stars "come out only at night"—but we'd be wrong. We simply could not see them in the glare of the sunlight. Now, with sun having set and our night vision slowly activating, we soon see the visual symphony of a universe unfolding before our eyes—a galactic kaleidoscope there all along but invisible to the eye.

Perhaps the betweenness of mind is like subtle starlight, there but lost in the glare of the daylight intensity of our interior constructive mental lives. We know the brilliant stars are up there in the daytime sky, but the sun's glare obliterates our ability to detect them, just as our thoughts and top-down constructive filters may blare-out the more subtle signals of the betweenness of our lives. Perhaps night vision's equal is a mind vision for the more subtle flow of energy, perhaps even of information, that connects us to one another in patterns that fill our interconnected world. This world of betweenness may be there, but hidden from view.

The patterns of energy and information flow that connect us have a texture to them too, something we can perhaps feel in the moment. With the mind-vision subtle sensitivity of mindsight, we can begin to sense this flow between. We often perceive this as one mind sending signals to another, a glance from you, a sigh from me, a shared smile between the two of us. Perhaps this is all there is—separate minds from separate bodies emitting signals to one another. But perhaps that view is more a part of a mentally conceived reality, perhaps a constructed top-down interpretation. One view from the field of social neuroscience, for example, suggests that separate brains are simply responding to signals from other brains. Perhaps that view could be broadened to one body sending signals to another body. Perhaps this single body view is the one source of subjective life: We feel from the locations of energy and information flow in the brain, or even the whole body. In this view, the mind remains within and our interconnections are simply signals shared. This may be the whole situation—mind within, communication between.

This would keep us within the perhaps accurate, and certainly more common stance, that the mind is only located within a person, within a single body, even in one brain. This personal solo-self view of mind may indeed be the whole truth.

But recall from our opening to this journey that we've named several facets of what the mind includes: consciousness and its subjective textures of lived life, information processing, and self-organization. To offer a brief review of the fundamentally *un*known, no one knows how consciousness and subjective experience arise from the brain, or the body for that matter. Information processing may be a bit easier to discuss, looking at the neural calculations of a parallel-distributed processor of the head-brain, and in our discussion above, perhaps even the gut- and heart-brains. In this view, information arises from energy patterns arising from within the body.

And self-organization? Well, we've already tipped our hand: We've stated the proposal that one aspect of mind is the embodied *and* relational, emergent, self-organizing process that regulates energy and information flow. As you read these words, information flows from this book to you as patterns of energy shared between us. Look at the Internet and how it helps you with information retrieval and storage. Energy patterns stored as zeroes and ones—digital processing. We can see that information flow is not a solo job. This places information processing beyond just your body. Information flow is not only within; it is also between. Self-organization is influenced by the complex system of energy and information flow, and that flow happens where? It happens within your body, and between your body and the world.

We've placed information processing and self-organization in what on the surface seems like two places, locations we can name as within and between. These two facets of mind, information processing and the regulation function of self-organization, are part of the system of energy and information flow that happens within us *and* between us. It seems like two locations, but this flow is not blocked by skull or skin, as we've discussed. These two facets of mind are part of one system, energy and information flow. This proposal to locate mind both within and between may be a big stretch for many people to accept, as it seems at first implausible. Two places at once, how could this be true? Yet if we consider the suggestion that mind emerges from energy and information flow, and that this flow is continuous in what we term "within" and "between," then we can see how one mind can be emerging from one flow. This is not a separating of realities or locations; it is opening our minds to the unity of these locations, these contexts, as the single setting of one process, one emergent flowing.

We return now to our two other facets of mind, subjective experience and the consciousness from which it is experienced. Could these other fundamental facets of mind also arise not just within, within the body and its brain, but also between? Sounds odd, I know, but let's see where this might take us. Let's see if we can relax all the top-down beliefs of consciousness being a cortical construction that would have us dismiss this notion outright, and remain open to exploring this possibility.

Patterns of energy and information flow that interconnect us to each other, and to the larger world in which we live, can be seen as a real process, even without our ability to be aware of them. Where does this flow occur? Within us, and between us. That is a fact. This would put the non-conscious mind—mind emerging from energy and information flow without awareness—fully in the location of not only within, but also between us. So we might say, just as with this flow within us, that when energy and information flow is not in awareness, we don't have a subjective sense of it. Okay. So the betweenness of mind may not have a subjective texture if we are not aware of it. In other words, non-conscious mind is both between and within. In some ways, this is simply reformulating the statement that both the information processing and the self-organizational facets of mind, as emergent properties of energy flow, are within us and between us.

Perhaps some of these patterns of energy and information flow that are between us, as with those within, may also be available to consciousness; these may have a subjective feeling to them. Here is where a bigger stretch within our common views may be needed.

When we let bottom-up fill us, when we extend our perception beyond top-down conceptions of a private solo-self, a personal inside-only mind, we may be able to gain a sense of something more. In the last entry we saw that personal identity could be suspended after an accident, illuminating the constructed nature of a personal, body-based identity. There we saw that a conduit of sensory flow was as real as a constructed sense of a *me*. In a parallel way, could our sense of self also include a betweenness? Doesn't our sense of belonging shape who we are? When you walk into a room of close friends, can you feel that connection? When you sit with close family members, can you feel the history that binds you together? When you've made a new relationship with someone that feels authentic, the next time you greet this new friend can you feel the difference now that you've made that connection? Can you feel not only a sense of connection, but also a sense of a larger *we* within the relationship, in the betweenness that arises, literally, between you and these others?

I am readily willing to acknowledge that this notion of a we-identity might be a projection of my own personal mind, a wish perhaps, a longing, to experience something more than just my solitary self. As is sometimes said, "We're born alone, we live alone, we die alone." Maybe that's all there is. But I long for something more than this. Perhaps my longing may be leading to a constructed view that something really exists, some mind-between, that truly connects us. But maybe that's just a wish, not founded in reality. My doubting mind constructs a scenario in which this is ridiculous: the mind is in the body, that's it. Just settle with that expansion beyond the head-brain, that's enough. But my doubting mind takes it further. Perhaps what many scientists say, that mind is only in the brain, the brain in the head, is indeed true. Perhaps the mind is always just a solo act, even an enskulled act. Perhaps the saying, "That's all there is, folks," is true.

But in my training in psychiatry in the 1980s, there was this odd sense that putting the mind only in the brain, or even only in the body, didn't feel right. There seemed to be something more, something that wasn't talked about, but something you could sense in the room sitting with a family in crisis, or in a group of my fellow residents trying to make sense of it all. There was a family mind, and there was a group mind. Our sense of who we are is created in this collective mental emergence. Not only shaped by it, influenced by signals from others, but living in some sea-between, not only within.

What if this "something more" were more than a wishful mental construction, and actually real? What if the mind as we are exploring it—perhaps with subjective textures in consciousness, and certainly energy flow and information processing sometimes not available to awareness, as well as self-organizational emergence—is actually not only within us, but truly between us as well? What if like fish, who may not notice the sea surrounding them, we also cannot see this mental sea surrounding us? What would this mean? I couldn't quite articulate it back then, but I felt a deep sense of this something more being something important. What are the implications of mind being between as well as within?

To feel this interconnectedness, we may need to use a different set of lenses than those often used by our analytical, constructive minds. I think of my experience away from medical school when I learned to harness a different mode of perception. Whichever side or part of the brain that is utilized, we clearly have differentiated modes of processing information. As we've seen, the left mode is dominant for logical reasoning, looking for cause-effect relationships in the world. The left mode is also active in using linguistic language, like these words, in making linear, literal statements. Yet I had learned in my time off, which was more like time on and in, that a different mode, one we can simply call a right mode, whichever side or part it emanates from in the brain, senses things differently in this right-mode conduit function. The interlinked perceptions of the right mode, too, have a differently constructed texture. In this right mode, the interconnectedness of things as "context" is grasped as relational meaning rather than merely perceiving the specific individual details, the "text," without sensing their interrelationships; the spirit of the law is sensed, rather than the literal interpretation of the law the left mode tends to emphasize. While these distinctions are sometimes hotly debated, Roger Sperry just a few years earlier, in 1981, had been awarded the Nobel Prize for discovering some of these hemispheric differences. Right mode perception sees between the lines and text to envision the interconnections shaping the whole; left mode constructs lists and labels to decompose the whole into constituent parts, to analyze ("down-break") rather than see the "big picture," often missing their interconnectedness (McGilchrist, 2009). Could it be the situation that the left mode may be able to readily study the withinness of mind and its origins in neural functioning, but is perhaps more challenged to perceive the reality of the betweenness?

Seeing the mind fully with its many facets may require both modes of perception. We need both day- and night-vision to see in this world. It may be that the more subtle patterns of energy and information flow interconnecting us, the full betweenness of the mind, are best perceived by the contextual right mode.

The scientist Michael Faraday, back in the 1800's, was one of the first to suggest that "fields" might exist connecting separate elements in the world to each other. From this view we can now use many aspects of electromagnetic fields in modern technology. Once my father-in-law, a farmer, was speaking in Los Angeles with his traveling grandson in the Ukraine over a video link using my smartphone. He looked astonished at seeing his grandson's face on the phone. I passed my hand around the wire-free object, visually demonstrating what Faraday's wonderful constructive mind had proposed: Electromagnetic fields in the form of patterns of waves are invisible to the naked eye but nevertheless real.

Before Faraday, we would have thought this impossible, if not some form of witchcraft. Today, with our smartphones and so many other electronic gadgets, we don't think twice about it. Usually we don't even think once about it, we simply use the devices. Couldn't it be possible that the human body and its brain or brains have some sort of send and receive functions parallel to those of the devices we've invented? Even if they don't have such abilities, the five senses we do have may pick up a range of signals that fill the air between us, letting us connect in ways that are real, though not always available for conscious perception. Sensing can be without awareness; and perceptual images we construct from those sensations can also be present without awareness.

But perception, with or without consciousness, can be trained. If the mode of this potential perceptual capacity is real, perhaps it could be developed in each of us to various degrees so we can sense this betweeness as well, or at least develop it better? As we've discussed earlier, some call this essense of what connects us a "social field," (for example, see the writings of Otto Scharmer and Peter Senge, or the related work of Nicholas Christakis and James Fowler), one that embeds us fully in the complex system of our interconnectedness.

Some of this connection is revealed in direct visible or audible communication, but we don't yet understand whether some of the other "stuff" of this betweenness might be electrical fields that could one day be measured and picked up by our sensory systems directly, or by some other yet to be identified factors.

One of those other factors might be the empirically proven physical aspect of the universe called "non-locality" or "entanglement" in quantum physics terms. Some scientists roll their eyes and say, "Oh, we shouldn't evoke quantum physics in discussing the mind." They've also said we shouldn't evoke discussions of energy and apply that to mind explorations. Maybe they are right, mind has nothing whatsoever to do with energy. What do you think? I personally don't think it makes scientific sense to limit our exploration in those ways, to avoid scientific discussions of energy in inquiring about mind, but maybe that's just my mind thinking something it shouldn't. So we'll just say here, as a reflective statement, that if mind is indeed an emergent property of energy, if that is true, then the science of energy ought to be brought into the discussion of mind. Careful research studies over the last century have revealed that energy, and now in this last year even matter (which is condensed energy, after all), have properties linking them that exert measureable effects on even physically separated locations (Stapp, 2011; Hensen, et al., 2015; Kimov, et al., 2015). In other words, elements (such as electrons) that were once coupled can become physically separated and then directly influence each other faster than the speed of light, even over long distances. That is a proven finding. We need to be cautious while also scientifically open to applying these non-local empirically established aspects of entanglement to questions about mind.

In this case, we can propose that it is logical, empirically supported, and a reasonable approach to say the following: 1) Energy and matter have a betweenness to them, a fundamental interconnectedness called entanglement, that is not visible to initial measurement or the five senses; 2) Given that this is empirically established, the question about mind experience having aspects of this interconnectedness is a reasonable, scientifically grounded question, not a conclusion or assumption; and, 3) We don't have a definitive answer if this is true—it is simply being raised as something science would logically suggest we consider, no matter what some scientists may say. Science supports this approach of inquiry in our exploration. We take this conservative approach, emphasizing as strongly as possible that such broad and in-depth questioning is reasonable as a part of inquiry, not with foregone conclusions of what the answers to these questions may ultimately be.

Might we develop a more subtle vision to sense the nature of our interconnectedness, such as fields of energy flow or other aspects of reality that link us, such as entanglement, those that may shape our subjective experience and other aspects of our minds yet are not often seen by our daytime vision? Wouldn't those non-local connections and fields of energy, real aspects of the physical world, have impacts on us even if we are not consciously aware of them at first, or perhaps even ever?

A case in point is light. Light varies along a spectrum of frequencies. Our rods and cones in the retina pick up a small range of those frequencies. But some light, such as ultraviolet (UV) waves, cannot be seen. Still, too much UV and you burn. Wasn't the UV there all along even if you didn't see it with your eyes? We are just beginning to learn about the effective use of subtle transmission of photons, beams of light, for healing (Doidge, 2015). Sending photons toward the surface of the skull has been demonstrated to alter the pattern of measureable firing in the brain. Photons going through the skull? Yes, energy is not constrained, as we've said, by skull or skin. So what we once thought impossible is now being revealed as true: Light can pass into the body's skin and bone and influence the brain. That's one example of how energy patterns may directly shape us—in this case, light that's not being taken in by the eyes. Why wouldn't there be many other energy fields, as Faraday suggested and we now know exist, and other properties of energy and matter that might also be influencing these bodies we live in? If mind is embodied and relational and an emergent property of energy, how could or why would the mind be independent of these factors?

As we will explore in detail in our entry on the when of mind, the scientific investigation of time also invites us to imagine how our mental lives may live at one level of observable reality—the quantum level—whereas the large-scale bodies and larger object-filled world we live in are often most readily observable at the classical Newtonian level. Quantum effects have been found, as we discussed, even at these larger size-scales, but these effects are often hidden beneath the glare of classical physics' properties. This may be another way in which mindsight empowers us to perceive our interconnected mental lives more than physical sight's view. It may just be, as we'll explore in the entries ahead, that we live with a quantum mind with all its various quantum properties readily active but in a classical Newtonian body and world where those more subtle but nevertheless real properties are often obscured.

I think back to my debate opponent who said that feelings were not real. Perhaps his perceptual system made his conscious mind devoid of emotion. He seemed to truly believe what he said, he told me later over some beer. We can honor the universal reality that perception is a constructed skill. He may have had emotions, even sensations of these internal processes, but, for whatever reason, they did not enter his conscious experience, so he never was aware of perceiving them. In a similar way, perhaps many of us don't often have the energy flow or other aspects of our connectedness, the mechanisms of our betweenness, enter consciousness. Perhaps we don't perceive this in-between interconnectedness. But if the sensory flow of those various possible mechanisms may be there without perceptual construction or conscious reflection, couldn't it be possible to develop and strengthen some kind of skill of accurately perceiving the world between?

It may be helpful to keep an open mind about what we think we may know and what reality may hold as discoveries unfold.

If subjective experience, somehow, is an emergent property of energy and information flow, and that flow happens both within and between, could we have betweenness as one source of subjective experience? Subjective experience might be within and between, a function of awareness of that energy flow, just as self-organization is within and between, but in this case with or without awareness.

What this might mean, though, is that if subjective experience is only felt within awareness, and awareness is also some kind of emergent property of energy flow, could the betweenness of energy flow give rise to some form of consciousness? The within source of energy flow that "gives rise" to consciousness is explored with the many neuroscience studies examining the neural correlates of consciousness. But we don't currently have the same sort of science to support the notion of consciousness arising from outside the body, or even outside the brain.

Could there be a betweenness source of consciousness, a form of awareness that arises from between? Many scientists would simply say no. But the truth is, we just don't know yet. And wouldn't such a between source of awareness then have a subjective texture to it? Even if such a source of consciousness is not identified, even if it actually does not exist, wouldn't we still have a way to feel, within awareness, the betweenness aspect of mind? Just as we can subjectively feel the wind on our face, the water surrounding our bodies as we swim in a lagoon, or the twirl of our synchronized bodies while waltzing with our partners, couldn't we subjectively experience the betweenness of mind? In other words, could we use our internally generated awareness—if that is its only origin—to feel, to subjectively sense, the betweenness of mental life even if consciousness itself turns out to be primarily an inside job?.

This betweenness of mind may actually be more shareable than the withinness. These were deep concepts that possessed me during my initial training to be a psychiatrist. I learned firsthand about the importance of sensing the mind between as a young therapist at the beginning of my psychiatric residency program.

One of my first patients as a trainee was a graduate student who came in with depression following the death of a colleague. We dove into psychotherapy, me not really knowing what to do, she open to seeing what happened. At the end of her year and a half of treatment, she was done with therapy, her depression and grief resolved, her life feeling full. Soon she was ready to move on to her postdoctoral position at another university and we were preparing to say goodbye.

I was curious to know what had been the exact aspect of the experience that seemed to have helped her, so at the end of treatment I said we would have an "exit interview" to review what had been most helpful and what could be improved. "Great idea" she said. I asked her, "What was most helpful to you?" "Oh, that's obvious," she replied. "Yes," I said, "I know, but if you had to put words to it, what would you say?" She paused for a moment, looking at me with moist eyes, and said, "You know, I've never had this experience before. I've never had this experience of feeling felt by anyone. That's what helped me get better."

Feeling felt.

I had never heard of such an eloquent way of expressing the connections we have with another person when we are felt, understood, and connected.

Between us was a sense that I was focused on her inner, subjective experience of mind. In that joining, she felt felt. Now I can imagine that feeling of resonance, of me being attuned to her inner world and changed by it, how that might have given her a sense of trust.

And with that trust, she and I could explore the inner world of her mind that was troubling her so deeply. The mind that emerged in her as we worked closely together—the resolution of her traumas from a painful past with her family, her feelings of helplessness in the present with the death of her colleague, her experience of hopelessness for the future that gave her a sense of despair—these were now resolved. Trust was the gateway to our journey to heal those wounds. What was the healing action? Feeling felt. I could be present for her, attuning to her inner life—her inner subjective reality—and then resonate with that reality. I might even attune to that inner world and connect with information processing that she was not in touch with, aspects of her non-conscious mind. And that, too, could shape my internal world even if it was not in her awareness. I was changed because of our connection too. This is the experience of resonance.

Sometimes that resonance can be consciously recognized; we are overtly aware of feeling felt as we sense we have become connected as a *we*. But such connection may be registered inside us even without participation of our conscious mind. Non-conscious information flow, a part of something called non-focal attention, can take in signals and create representations of experience without accessing awareness. These conscious and non-conscious experiences of our connection can create trust.

We'd learn later that trust facilitates learning as it harnesses neuroplasticity—the ways the brain changes in response to experience, which we'll soon explore. I am an acronym nut, you now know, so you can see that the essence of the therapy that created the experience of feeling felt for my patient was the *part* I provided in our relationship—*p*resence, *a*ttunement, *r*esonance, and *t*rust—PART.

This experience also gave me the sense that the mind is not just within us—it is also between us. But at the time in my psychiatry program, after having finished medical training, I wondered how in the world that could happen. If the mind were simply brain activity, or simply personal and private, how could a relationship with its deep sense of connectedness involve such a powerful sense of the betweenness that was there? As we've just discussed, perhaps this sense of connection was simply a mirage of our individual minds. Was

I just imagining this, projecting onto a relationship something that is really a solo process, a mind in isolation? Aren't we really all alone? Or could the mind's location actually be both personal and within while also interpersonal and between? Is there an actual social field in which we are truly embedded? Just like Faraday's fields, could this social field be real and created by shared energy flow patterns that happen when we are in proximity with each other? Was I simply attuning to this field? Could other couplings be arising beyond those that arise in the between of proximity, such as the wild, theoretically-sound but open question of entanglement? Could the focus of one mind on the internal world of another, that betweenness of feeling felt, additionally involve some entrainment of energy patterns such that they then later exhibit this aspect of non-locality that would occur even when we are not in physical proximity with each other? Is this why and how we come to feel so in-sync with our life partners? If mind is an emergent property of energy, the empirically-proven non-local nature of coupled energy and even matter would suggest at a minimum the potential reality of such entangled connections that occur at a distance. This range of possible ways of feeling felt is open for exploration and study.

And so at this point in our journey, we've taken on the question of the where of the mind by opening our own minds to the possible between location of our mental experience. The within origin of mind, while located in the brain in our heads as a notion for over 2500 years now, ought to be considered with an open mind as well. Is the mind truly within? Or is this an illusion, just as my sense of a personal self was knocked out after my horse accident, that the mind itself creates? As a young trainee, caring for my patients made the betweenness of mind come alive; and just as real, the importance of the internal locus of mind came into sharp relief.

Around the same time as caring for my patient who said she "felt felt," another patient's experience with his trauma from fighting in Vietnam also helped me understand the power of therapeutic relationships and the importance of seeing how the mind is both embodied and relational. Bill was filled with flashbacks and other intrusive symptoms of posttraumatic stress disorder (PTSD). None of my supervisors could help me understand how trauma might impact a person so as to give rise to PTSD. So, having been trained in biology and loving my neuroscience training in medical school, I turned to the biological research on memory, and found emerging scientific studies on a part of the brain—the hippocampus—which integrated earlier layers of implicit, sometimes called non-declarative, memory into more explicit or declarative forms. Learning about the embodied ways in which our neural circuitry mediates memory enabled me to make sense of the otherwise confusing profile of PTSD symptoms in my patients.

I would later meet with researchers who discovered these distinctions, and would propose the following:

The part of the brain that integrates memory from initial input to accessible and flexible reflections on a past event is the hippocampus. The hippocampus, when damaged, blocks the transformation of memory from initial layers to more elaborate mental representations that can be reflected upon later on. These were the constructive mechanisms of memory—from encoding to storage to retrieval—which then move memory from early, basic forms into more complex aspects of autobiographical recollection. Cortisol, the hormone released with significant stress, shuts off the hippocampus in the short run, and, if prolonged in its secretion, can even damage the hippocampus. Since trauma leads to a secretion of excess cortisol, could the hippocampus be temporarily shut off during an overwhelming event and in some cases damaged following the event?

There was other experimental data that seemed relevant to what many patients were describing. If the focus of attention were directed away from one aspect of the environment, the hippocampus would not be engaged in the encoding of that part of the experience and the explicit, flexible forms of memory would not be created. So if people could divide their attention, a process of dissociation in which they paid conscious attention to a non-traumatic image or aspect of the world, then they might also distract the hippocampus that needs focal, conscious attention to be engaged. If that dissociative process were initiated, hippocampal processing could also be blocked.

Yet non-hippocampal implicit memory does not require we focus our conscious mind in order to encode those elements of memory. Might, in fact, the excess secretion of adrenaline during a trauma lead those non-declarative aspects of memory, the implicit components of emotion, perception, bodily sensation, and behavior, to be even more strongly encoded?

The reason these questions seemed so important was that without hippocampal function, the implicit layer of memory would remain in a pure form. The research I was reading revealed that when pure implicit memory was retrieved from storage, when a cue elicits its activation, the emotion, perception, bodily sensation, or even behavior is engaged, but is not tagged with the sensation of something being recalled. That blew my mind. In other words, the basic science of memory might be useful to clinically explain how a traumatic event could be encoded only into implicit memory, and later on come back when retrieved as the wide array of symptoms of PTSD, including intrusive memories and emotions, avoidant behaviors, and even flashbacks. A flashback might be an array of implicit-only memory retrieved fully into awareness but without the explicit tagging that something is coming from a time in the past.

The researchers I met with affirmed the theoretical soundness of the hypothesis. In my clinical work, I found that with these new findings from the science of memory and the brain, there was the possibility to explain the mechanisms beneath the pattern of suffering for those with PTSD. Even more, this framework could guide a new approach to therapeutic intervention. If I could play the proper PART with my patient—being *p*resent, *a*ttuned, and *r*esonating so that *t*rust was cultivated, then the conditions for change could be created. But something more had been needed beyond the relational connection—this betweenness of the mind in therapy. I also needed to understand the inside of the body, to know how to effectively work with the circuitry of the brain, in order to get to the withinness of the mind. The mind was in two places—between and within.

Our communication and connection allowed a betweenness of our minds to emerge. With individuals or in group settings, with a couple or larger family, being present felt like coming with an openness to energy and information flow that invited something to emerge in therapy that could easily be shut down or warded off. That shutdown seemed to emerge from a reactive state of threat, something you could feel in the social field, the spaces between. You could feel it at times when you entered the room. The betweenness of mind for me as a young therapist felt palpable. How we as therapists set the tone of communicating would facilitate, or inhibit, the emergence of energy and information flow that filled the room. If I had not resolved some aspect my own life's history, my patient's experience would be hampered. You could feel such a conflict in the room. That room-filled flow was something that had a texture to it, a feeling to it, a substance to it you could almost touch. It felt real. You could sense the social field with mind-vision. The mind, emerging between therapist and the individual, couple, family, or group, could be felt.

No one was talking about it; but that did not make it unreal. I recalled my days of dropping out of medical school when the mind had been absent in the clinical instruction I was receiving. But the concept of mindsight helped me recall that the mind was real, even if a training program, a medical socialization process, acted as if it was not real. Could betweenness be a place of mind as much as withinness? I wondered about this as a psychiatric trainee back then. There just weren't any conceptual notions that could anchor that question in some framework of inquiry.

The betweenness that was shaped by my presence in psychotherapy seemed to open a gateway to focusing more freely and fully on the withinness of the patient's mind. How could I work to change the circuits of a given patient's brain, ones that were negatively imprinted by experience, in a way that created and consolidated positive changes in their condition? If lasting change were to occur within therapy, wouldn't the brain of the patient have to change?

Since a therapist cannot perform neurosurgery, what could be done to change the patient's brain in a positive way?

Experience. Experience means the streaming of energy and information flow—between and within us, shared among people, and through the body, including its brain. Therapeutic experience would mean streaming that flow in a particular way. The patient could feel felt between us, but what would I do for the withinness of that flow? Even if I knew how that should be done, how could it be achieved?

The answer seemed to be attention. Attention is the process that directs the flow of energy and information and can be created in communication between people as well as within people. So if I write to you the words "Eiffel Tower," I just invited your attention to be directed to that architectural icon of Paris. Communication within relationships shapes attention interpersonally, and it guides attention internally. With attention we direct the flow of energy and information. In consciousness, when attention is coupled with awareness, we call this focal attention. Research revealed that focal attention is what is needed to activate the hippocampus.

The between and the within of mind are deeply interconnected. Neither the skull nor skin is an impermeable boundary for energy and information to flow. That flow happens between and within us.

My clinical hypothesis was that if I could bring my PART to a patient, trust could develop. And within that trusting relationship, an open sense of presence and receptivity would emerge in the betweenness, then in the withinness of mind. I could guide the patient's focal attention toward aspects of implicit memory—emotions, perceptual images, bodily sensations, and behavioral impulses—so they could be linked together by the hippocampus to achieve the integrated state of explicit memory. I could even work with the mental models or schema, and the priming processes of implicit memory as well. When focal attention rests on these elements of implicit memory, the possibility arises to transform them into the more integrated explicit layers of memory. Explicit processing by the hippocampus creates both factual (knowledge of the fact of something from the past) and autobiographical (knowledge of oneself at a time in the past) memory which, when retrieved, would have the clear feeling that this is coming from the past. I called that an "ecphoric sensation," the subjective sense that in this moment, what is being experienced is something coming from a previous event, not something happening now. It is this ecphoric sensation that was missing with pure, implicit-only memory retrieval.

The trusting relationship would create an increased tolerance to "be with" the emotions, images, bodily sensations, and behaviors that before may have felt too frightening, too overwhelming, to be taken into awareness. In that social field of acceptance and receptivity, what before was unmentionable and unbearable became nameable and tamable, like the widening of what seemed like a "window of tolerance" for that particular emotion, image, or memory. Therapy would widen windows of tolerance so that focal attention could integrate previously blocked implicit memory into its fully integrated explicit forms of facts and autobiographical knowledge. Within the boundaries of the window, energy and information flowed in harmony with flexibility, adaptability, coherence, energy and stability—the FACES flow of integration. With a narrow window, an individual's flow would tend to move beyond the boundaries into chaos on one side, or rigidity on the other. Widening the window meant bringing more harmony into a person's life by enhancing integration. All of this felt to be happening both between and within.

Further, implicit memory has created generalizations of those events into a schema or mental model, such as "all dogs are bad" emerging mentally if a dog has bitten you. These are the constructor's means of making more efficient cognition, as we've explored in the last entry. The readying for a future event, called priming, is also an implicit element that could be dissolved with effective treatment. These aspects of implicit memory, when isolated and un-integrated, might be the neural foundations for the mental suffering of PTSD.

I was eager to try out the hypothesis and see if such an approach could help. While it was not a controlled study of hundreds of patients in a double-blind set-up, these clinical observations came from many single cases without a research control group. I knew as a science-trained person that these clinical applications and findings were simply anecdotal, not empirical data. More like the initial findings of a pilot study, the therapeutic effects were positive and robust, lasting for months and years, and later on, in some cases, decades after treatment. It was exhilarating to see how science could be applied to help reduce suffering in new and effective ways.

This journey combining scientific, objective findings with the subjective reality of the mind implied that there was likely a direct linkage among mind, relationships, and the body's brain. Mind and matter shaped each other. There seemed to be no need for the frequently stated separation that these linguistic packets imply. Energy and information flow connected all three into one whole of mind, body with its brain, and relationships.

But how could a broad conceptual understanding combine molecules and mind, neurons and narratives? What was the essence, I wondered back in those early training years, which linked mind, brain, and relationships? This was all happening before the Decade of the Brain, before that experience of being with that group of scientists who revealed this lack of a shared definition of mind, and before imagining that the mind might be an emergent process of self-organization arising from and regulating energy and information flow, within and between.

There was something about these questions, the relationships I had with my patients and the researchers, as well as the research ideas, that felt like a sea of interconnected information, some energy that was being shared in each of these relational settings. I know it may sound odd, but there was a kind of energy and information atmosphere, a *mindsphere*, I would laugh and call it, that surrounded us and suffused our lives with an often invisible atmosphere of energy flow that shaped how we experienced our worlds. A consilient idea I'd learn of years later is the notion of a "noosphere" (Levit, 2000). Like fish swimming in the sea or birds in the sky, which may pay little attention to the water or aquasphere and air or atmosphere surrounding them, we may often be unaware of the information bathing us, our noosphere. Perhaps the mindsphere may not just be something that affects us, something we breathe, but instead something from which we emerge, if indeed mind emerges from energy and information flow. Could our experience of mind be emerging not only from within us, but from aspects of this sea of energy and information, perhaps creating our minds as the emergent source of our mental lives? More than information, the mindsphere would include our shared exchanges of energy and information, linking us to each other, not merely serving as sources of stimuli. The mind, then, might be truly relational, not just sensitive to these forms of input.

As I walked along the paths of the University of California at Berkeley this morning, now thirty years later as I write these words to you in this second decade of the twenty-first century, I could sense how the bustling students were living in a mindsphere that filled them with ideas. A place of learning like this, and even our larger society, surrounds us in a sea of information created by energy flow patterns that shape who we are and how we unfold. My daughter invited me to attend a class on ecological systems, and I learned a ton—both from her letting me know what was going on, and from the teacher. Professor Paul Fine offered a definition of an ecosystem as a complex system with a topographical unity of a volume of land, water, and air that extended over a portion of the earth for a certain period of time. Part of the class's task was to memorize the woody plants in a range of local ecosystems they were visiting. The assignments were apparently daunting, as the professor elected to quote his own teacher, Burton V. Barnes, in reflecting on how hard the emotional experience was, initially, to commit these names to memory. Here is the list that perhaps may illuminate how this journey into the mind may be unfolding for you too: denial, anger, whining, confusion, acceptance, resignation, and joy. I know this deep dive into mind may at times feel daunting, too. I hope we are in—or at least moving toward— the joy stage.

Our minds live in such mind-systems, though they span more than topographical space. Just as the atmosphere influences local landscapes within various habitats of the ecosystems it envelopes, so too, the mindsphere influences our individual mental states, our personal inner landscapes, what we can call our *mindscapes.* We all have localized personal habitats with mindscapes shaped by the larger mindsphere in which we are all collectively embedded. In many ways, these notions of mindsphere and mindscape reveal the inter and inner nature of the mind, the between and within.

But even if we could state that the *where* of mind was both within and between, what was the scientific foundation, I wondered thirty years ago, that could link these two seemingly distinct locations of one singular mental entity? Within and between ought not be separate entities, but two aspects of one reality, the reality of mind. I was desperate to find ways to join with others to work these fundamental issues into some useful conceptual framework for all concerned, from patients to clinicians and scientists.

After training in adult and then child and adolescent clinical psychiatry, struggling with the realities of a field trying to find its own clinical identity, I decided to become a researcher and obtained a National Institute of Mental Health (NIMH) Research Fellowship to study family interactions at UCLA. I was urged to focus on some specific disease or medication treatment, and my academic mentors saw little merit in studying attachment relationships. But I was intrigued by the findings in attachment research and how they illuminated these questions about the *where* and *what* of the mind. I wanted to study the ways in which a person's life narrative had been shown to be the most robust predictor of how their own children would become attached to them. In other words, it turned out not to be what happened to us as parents that mattered most; it was how we'd made sense of what happened to us that was empirically shown to be the strongest factor that influenced how our children were attached to us. And that attachment would, in turn, lead to positive or negative influences on the development of that child's mind—how she regulated her emotions, interacted with others, made sense of her own life. Something about the making-sense process of the parent created a mindpshere in the family that supported secure attachment. What could that be?

I was fascinated with the field of relationship science, and saw deep potential in starting with healthy relationships and how they impacted the development of the mind before exploring more about how a mind might not develop toward health and resilience—to push its growth toward dis-order and dis-ease.

My own teacher of neuroscience, David Hubel, had, along with

Torsten Wiesel, received a Nobel Prize when I was in school back in 1981 for demonstrating how experience shapes the structure of the developing brain of kittens. Experience during early development had a major impact on the structure and function of the cat's brain. Those findings and his teachings impacted my professional development, imprinting in me the reality that our brain is shaped deeply by experience. Surely there must be ways of taking the advances in our understanding of brain structure and function and weaving them with how relationships, as a form of experience, shape the mind.

During these years of my clinical training, I wondered if some of Hubel and others' discoveries might be useful in helping patients heal from trauma or find strength to face their own inherent difficulties in mood, thought, or behavior. How patients recovered from trauma seemed directly related to how they made sense of their lives and created a coherent narrative of their life experiences. Could the hippocampus' role in integrating memory play a part in this making sense process? This was exactly what attachment researchers had independently uncovered: Someone who had made sense of their trauma would not pass that trauma on to their own children in the form of disorganized attachment; someone who had been traumatized but did not make sense of its impact on them would be, sadly, very likely to pass that disorganization on to their own children. Could the resolution of trauma be connected to some core making-sense narrative process that was related to a fundamental process in the brain? How would this making-sense influence the social field of the family to support the development of secure attachment and a healthy mind?

The first attempt to use the discoveries of the hippocampal processing of memory had been useful, at least conceptually, in approaching an understanding of and developing a treatment strategy for those with PTSD. Although we were not taught much, if anything, about the brain in our clinical training back in the 1980s, as surprising as that may seem, my first therapy experiences and prior training as a biologist brought me back again and again to the question of how my patients' minds might be best understood by some interaction of the brain and our relationships. Both were important. The mind was within and between, and that is where mind-therapy, psychotherapy, seemed to need to be focused to work well.

If experience did change the activity of the brain, and that change could alter developing brain structure as Hubel and others had revealed, perhaps parent-child interactions molded the child's brain. Might the relationship between a therapist and patient also shape the brain's function and structure? If so, how could that effect be understood? If these relationship experiences were shaping the brain's structure, could that lead to lasting therapeutic improvements in a patient's life, with or even without medications?

These questions filled my own mind each day of working in science and working as a therapist. Was there a way to think of what a healthy mind, brain, and relationship were all about? Was there really a way to consider that the mind could be both within and between?

Neuroplasticity and Cultural Systems

An old teaching in neuroscience was that the brain had very localized functions and basically stopped growing with adulthood. We now know that the extremes of both views are not true. Functions in the brain—such as memory, emotion, or even motor activation—are widely distributed, not just found limited to one small area or another. Even strict divisions between sensory input and motor output appear not so fixed. The distribution of maps in the brain, those activated regions that work together to shape mental functions, seem to continually change in a dynamic way as we go through life's experiences.

The brain also continues to grow throughout the lifespan. Yes, there are early periods of important growth when the brain is vulnerable and needs certain input to shape its development in healthy ways. But the brain does not stop growing after childhood, or adolescence. Changing in the long-term in four fundamental ways, the brain takes in experience with the activation of neural firing. This firing, at a minimum, can lead to temporary, short-term chemical associations among neurons revealing immediate or short-term memory. But longer-term impacts on brain structure can happen even in adulthood. These changes can involve: 1) the growth of new neurons from neural stem cells, at least now documented in one area in adults, the hippocampus; 2) the growth and modulation of synaptic connections among neurons, changing their ways of communicating with one another; 3) the laying down of myelin by the supportive glial cells, enabling action potentials of ions flowing in and out of the neurons' membranes to stream 100 times faster and the resting or refractory period between firings to be 30 times briefer (30 X 100 = 3,000 times not only faster,

but more coordinated in timing and distribution); and 4) the alteration of the non-DNA molecules that sit on top of the DNA called epigenetic regulators, such as histones and methyl groups. Epigenetic changes are induced by experience and then alter how future experiences will allow the genes to be expressed, proteins produced, and structural changes to unfold.

These ways the brain is altered by experience are a part of what is called the neuroplasticity of the brain. The discoveries of how moldable the brain is in response to experience would soon create a revolution in how we can change with experience—and open the doors to how relationships and the mind might be understood to alter the brain itself.

These findings can be woven with our discussion of mind as emerging from energy and information flow, within and between. How attention directs that flow will activate certain neural pathways and activate certain interpersonal experiences. Within us, attention drives the activation of neurons in the brain, at a minimum. Perhaps this inner attention drives energy flow throughout the whole body. When we communicate with one another, too, as I write and you read these words, we are also harnessing the power of attention to direct the flow of energy and its symbolic forms we call information. Energy and information flow between us as well as within us. With shifts in neural activation, the opportunity to change the structure of the brain is created. With shifts in external attention, the opportunity is created to alter the internal neural firings that shape not only the activity in the brain in the moment, but also alter the structural connections in the brains of those engaged in the interactions, in the communication, among people in the world. What this suggests is that the mind, within and between, can change the structure of the brain.

The way I have come to recall this connection of attention shaping energy and information flow to neural activity and growth is this way: Where attention goes, neural firing flows and neural connection grows. This not only helps us understand how psychotherapy and parenting work, but also how our societies shape our minds as well.

From a systems science perspective, in the discussion of the interlinked connections in which we are embedded, we can use the term *culture* to focus on several features of our lives. Peter Senge describes three layers of how this system can be experienced: events, patterns, and structure (Senge, 2006). On the surface we see events as visible results of the system, what he describes as the "tip of an iceberg." Just below that tip are the system's patterns of behavior, not directly visible within a single event, but nonetheless patterns that are present, impactful, and detectable with recognition of hidden patterns within events. Further beneath visible events and patterns of behavior is the system's structure, which can be described as having these further three components: habits of thought, habits of action, and artifacts (the physical aspects of our culture, such as diaper changing tables in a men's restroom in modern societies, reflecting and reinforcing the system's stance that men should care for infants, not just women). The system's structure is not often seen at first glance, but can be perceived when we look deeply at these habits of thought and action and artifacts underlying the system's patterns and events.

Similarly, we can never consciously "see" Faraday's fields with our bodily apparatus, but those energy patterns are real. We do not see non-local entanglement of some elements with others, but this entanglement, for particular pairings such as energy and matter, is real. Likewise for systems, we may be able to see events, compile these events into a perception of a system's patterns, and even detect aspects of habits of thought and action, but many of these aspects are hidden from plain sight, especially at first glance. Artifacts of a system's structure are indeed visible, and the ripples they create in how we think and act, especially how we interact with one another, can be ubiquitous and visible in our culture, yet often not appreciated for the ways they shape our mental lives. They can be like the sea that surrounds us but often goes unnoticed. The various aspects of our culture's features shape us with and without our awareness. This is our mindsphere.

Imagine this scene. Energy and information surround us. Their flow enters us. As energy enters the nervous system, neurons fire. As these neurons fire, any of the four changes of neuroplasticity can be evoked (neural, synaptic, or myelin growth and epigenetic modifications regulating gene expression). In turn, as neuroplastic changes happen in the brain, the energy and information emitted from those changed nervous systems also change. In other words, energy and information patterns sent from an individual are shaped directly by the kinds of neuroplastic changes induced by the mindsphere. This change in energy and information flow emitted from the brains, and the bodies, of all the individuals in that sphere then shape the social field. The system's events are changed and can be seen by the eye, but the system's processes, and even its structural components of habits of thought and action, may remain inaccessible to everyday surface vision. As events and artifacts change, we more readily see what is shifting in the mindsphere. This social field is influenced by the invisible neuroplastic changes arising within the individuals comprising that system; and the social field is then inducing shifts in the mindsphere—the flow of energy and information between us within a culture—that in turn change how neurons fire and rewire. This is the recursive, self-reinforcing nature of the mindsphere: It induces changes in the very neural structures that maintain and shape the form of that flow.

This perspective helps us understand the nature of how our relational minds shape our ways of communicating and connecting to one another in what is called *cultural evolution*. Evolutionary views of how we've evolved over the last forty thousand years suggest that while our brains may have achieved a certain state of genetically determined anatomic evolution about ninety thousand years ago, some shift in how we began to have the capacity for symbolization dramatically altered our cultural lives as revealed in how we created tools and visual representations in this Upper Paleolithic period, such that the distributed mind became the driving force for our cultural evolution (see Dunbar, Gamble, & Gowlett, 2010; Johnson, 2005). We can view these suggestions as how shared symbolic forms within the mindsphere, such as how we use language, then stimulate changes in our mindscape that can induce neuroplastic changes that in turn support a more complex mindsphere.

The philosophers Andy Clark and David Chalmers suggest that language

> is not a mirror of our inner states, but a complement to them. It serves as a tool whose role is to extend cognition in ways that onboard devices cannot. Indeed, it may be that the intellectual explosion in recent evolutionary time is due as much to this linguistically-enabled extension of cognition as to any independent development in our inner cognitive resources. (1998, p. 17)

The key is that energy and information flow occurs throughout the system: within the body, the internal, embodied, personal mind we have called the mindscape, and between the body and the world surrounding it, our collective, relational, mindsphere. To deeply understand the *where* of mind, we can embrace the powerful insights of neural science and systems science. We can see the common denominator of these seemingly separate views as energy and information flow. This flow is not limited by skull or skin, but suffuses our lives in systems within and between.

Even if consciousness is ultimately affirmed to be an emergent aspect of our mindscape alone, the personal workings of our brain and perhaps the body as a whole, we can become aware of the mindsphere and have a subjective sense within consciousness of this betweenness of mind, just as we can be aware of the mindscape and have a subjective sense of the inner aspect of mind. Within awareness or without awareness, our mindscape and mindsphere depict the embodied and relational mind that shapes and is shaped by the neuroplastic changes of the brain and the culture in which we live.

Reflections and Invitations: Within and Between

When the notion that the mind might not be simply a product of brain activity alone first arose for me as a student, it seemed like heresy in scientific conversations in the professional world I was living in at

Photo by Lee Freedman

the time. But as a trainee in psychiatry with a research background in biochemistry, it seemed that the systems being explored ought to have some universal aspects to them, and some reciprocal influences on each other. The salmon I studied in college might be responding to changes in the salinity of the surrounding water by altering an enzyme that changed the density of particles in the fish's bodily fluids. The fish adapted to changes in the environment, as the fish's own nervous system was programmed to have it swim from freshwater stream to salty sea. Likewise, a person in crisis could respond to ways in which the receiver on a suicide crisis helpline communicated with compassion and connection, instilling a sense of not only trust and feeling felt in the moment, but hope for the future.

I had learned, too, that starting a medical school experience and finding the socialization process unhelpful didn't mean I had to passively adapt to my environment's demands. I could take time away, taking time-in to reflect on what had been happening, develop new modes of perceiving the world, then return with new concepts and perspectives. The mindsphere surrounding us does not have to be the one drowning us; we can change the world around and we can change the world within. Knowing that the mind was real, using the concept and process of mindsight as a protective shield, helped my internal world develop enough perspective and resilience to re-enter that medical world and try, as best I could, to hold on to what felt real, like the mind and its subjective reality. Now in psychiatry training, the issue being raised deeply in my mind was the following question: Could the mind be both within us and between us?

Enzymes and empathy. Brains and relationships. Matter and mind. I was driven to understand the connections among these seemingly distinct but equally real aspects of our lives.

Back then it felt that there ought to be a natural way to begin to imagine how we could understand the flow from neuronal activity to narratives. These stories of our lives within us would also be shared in our relationships with each other, in our communities of caring, in our families and friendships, in our neighborhoods. The mind seemed to be located both within and between, meaning we live within these bodies, within these relationships, and each of these locations of mind shape our lives.

The mind is within and between. That is the *where* of mind.

If you believe that your life is simply what it is by way of genes that shaped your brain's structure and therefore its function, you may also imagine your mind is a passive product, almost a side effect, of neural firing. You might naturally place the mind in your single skull alone, emanating from your gene-produced brain. That gives the impression of a unidirectional flow, from genes to brain, brain to mind. In that flow, you are just an outcome of the inevitable, along for the solo ride of your life.

But back then it was becoming clear, and now we know from repeated empirical studies, that the mind can drive energy and information through the brain in new ways, ways that may not happen automatically. This neural activation with mental effort—with the intentional focus of attention in particular ways—may be creating a different pattern of brain firing than would happen naturally. We now know, too, that this mental initiation of brain activity can activate genes, change enzyme levels that repair the ends of your chromosomes, and even alter epigenetic regulation of gene expression, as we've discussed. You can see how your mental intention and mental belief shape your internal experience with your mind, and actually change your brain and the various molecular mechanisms underlying neural functioning and bodily physiology. You can intentionally shape your mindscape.

With communications we have between, in our relational communications we have with others, we have the focus of attention directed in a way which creates new intentions and beliefs within. In this way, too, the relational location of mind can even drive changes not only to the mindsphere, but also to the brain by way of the creation of new mindsets with their beliefs and intentions. The mind—both within and between—can change the brain.

If you believed that you were just a product of your genes, that the mind is simply an outcome of a brain unchangeable, such a view might understandably also give rise to a feeling of helplessness. Beyond despair, in turn, you might feel quite alone. In this view, mind is an outcome of brain, brain is in your head, you are not able to influence much of all of that, you might think. But in Carol Dweck's wonderful work on mindsets, this perspective might be part of what she calls a "fixed mindset" in which you believe that what you have is what you have, based on innate characteristics that cannot be changed (Dweck, 2006). Instead, a "growth mindset" is the mental view that what you have can be changed by effort. Academic achievement, intelligence, even personality traits can be changed if one puts in effort. Try a little harder and you can achieve all sorts of things, research reveals, from success in school to changes in habitual ways of behaving. Dweck and others have even revealed that teaching people about the power of the growth mindset and the ability to change your brain with mental effort lead to significant enhancements in school performance across a range of situations.

Even the way we believe stress impacts us can change its impact on us. If we interpret the increased heart rate and sweating of a "stressful experience" as excitement and not helplessness and terror, for example, we can have a different way of performing, say, a public presentation (McGonigal, 2015). Such a change in our belief, too, can alter our own physiology from a damaging state of threat to an empowering state of challenge. What we do with our minds changes our experience, including the activity of our brains and the physiological response of our bodies.

If you reflect on what you believe is unchangeable in your life, the "story of you" that is your personal, constructed narrative, you may find some areas that are not alterable. Not everything can be changed—my five-foot-eight height is not open to change so I can now become a professional basketball player with wishful thinking. But if that sport were my passion, with some reflection I might be able to find many ways to express it.

But you may also discover that patterns you thought were fixed are actually things that with mental effort can indeed be changed. Knowing what we now know, without a doubt, that the mind can actually change the function and structure of the brain itself, may help you see how mental effort can change long standing characteristics we thought were permanent (Davidson & Begley, 2012).

If you reflect at this moment on this notion of an inner and an inter location of mind, how do you feel? Can you sense some of the internality of mind, your mindscape, how energy and information stream through your body, including your brain? At this location, keep in mind that you can, with intention, shape the way your body functions. That is not hyperbole, guessing, opinion, or wishful thinking—that is science. Whether you use your mind to focus on this book, do reflective practices like meditation, or in many other ways develop an open awareness called presence, you can intentionally shape the well-being of your internal life, your brain, and its body.

If you reflect now on the inter-location of your mind, on the mindsphere, how does that feel? If you relax top-down conceptualizations of mind within you, can you begin to sense, even with the emerging of this mind-vision, this more subtle sensation of something between, some "thing" in between you and significant others in your life? These others might be people close to you, a pet, maybe even a tree outside. The cover of this book has a photo of me on a cliff in Norway, and you might feel, even from this visual image, what I may have been sensing in the awe of that scene, open to the vastness of the fjord, feeling both small and fragile, and immense and strong all at the same time. Within and between, the mind is both.

Opening to the sense of connection to other people, pets, and the planet, or to any other entity outside your body, is what we mean when we speak of the betweenness of the mind, of your mind. You may feel this readily once invited to reflect on these connections. Or you may feel this as a vague sensation at first, or as nothing at all. For many of us, this is not a sense we are familiar with, nothing that was developed in school or at home, nothing encouraged in our society, and therefore it isn't necessarily easy to gain access to at first. It can feel like an untapped potential, something we have a birthright to experience, but one that has yet to be cultivated.

Soon we'll discuss in detail a reflective Wheel of Awareness practice and explore what can be called our *eighth sense*, a sense of our connections to things outside these bodies we inhabit. You can try the Wheel and see how that feels. For now, simply know that if this scientific reasoning and close observation of mental life are true, then you may find exploring this eighth sense an exciting and rewarding adventure to develop your sense, perception, and perhaps ultimately awareness, of this betweenness of your mind.

Seeing the mind as both within and between gives us new insights and inroads into change. We not only work on the important inner life of the mind, including our beliefs of growth or fixed mindsets, but also work on the outer life of the mind. The mindsphere is filled with energy and information flow that can be shaped to promote well-being, or inhibit it. The ways we give dignity to all people, how we differentiate our many ways of being, and then link with compassionate connections is a simple framework of integration that applies to the important internal, personal aspect of the mindscape, and external, relational aspect of the mindsphere, the social field in which we live.

The friends we have, activities we engage in, the way we spend our time with health-full or health-less experiences, shape our mind in direct and impactful ways.

With our journey we have affirmed our fundamental proposition: The where of the mind is both within and between. How we cultivate a sense of meaning and purpose, the *why* of the mind, within these important two locales is the focus of the next entry of our journey.

CHAPTER 7

A Why of Mind?

WE ARE MOVING MORE DEEPLY INTO THE TERRITORY OF THE meaning and purpose of mind with this entry. Having explored the mind's emergence as a self-organizing process that harnesses integration, one filled with conduit and constructor functions, one located as a mindscape within and a mindsphere between, and having seen the importance of sharing our own subjectivity in creating a "feeling felt" experience that links two as an integrated whole, we now ask about the why of mind. That this journey would lead to the world of spirituality and religion in my own life was something my own mind was surprised to discover.

Meaning and Mind, Science and Spirituality (2000-2005)

The turn of the millennium opened new immersions into mind that I never could have anticipated. Since the first edition of *The Developing Mind* had just been released months earlier, new invitations to teach outside of UCLA presented themselves as a range of people absorbed the proposals in the book about interpersonal neurobiology as a way of looking at our lives. Offering a definition of mind as more than brain activity, as the emergent self-organizing process that was both fully embodied and fully relational, opened the way to move beyond the small discussion group setting and connect to a wider audience. I couldn't have expected that one of the first teaching invitations I'd receive included an audience with the Pope.

The Vatican's Pontifical Council on the Family sent an email stating that Pope John Paul II would like me to come to Rome to discuss with him the importance of the mother in the development of the child. I wrote back that I'd be honored to speak to the Pope about the importance of attachment figures, including the mother, father, and others, in shaping how the child's mind developed. I inquired about what Pope John Paul was specifically interested in, and they wrote that he wanted to know why the mother's gaze was so important in a child's life.

Preparing for that trip to the Vatican required that I write a paper, and I chose to write an article titled "The Biology of Compassion." Embedded in that document was everything you and I have been exploring, how we can see the mind as beyond the activity of the brain, how subjective experience may arise from our neural firing but is not the same as neuronal activity, how the mind may also have an

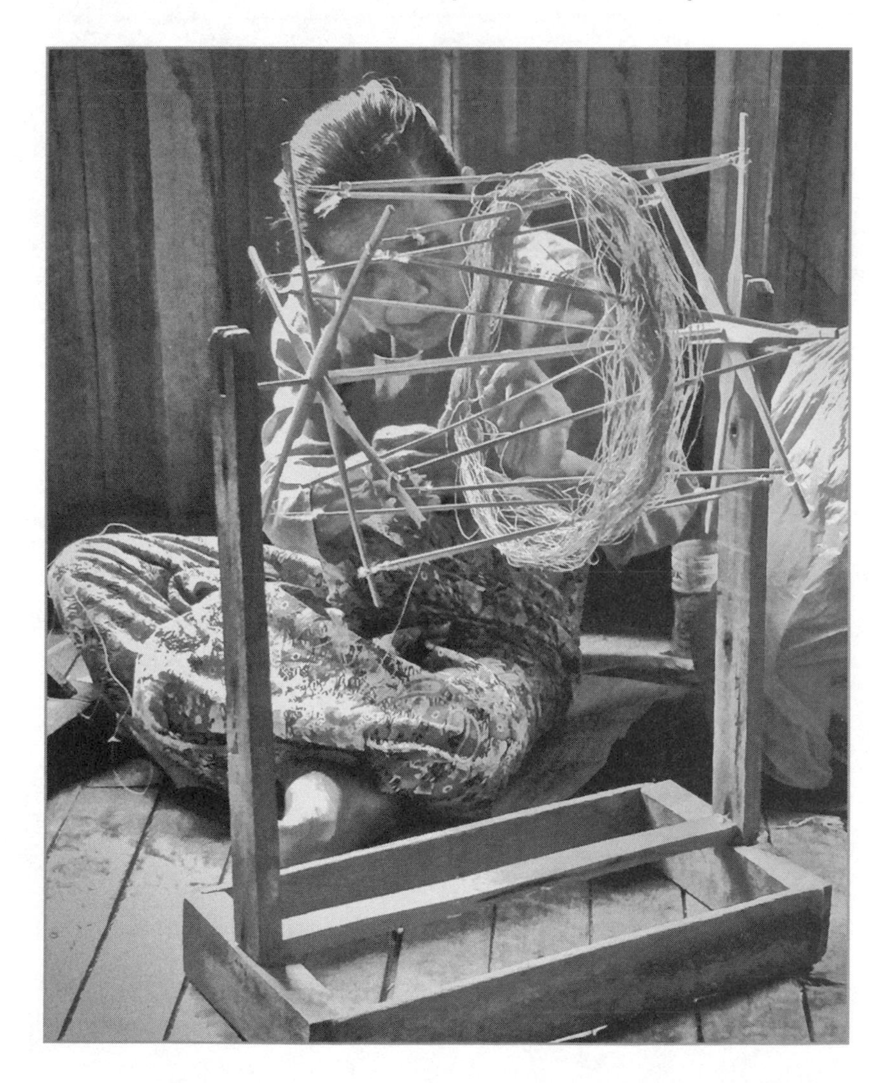

Photo by Madeleine Siegel

emergent process of self-organization, and that this process is both embodied and relational. In seeing integration as the basis of health, we could then sense how the embodied and relational mind moves toward integration in the process of cultivating well-being. One fundamental expression of that integration is compassion and kindness.

When my wife, two children, and I arrived in Rome, we were escorted to the Vatican and settled in for our nine-day visit. We received a "back-stage pass" that enabled us to comb through the back hallways of that intricate institution, and we soaked in the complex architecture and artful adornments of this center of the Catholic faith.

I was raised without any formal religion, but as a child and adolescent my family and I were members of the pacifist Unitarian and Quaker churches. Like many others during the Vietnam War in those groups, our historical origins were Jewish. You may know the basic history of Jews beginning as a wandering tribe of nomads who had two qualities that made them standout from the settled city dwellers of the day. One was that they saw time as a line, rather than a never-ending, unchangeable circle. The other was that they told stories that were filled with real facts, the positive and the negative, rather than the idealized narratives of the city culture. These two factors led this nomadic group to see that they could change the course of how life unfolded, for them, or for others. The sense that arose of "being the chosen people" was their belief that a single god had chosen them to be responsible to heal the suffering of others (Cahill, 1998; Johnson, 1987).

The history of the Christian faith began in the region now called Israel. A Jewish man of Nazareth found his way to Jerusalem and taught a variation of common Jewish values. He sadly was arrested by the Romans who, as was the practice in those days, put him on trial and chose to have him executed by crucifixion. Over the ensuing nearly three centuries, those who had followed Jesus' teachings, including the apostle Paul, came to see the teachings of Christ, whom they considered in Jewish narrative to be the messiah, the son and messenger of the one and only God, as not a variation of Judaism, but something separate. Some believed it was the Jews who elected to have Christ killed, not the Romans.

A revolt against the Romans by a small group of Jews around 70 AD precipitated the destruction of the Jewish temple in Jerusalem and the banning of the Jews from that region. This was a banishing of that old group of narrating nomads, those that had become Jewish people, who were now on a journey of wandering called The Diaspora." For two thousand years, this wandering has been a part of Jewish identity, a part of the narrative theme of being a member of this cultural group, these nomads gone global. But what does that really mean, I wondered as we wandered around the Vatican grounds, to be a "member of a cultural group"? Isn't the true group we belong to composed of all those beings we call human? And why have a species-limited identity? Couldn't our group include all those we call living beings? Where did group membership begin and end?

Who are we? Why do we have a sense of personal or cultural identity? At the Vatican I wondered about my historical origins, about how the cardinals and bishops, and the Pope himself, had come from their own backgrounds and found a life devoted to the Catholic faith.

Inside I felt like I was first a human being, a member of a large human family. Moving outward, I felt as a member of the living beings of our planet. If I let my mind move even more broadly, I was a conduit of things on Earth, a living being, yes, but also a part of the whole physical world around me.

If I let myself focus more narrowly, what arose was a sense of belonging to layers of groups, some of which might not even know I was a member. In these belongings, sometimes I had a role, like being a parent, or more specifically, a father. I was a father to my two children, who held my hand as we strolled the broad avenues and back alleys of Rome. And I was a member of all fathers, and could relate to fathers I saw on those streets. I was a citizen of a country, like other citizens, and this country, right now in this lifetime, was the United States.

But not long ago, those who are my ancestors just two generations in the past, were here in Europe. The Diaspora had forced them on a journey, I am told, for two thousand years of wandering, when they were periodically extruded from whichever land they had made home. My grandmother would tell me of her life growing up in a *shtetl*, a village of Jewish people who were not allowed to live in the cities of Eastern Europe. This village was in the Ukraine, just as my other grandparents and great-grandparents had lived in such shtetls in Lithuania and Russia. I never met them but I know them, inside somehow, in the stories I've been told, in the genes I carry shaped over these centuries of running, and in the epigenetic controls that regulate those genes.

Epigenetics, as we've discussed, refers to how we have regulatory non-DNA molecules that sit atop our genes and regulate their expression. When genes are activated they lead to protein production and shape the function and structure of their cells. In the brain's case, we are learning that experiences can shape our epigenetic controls, and with those changes, the brain's activity and structure modify to adapt to prior experience.

In some cases, research has now revealed that if the timing of an experience is right, we can actually pass such changes on to the next generation by way of our eggs and sperm (Yehuda, et al., 2014; Youngson & Whitelaw, 2012; Meaney, 2010). For a female, for example, if she is in the womb when her mother experiences a stressful event, like a famine, she will acquire epigenetic changes her mother had to endure with the lack of food and then pass on to her own children, through her eggs that were maturing while she was in utero, adaptations that make her metabolize a food as if there were a scarcity. If food is plentiful, though, her body will hold on to those calories and she'll be more likely than another child to develop obesity and diabetes.

In my grandmother's situation, it may be that since she was in the womb at the time her father was murdered in a pogrom, a common action to kill Jews in the shtetl by, in this case, the Cossacks of the Russian army, then it is quite possible that the trauma my great-grandmother experienced may have influenced how my grandmother—in the womb as her own ovaries were developing—would then pass those adaptations to danger on to my father, and then he on to me. Such adaptations might include the beneficial neural processes that would make our offspring more likely to survive by being more vigilant for danger, what Steve Porges calls "neuroception" (Porges, 2011)—scanning the environment more intensely, responding more rapidly, reacting more intensely.

For our journey into the mind, it is important to understand that we are influenced by more than our experiences and narratives we hear as part of those experiences in relationships with others. We are also shaped by our bodies. We have an embodied mind. This means that genetic factors shaped by evolutionary selection and epigenetic factors shaped by our immediate ancestors' experiences will contribute to the formation of our bodies shaping the function and structure of our brains.

It would be natural to think that running from danger would select for the survival of those who could detect it in their environment and choose to escape it. That might be some aspect of my own feeling of not accepting the socialization of medical school, for example, as something I should trust. The epigenetic shifts in my immediate ancestors, in my grandmother, may have furthered that adaptation to not trust, to be wary, and always look for a deeper truth beneath the perceived façade of what is acceptable or what seems on the surface to be safe. Not trusting the world around you could keep you alive.

I can see now that these ancestral origins embedded in my epigenetic and genetic legacy combined with my personal experiences, of losing my identity after that horse accident in Mexico and training to understand the biological mechanisms of things underlying life processes, may have contributed to my senses of identity in Rome. I felt a kinship to all living beings, connected to our larger human family, yet also felt there were things afloat in our world—in how we drew lines that separated us, boundaries of our minds, cultures, and belief systems, that were perhaps true but also superficial separations. With integration seen as the honoring of differences and cultivating compassionate linkages among those group-identity entities, then integration could help reveal that while differences were fine, the connections were essential. I was not raised Catholic, but I was a human being. I may have had a different cultural background being raised as a Unitarian, a different ancestral background as someone with Jewish heritage, but in the end we all began in Africa (see John Reader's 1999 account of this important and fascinating finding). Whether we stayed on that continent or our ancestors were part of the proposed small group of less than a few hundred individuals who migrated about a hundred thousand years ago through the desert and up into Europe to populate the rest of the non-Africa world, we came from the same origins. We are each living beings; we are all human beings. The heart of this being human is the mind we all share.

Reflecting on all of this as I sat in the Vatican helped as I gave my public talk in the Synod of Bishops, a half-millennium-old structure where major meetings were held. Cardinals, bishops, and priests from all over the world were assembled. It was the first time a talk I was giving would be translated, and was done so simultaneously so I didn't need to pause. Still, the timing of the response to anything funny I shared revealed the speed of the various translators, where ripples of off-sync laughter would peel from different areas of religious figures from around Europe, Asia, and Africa. Each one of us came from different tribes that had found a differentiated life of language and location around this planet, but each of us were capable of love and laughter.

What could I say? Beyond presenting the approach of interpersonal neurobiology (IPNB) and everything you and I have been exploring together, there was an opportunity to directly address the finding of *alloparenting*—that we as a species evolved to have more than one caregiver. This sharing of child-rearing responsibilities may be a basis for our collaborative nature (Hrdy, 2009). This also meant that the responsibility for raising children should extend beyond the mother's shoulders and be shared with the father and others. This didn't go over so well, as many of the responses stated the equivalent of this: "Dr. Siegel, don't you think that the reason you have so much violence in America is that you let your women work?" We discussed the issue, and I stood firm with the science: We humans have the capacity to have more than one attachment figure. Selective attachments, yes, but not just with one person.

We also discussed the connection between science and spirituality. The Pope seemed quite interested in this topic, having recently pardoned Galileo, who was condemned by the Church for suggesting the earth was not the center of the universe. There was a fascinating openness to John Paul, and though he was on medications for his worsening Parkinson's disease, he seemed genuinely interested in reaching across the divide and finding some kind of bridge between the two worlds. Could religion and its traditions find some common ground with empirical science? Could there be a link between spirituality and science?

These questions filled my mind during that visit, and have stayed with me to this day. As I walked with my family around Rome, I remember one powerful moment in front of the Vatican. We had just visited the ancient Roman Pantheon, a religious architectural wonder of the world with its imposing dome and massive interior that had representations of the many Roman Gods. We had also visited the old Jewish Synagogue, also across the river from the Vatican. Prior to this trip, having been a part of the Unitarian Church which embraces all religions and teaches the importance of seeing the good in all traditions, my own children had visited Islamic Mosques and Baptist Churches. As a family at home, we would go regularly to a Hindu center in town, strolling through its peaceful grounds home to a shrine that holds some of the ashes of Mahatma Gandhi.

With my 10-year-old son on one hand, and 5-year-old daughter on the other, we walked slowly in Saint Peter's Square, the pigeons searching for food on the blustery December day. My son looked up at me and asked me to explain how it was possible that there was so much belief in so many things that contradicted each other. How, he wanted to know, could it be that the believers in many gods could construct the Pantheon and believe in what they believed, and the Jews could have their teachings and believe what they believed, and the Christians could construct this massive place, the Vatican, and believe what they believe. He then asked that since they can't all be true, why would they have such conviction that their story is the right one? Which is correct? My daughter listened attentively to her brother's questions, and then they both looked at me, waiting for my response.

I will never forget that moment. I paused and wondered what to say. I was raised to believe in being human, to defend the rights of all people to find their way to their own truths. And with the IPNB view of integration, it made sense to differentiate our cultures, religious beliefs, and ethnic identities, and honor those differences and promote compassionate linkages. That would be an integrated world, a world of compassion, a world that enabled people to belong and thrive not only in spite of differences, but because of those differences. Integration could be envisioned as the source of kindness and compassion. That would be an integrated world, a world that flourished, a world in which kindness and compassion were signs of well-being. And that's what I tried to say.

My son asked why people would need to have these different beliefs anyway. Why did they need religion? He asked again, which religion was 'right'?

Once again I paused as we continued to walk through Saint Peter's Square, looking at the statues of the apostles atop the massive Vatican Church, seeing where we had climbed the stairs the day before, wondering where we'd be tomorrow, and in a week, and in a month and year and decade from now. And then I thought of what to say.

I said that people evolved to have a brain that can make a map of what we call time, to imagine what might happen in the future, to recall the past. Once that happened, I said, we had an important change in our lives. We now could realize that life does not last forever, that we all die. As social creatures, we share stories with each other. In those stories, we try to make sense of the world. How can we make sense of the meaning of life if life ends? All the world's religions, I said, in one way or another, try to address this question. Why are we here? What is the purpose of our lives? What is the reason for living? What happens when we die?

My son and daughter just looked at me as I spoke, and then as we walked on through the square we were all just silent. The pigeons flew off as we walked by, people strolled, and the marble apostles looked down.

About a year later, I recall my then 11-year-old son coming to me with a follow-up question. "Dad," he asked, "if we all die one day, and know we are going to die, what is the reason to do anything? Why are we here?" I just looked at him, his pre-teen face filled with innocence and concern. What would I say? I told him that finding what has meaning in life is something that will emerge in his own journey of discovery. I could tell him what has meaning for me, and if he ever wanted me to tell him what my meaning was, I'd be happy to. But finding his own way to that sense of purpose, of addressing the question of *why* he is here, would be his own discovery.

He looked at me with a sense of understanding, paused, and asked, "Well, Dad, if we are all going to die, and we know we are going to die, do I still need to do my homework?"

As the first part of the decade unfolded, I reflected on these questions of the *why* of our minds not only with my family, but also with friends and colleagues in science and clinical practice. With the positive reception following the publication of my first academic book, and then the release of my first parenting book, I began to feel more at ease in my doubting mind that there might be something true, and perhaps even useful, about these ideas and this approach.

I was asked by a number of local clinicians to offer a study group to help them apply the ideas of IPNB to psychotherapy. Soon that one group grew into seven groups, and I found myself moving from the privacy of my psychotherapy practice where I'd been applying these ideas for over 10 years to teaching other professionals how they might understand the mind and cultivate integration in not only their own lives, but also the lives of their clients and patients.

Some of these study group participants were new and young therapists just entering the field of mental health. But most were seasoned professionals, often older than me and with decades more experience. I was intrigued that they found this material of interest. As months folded into years, with more and more therapists participating, I began to receive their input that using the framework of integration as the core basis of health transformed their work. People working with them, many therapists informed me, were able to change and reach new levels of well-being with integration as the goal.

Seeing the mind beyond merely descriptions of mental activities like feelings and thoughts, viewing the mind as not just brain activity but fully embodied, and conceptualizing the mind as a self-organizing, embodied and relational process proved useful in clinical practice. Whatever the background of the therapist—from specific cognitive behavioral approaches to body-centered work, from narrative-based therapies to psychoanalysis—these courageous individuals became the informal vanguard of an open pilot study to see if integration worked.

At the same time, invitations to teach outside Los Angeles were increasing. I began to have a travel schedule of teaching that enabled me to reach people in the fields of mental health in other states, on other continents. What was striking was that everywhere I went, from Asia to Australia, Africa to Europe, there was this finding we discussed in the very beginning of our journey: Very few professionals, or scientists, were ever offered a definition of what the mind was.

That period of time was filled with puzzlement for me. Were these ideas accurate? Could a definition of the mind be offered, at least as a means of beginning a discussion? Would this view of integration as the basis of health benefit anyone beyond my patients? Locally, nationally, and internationally, the feedback from participants in seminars in person and online was clear—this approach not only made sense, it could be practically applied to help reduce suffering and move people toward well-being and a more meaningful life.

Whether one is suffering from an experience-caused disorder like post traumatic stress or a non-experientially-caused condition like bipolar disorder or schizophrenia, the fundamental mechanism in the brain appears similar: impaired integration.

For evaluation, what became important was not categorizing a person into a *DSM* grouping that might limit our understanding of who they were and thus who they could become, but rather seeing the chaos and rigidity that emerged in their life as coming from impaired integration. Moments of harmony with that FACES flow of being flexible, adaptive, coherent, energized, and stable revealed times of integration.

The first issue was to not pathologize and categorize. If there were seven billion of us on the planet, then there were seven billion ways we could be. In addition, each of us, regardless of culture or history, is human. As humans, our minds emerge as energy and information flows within us and between us. Sensing that flow as chaotic or rigid when outside of integration or in harmony with integration became the task of assessment, of getting an open picture of what was going on in a person's life.

Those nine domains of integration, described earlier, became apparent as the months and years then rolled into decades of doing therapy this way. Many professionals requested regular workshops to learn more about this mindsight approach that was built upon the consilient view of interpersonal neurobiology. In that teaching my doubting mind often urged me to stop right there and then. It was one thing, this inner critic said, to simply connect with my patients and work closely with them. That they were mostly getting better was enough supportive evidence that I should keep on privately helping people in this integration based way, but it was likely best to "keep it to myself"—which I did for years.

Now I faced ever more professionals, many of them quite experienced and older than me from this wide range of approaches. I taught them about the notion of integration, of the mind as a self-organizing process, the fully embodied and relationally embedded nature of the mind, ways to use the focus of attention to cultivate integration and stimulate not only healthy functioning in the moment, but likely changes in the structure of the brain in the long run.

I was nervous. Would they find trying this approach with their own clients to be effective? I felt I could only do my best to articulate the ideas and clinical strategies. When the first set of feedback began to emerge at the turn of the new millennium, I was shocked. It was working. Many of my students, clients, and patients of my students were moving into new areas of growth that weren't imagined likely to be achieved prior to this new approach.

The basis of a mindsight approach to therapy is that we all have a natural drive toward integration, toward optimal self-organization. As complex systems, self-organization is an emergent property of who we are—it is, as we've suggested, a fundamental facet of our minds. But "stuff" sometimes gets in the way. For some of us, this "stuff" that challenges well-being arises because of our sub-optimal experience with caregivers early in life. For others, random events, genes, epigenetic factors, or toxic chemical exposures or infections may negatively influence how the nervous system achieves integration early in life, or during the formative period of adolescence.

Whatever the cause, the intervention in the face of the rigidity or chaos emerging from impediments to integration still involves the power of relationships to inspire us to rewire our brains toward integration.

As my clinical colleagues continued to use these new approaches to treatment and to understand in new ways old approaches that worked, they'd tell me they found new ways to deepen and broaden their clinical impact.

I felt so moved by these teaching experiences that I could finally begin to soothe my doubting mind. I wrote many case histories revealing the nine domains of integration and how clinical evaluation, treatment planning, treatment implementation, and outcome assessment could be approached. The experience of writing and releasing that book, *Mindsight*, still surprises me. Even people who are learning this material though the written (or spoken) word of a book seem able to take in these ideas, move their lives from the chaos and rigidity that were imprisoning them, and enter a new state of flourishing that integration empowers us to create in our lives.

Integration as the "Purpose of Life?"

It's audacious to even attempt to address the question of why we are here, of what the "purpose" of life is, I know. The part of me that doubts so much, my doubting mind, urges me to back off from continuing this part of our exploration of the *how, when*, and *where* of the mind—just skip over the *why* part! But if, at this point in our journey together, we can continue with the same stance of asking questions and not presuming absolutes or final answers, perhaps moving forward with this direct discussion of the *why* of mind is not only okay—it's important. So much in modern life is confusing and often leaves us feeling lost as we are bombarded by an infinite world of information flow in this digital age. (I say silently in my internal mind that the Internet ought to really be called the "infi-NET" as it never ends and never gives a sense of completion.) Never done, endless in its streaming of energy and information that distracts and distances at the same time it connects and constrains our minds to certain ways of being. There is so often a feeling of insufficiency and urgency. With all these aspects of modern life preoccupy-

Photo by Madeleine Siegel

ing us, a feeling of being overwhelmed with data can make a question like "Why are we here?" seem superfluous. We are here to consume and share. "We share therefore we exist," is a habit of thought, a habit of behavior, shaped by the artifacts of the digital culture in which so many of us in modern life are immersed. Yet exploring the question of the *why* of mind brings up powerful issues about our purpose in life that may be just what we need, especially in these overwhelming times. Why *are* we here? Why does the mind work as it does? Why are there so many different ways of living, different beliefs, differing stories?

Over the years of studying and exploring, teaching and practicing, living and reflecting, a view has emerged from patterns I've observed that suggests the following direct, outrageous, and surprisingly simple statement: integration may be the *why* of the mind.

From our discussions of the mind as being more than merely brain activity, and fully embodied, we've come to consider a larger perspective on what mind entails. By viewing subjective reality as not identical to physiology, even to neural firing in the head, we've come to realize that mental life is not the same as activity encased in the skull. We've also come to a place in our journey to consider that mind may be more than merely subjective experience and the consciousness that permits our awareness of that sense of lived life. At a minimum we've seen that energy and information flow, when in awareness, has a felt texture, something we are calling a prime, something not reducible to something else, like neural activity or other forms of energy flow.

Subjective experience may emerge from this flow, but it is a feature of it, not the same as it. And sometimes, perhaps even most often, that energy and information flow happens without awareness. Mental life, including its thoughts and feelings, may be dependent upon energy flow but not always in awareness.

When we've taken the next step and suggested that mind is possibly more than energy and information flow inside the body, that it is more than embodied as our mindscape, we've considered that mind happens in relationships with other people, and in our interconnectedness with other aspects of the world outside of these bodies in which we live. Here we've seen that we have an equally important mindsphere. The common denominator that links our inner bodily experiences with our inter-relational experiences is energy and information flow. This is what our mindscape and our mindsphere share. When we see that this flow happens as a fundamental part of a system bounded neither by skull nor skin, we've come to embrace the notion of an embodied and relational mind. When this system of mind can be viewed as having the three qualities of being open, chaos-capable, and non-linear, the mind can be seen as a part of a complex system.

This complex system, also called dynamical, has emergent properties, and one of them we are proposing is the subjective reality of lived life. This might mean, too, as we'll discuss in the next entries, that the consciousness that gives rise to subjective experience may be an emergent property of our complex mind systems. Another, possibly related, emergent property we've been proposing is self-organization. We've seen that this emergent, self-organizational process arises from energy and information flow and turns back and regulates that flow. That's the recursive property of self-organization: It arises from and then regulates that from which it continues to arise. That's recursive feedback. This is one aspect of how we can say "the mind often has a mind of its own."

We've further seen how self-organization moves a system, naturally, without programmer or program, toward maximizing complexity. That term is often taken as a bad thing to do, especially at a time when we want simplicity in this ever more complicated world. We've seen that this natural push toward maximizing complexity is actually quite simple, rather than being complicated, as it is created by how we differentiate and link elements of the system. When we do that, what is the result? Harmony. Like a choir differentiates their individual voices yet links at harmonic intervals, self-organization gives us a deep and powerful sense of vitality. This is the natural self-organizational movement toward harmony.

FACES is the flexibility, adaptability, coherence (holding dynamically together over time, being resilient), energy, and stability of a self-organizing system when it is linking differentiated parts—when it is integrating. What are the features of the resilience of coherence? Being connected, open, harmonious, engaged, receptive, emergent, noetic (sense of knowing of the truth), compassionate, and empathic are the COHERENCE features of an integrated flow. (My apologies, again, for all these acronyms. I just find them so helpful for assembling and summarizing key elements of what we are exploring.)

This innate push toward integration of the mind as the self-organizing aspect of our embodied and relational lives, of energy and information as it flows within and between, might be the *why* of the mind. Whether in our internal, personal mental lives of our mindscape or in the mindsphere that connects us to one another, integration could possibly be one of our central reasons or "purposes" for being.

You may imagine from this review and from where we've now come in our journey together how exciting this all feels for me. I wonder what you feel, think, and wonder about all this. Is going into the *why* of mind too outrageous? Is taking our discussions to this point pushing too far? My doubting mind is quite nervous at this moment, but let me share an experience with you that makes me feel that at least trying to articulate some of this issue of the *why* of mind may be worth the effort.

When I would present various aspects of integration to a range of groups, something surprising came up time and again. A clinical case I would often present (and later would describe in the first chapter of *Mindsight*) reveals that the midline areas of the top of the brain, the prefrontal cortex, enable nine functions to arise. These include: 1) body regulation (balancing the body's brakes and accelerator); 2) attuned communication with self and other (focusing attention on internal mental life); 3) emotional balance (living with a rich inner life of feelings); 4) response flexibility (being able to pause before responding); 5) soothing fear (calming fear reactions); 6) insight (connecting past, present, and future with self-understanding); 7) empathy (mapping the inner mental life of another); 8) morality (thinking and behaving as part of a larger whole, broader than a personal, bodily self); and, 9) intuition (awareness of the wisdom of the input from the body).

At various lectures around the world, people would come up to me and comment about this list. For example, when I was teaching in Alaska for those working with families in the northern islands, a leader of the Inuit tribes came to me and said, "Do you know what that list is, the one about the integration?" Yes, I said, it is about the prefrontal cortex's role in linking cortex, limbic, brainstem, somatic, and social energy and information flow to one another into a coherent whole. "Yes, that is what you said, I know," she went on. "But that list is exactly what my people have been teaching in the oral tradition for over five thousand years as the basis for living a wise and kind life." I was silent, and peered into her eyes for a long time. When words came to me, I tried to tell her how grateful I felt to her for sharing these reflections, but after speaking briefly I realized that words were not enough; a deep shared gaze and shared silence that unfolded afterwards seemed better.

As time went on, I heard similar responses from people who represented a range of ancient wisdom traditions, including the Lakota tribe in the Midwest of the United States, the Polynesian culture in the South Pacific, and those from the Buddhist, Christian, Hindu, Islamic, and Jewish religions.

What was going on here?

I reflected on my son's questions about how all these different belief systems could coexist. Could it be that what they all shared was exactly what that leader of the Inuit tribes of the Northern Alaska islands had suggested? Could integration be the basis not only of health, but also of the world's wisdom traditions?

If this were true, it might give us a useful bridge between science and spirituality that could deepen dialogue and promote collaboration across these various pursuits. Building on integration might help us integrate our common humanity—embracing differences while cultivating compassionate linkages. I felt deeply grateful for the journey, and ready to continue to ask questions about the *why* of the mind, and how we might move toward a more integrated world.

Reflections and Invitations: Purpose and Meaning

We've come a long way now in our journey. Would you ever have imagined that asking fundamental questions could not only be fun, but also take us to such broad applications, like seeing a connection between religion and research?

As I continued to explore the applications of these consilient ideas of IPNB in my clinical work with patients, I found that their impact in therapy was not only to alleviate the distressing symptoms of chaos and rigidity, but also move patients to a new sense of identity. That breathing-across all the domains of integration as we've discussed earlier—what I first called "transpirational," but now simply call "identity" integration—was something that seemed to simply arise as people worked on the other domains, from the domain of consciousness and vertical integration, to interpersonal and temporal domains. It felt as if the natural push of their lives as complex systems could be liberated with the right focus of our work. This approach meant that getting the "stuff" out of the way was the task—not so much doing something to someone, but instead permitting something to be released—so that the innate drive toward integration, the purpose of the mind, could be liberated from its prison.

One of those prisons seemed to be identity. A first level of personal identity is belonging to our own body, an individual self. Then we have our relationships with our family, our attachment figures and others, who are part of a close-knit unit. Sometimes this extended unit of personal identity is expanded to include those in our community, our neighborhood, or our membership in a religious group. As an old Jewish joke recounts, when a man stranded for twenty years on a deserted island was finally discovered, he asked if his rescuers would like to see the structures he had built. He showed them his modest home in a small valley, a library, hilltop temple, exercise area on the side of the nearby hill, and another temple near the beach. The rescuers asked why, since he was only one person, he would have built two temple structures. The man's response: "I would *never* be a member of that other temple!" In-group/out-group distinctions are in our DNA—in the "old days," other cave-members could harm us. Those of us from our Cave A were terrific; those from Cave B were rotten scoundrels. We need a group to belong to in order to feel safe and know whom we can trust and whom we need to be wary of in order to survive.

We also could have membership in a group with broad ethnic ties, religious beliefs, and cultural practices. Yes, we could feel "familiar" associating with those "like us," but when, I wondered, does this like-us boundary get set up? Aren't all human beings truly like one another? We have differing skin-colors, national origins, genders, sexual orientations, political persuasions and religious beliefs that tie us together while at the same time they pit us against each other.

As we've evolved to have an "in-group" or "out-group" distinction that helped us survive, it's natural to imagine that having group membership feels important, if not a matter of life and death. But where does that grouping process of our brains that shapes our embodied minds ultimately lead? As we discussed along the walk at the Vatican earlier, aren't we all a part of one humanity? What do these divisions of one "type" of human from another actually do for our well-being now? And furthermore, aren't we all a part of living beings, and even of our whole planet's ecosystem? When we remove ourselves from belonging to this differentiated but interconnected whole, where does it leave us?

Limiting one's identity in this time of global need seems to push against the importance of integration, especially the integration of a broad and embracing identity. But under threat, how can we promote such a widening of integration in our lives? Part of the answer may be in diving deeply into how our relational minds can be seen as not bound by fate, not inevitably tied to what our brains have evolved to create in our lives. In other words, the mind can rise above our inborn proclivities of the brain, genetically and epigenetically influenced propensities to impair integration, and move us toward a more helpful and healthful integrated way of being in the world.

There is a kind of being and of doing that are helpful in this liberation. It involves sensing chaos and rigidity, focusing on which domain they might be arising from, then cultivating differentiation and promoting linkage. That's the fundamental conceptual approach. Though there may be neural proclivities or social pressures that inhibit differentiation and block linkage, thus pushing us away from the harmony of integration and toward chaos and rigidity, we can use the pause of being present, of being aware, to intentionally create new pathways toward integration. For me, back then and to this day, the presence that is the portal for integration to emerge is the purpose of life, the meaning embedded within moment-by-moment living.

In day-to-day terms, we sense when things are off and instead of reacting impulsively, on automatic pilot driven by genetically shaped and culturally reinforced neural reflexes, we rise above these with our conscious minds to create choice. In many ways, these pathways toward integration are like the saying often mistakenly attributed to the psychiatrist and Holocaust survivor Viktor Frankl: "Between stimulus and response there is a space. In that space lies our power to choose our response. In our response lies our growth and our happiness." I spoke recently with Dr. Frankl's grandson, Alex Verely, who clarified that his grandfather actually never made that statement. Steven Covey has revealed that in fact, this quote which he made popular in earlier writings he actually first read in an unidentified book and is not from Dr. Frankl, but simply reflects Frankl's approach to finding meaning and freedom (Pattakos, 2010). Nevertheless, whoever actually penned these words, they reveal a universal truth: with conscious reflection, we can choose a different path from that which comes automatically. To find meaning in life, it may be necessary to cultivate this presence of mind, this pause from reactivity, this space of the mind, to enable the natural push for integration to arise. This is how, as we've discussed, we can let meaning and purpose unfold rather than make them happen.

When people ask moral questions about right and wrong, I find it helpful to reflect on the fundamental notion of integration. Is the dilemma being explored something that leads to chaos or rigidity—or does it cultivate harmony? Is there an honoring of differences and promoting of linkages? When integration is offered as a core principle guiding ethical investigations, the resultant discussions often open the way for respect across a wide range of backgrounds and beliefs. Integration connects us to a process of inclusion and empowerment.

As all of these notions of purpose and meaning were emerging with the experience of integration, it was crucial and reassuring to have the continual input from both seasoned and new therapists about the efficacy of this individualized approach to assess, plan, and implement clinical interventions based on integration. My colleagues and students, these fellow travellers along this journey, were charting new territory together to harness a working definition of the mind and promote integration. They and those they were helping were finding new ways to not only heal and reduce suffering, they were also discovering new meaning, wholeness and a sense of purpose in their lives.

A call from Deborah Malmud, a courageous and creative publisher from W.W. Norton & Company in New York, helped expand this work from primarily an oral tradition to a written one. Deborah invited me to start a professional series of textbooks, of which this book is a part, and I elected to have it be broad and emphasize the whole of interpersonal neurobiology. We decided to begin with a focus on psychotherapy, and I am honored to say that with my colleagues we have now published dozens of professional texts in the field. I can't describe the feeling of gratitude to all who have joined together to create this new approach to mental health.

More than simply a new set of books, this approach set out a new way for us to join in a journey together. We could offer ways to combine scientific disciplines, from math and physics, neuroscience and psychiatry, psychology and anthropology, and propose a consilient framework of what the mind is and what mental health might be. This new field of interpersonal neurobiology could remain a way to inform us, in fields such as mental health and education, parenting and organizational life, rather than serve as a specific approach. Interpersonal neurobiology could be a health-based view of human life. In this view, all of us, teachers and students, parents and children, therapists and patients, are interconnected members of a diverse human family and a deeply interdependent world.

There is a saying attributed to Mahatma Gandhi that comes to mind time and again. You may be familiar with it: "We must be the change we wish to see in the world." It appears that the closest statement that can be located in actual printed text is this: "The best propaganda is not pamphleteering, but for each one of us to try to live the life we would have the world live." (Johnson, 2005, p. 106). The empowerment of that stance is that we can inspire others by how we ourselves live our lives.

These statements misattributed to renown figures, this one to Gandhi, the one of the importance of mental space to Frankl, and the one regarding the notion that not everything that counts can be counted to Einstein, as we discussed in an earlier entry, invite us to realize that wisdom is not owned by a single individual. We do not need a famous person to coin phrases in order for them to have important truths deeply embedded in them. Coming to learn of these misattributions at first was jolting for me, but then gave rise to a deep feeling of our common humanity and our collective intelligence. After all, a human being created each of these phrases. Wisdom is something that can arise within each of us and be shared by us all. We can each illuminate reality and inspire one another to live a life filled with truth and connection. We do not need to wait for a designated wise one or leader; leadership can be within each of us. And don't these statements about being the change you wish to see, of that space between stimulus and response, and the importance of the immeasurable in life, each still have a crucial role to play in our lives no matter who created them? These are human statements illuminating the wisdom of being human. Each of these messages also reveals the importance of differentiating our mental lives by making the space to open to the immeasurable and often invisible inner world, and to create the presence that enables us to honor and link the differentiated aspects of our lives, within and between.

When we address the *why* of the mind, we come to the simple truths we've been exploring throughout our journey. Integration may be the reason we are here. Beginning with integration within, extending integration to those you are connected with, and moving integration into our larger world: these may just be the reasons we are here, the general guideposts of being on this journey, of being and doing in this life.

A common statement in wisdom traditions, now affirmed by careful research in a range of scientific disciplines is this: If you want to be happy, help others, and if you want others to be happy, help others (Vieten & Scammell, 2015). When we embrace integration as a central drive in our lives, we cultivate meaning and connection, happiness and health. Finding a way to differentiate oneself from others, and then link to them by helping support their well-being, is a win-win situation. In a way, that's how we can promote not only compassionate care, but also empathic joy, the interpersonal experience of enjoying others' successes in life.

Can you imagine a world that might be even moving in that direction? I invite you to consider how integration may be, or could be, playing a role in your life. Reflecting on the larger values in life, considering what has meaning for you as you live through these moments and minutes, months and years, has been shown to help reframe the impact of the inevitable stress we experience in our day-to-day lives. As you reflect on the *why* of mind, how can you enhance integration in your life? When life becomes filled with chaos or rigidity, how can you pause and reflect on what domain of integration may not be engaged? Permitting a pause and making the space in your life to reflect on these three states of chaos, rigidity, or harmony illuminates the flow of life. As we've suggested earlier in our journey, this flow is like a river. In the central flow is harmony, the harmony of integration. But when differentiation and linkage are not well honed in that moment, the flow moves toward one or both banks outside that flow, the banks of chaos or rigidity. Making mental space to pause and reflect on the river of your life can awaken you to what you may need to shift in your experience, helping you find a need to differentiate more, link more, and find the way toward harmony within and between.

The open potential of integration means that it is never done. We are always on an unfolding journey of discovery, finding ways to differentiate and link, to create more integration in our lives. Whether this is in small, personal ways, or large, relational ways, creating the space in our lives to move toward integration can feel empowering and filled with connection and meaning.

Reflecting on life's flow, you may find elements of your experience in need of a tune-up, and a tune-in. Taking care of yourself first, being sure you have the space of mind to use your awareness to differentiate and link, to move toward a more coherent way of being, can be deeply empowering. Being kind to ourselves may involve accepting that we are all emergent beings, moving sometimes toward chaos, sometimes toward rigidity, in this meandering flow of life. The journey of integration is never done; integration is an opportunity and a process, not a final destination or fixed product. Realizing that we must be the change we wish to see invites us to welcome integration, from the inside out. Accepting that being human is a journey of discovery, a verb not a noun, enables us to embrace integration as a principle, not a prison. We can set an intention toward integration without being committed to some particular outcome or result. Our lives then become an exploratory expedition embracing new learning, moment by moment. We may find this journey of life not only filled with moments of confusion and challenge, but also with surprise and humor, pleasure and delight, as we cultivate a deep sense of connection and satisfaction, and perhaps even meaning and purpose along our way.

Integration emerges across our lives within the unfolding of each lived moment. In the next entry, we'll dive more deeply into the notion of what a moment really means as we explore the nature of time and the when of mind.

CHAPTER 8

When Is Mind?

WHEN WE CONSIDER WHERE WE'VE COME FROM, AND IMAGINE where we could be going, we reflect on the deep nature of time. Perhaps what some physicists are proposing now, and what mystics and meditators have suggested for ages, is indeed true: Time, as we know it, may not exist. Time as something that flows, something we can run out of, something that we're pressured to hold on to, is a mental construction, a self-created stress, an illusion of the mind. All we have, from these scientific and spiritual views, careful quantitative research, and contemplative investigating reflection, is now. And if now shapes not only when we are, but also where we are, and how we are, then what is now really made of? How do we become aware of now? Why do we feel as if there is more than just now, more to preoccupy us with the past, to fear in the future? The question of the when of mind is then a query into the essence of the reality of each lived moment. How might this when of now relate to all the who, what, where, how, and why's we've been exploring? Let's dive in, right now, and see what emerges.

Exploring Presence in Mind and Moment (2005-2010)

A surprising invitation came to me at a time when all of these questions and ideas were being expressed across a range of talks and publications. I was then ten years beyond my full-time faculty role at UCLA, but busier than ever with academic pursuits. Harvard Medical School requested I give a keynote presentation about the importance of emotions and narrative in medicine. Knowing that I had dropped out of this very school because of the lack of focus on these inner mental

experiences, of seeing the powerful need for a focus on the feelings and stories of our lives, I was perplexed and exhilarated all at once. It had been 25 years, almost to the day, since I'd left the program in despair about these same issues, and here I was, back in Boston, arriving at Massachusetts General Hospital's Ether Dome.

Seeing the fundamental relational and embodied nature of mind made it scientifically straightforward why physicians need to heal through seeing the minds of their patients—not just their bodies. If only I had had this perspective and we had this knowledge back then. Looking deeply into the students' faces and later discussing these issues with them, I could see that science could help support the humanity of medicine.

The mind is the heart of being human.

This journey into the essence of life, of the mind into the mind, this journey you and I have been taking, could allow us to ask the fundamental questions needed to bring the mind to life, make it real, reveal its central importance in health and healing.

Seeing the mind was exactly what the insight and empathy of mindsight were all about. That very word that had been a life preserver as it enabled me to return to these halls and had been a guiding light, an intention and attitude bolstered by science but felt from within that could save my sanity. In addition to the important ways we sense our own or other's subjective mental life, mindsight included a respect for differences and the creation of linkages with compassionate communication. Integration was also inherent in mindsight.

Mindsight was simply this trio of fundamental and interconnected human experiences: insight for the mind within; empathy for the mind of others; and integration to promote the kindness and compassion of healthy and health-promoting relationships.

Words cannot describe how it felt to be there in the Ether Dome, exploring these issues in the very room where a quarter of a century earlier I was in despair. My 15-year-old son was in attendance, and seeing his eyes focus on all that was going on in the room around him amplified that moving moment.

Around this same time, a serendipitous use of the word *mindful* in a book, *Parenting from the Inside Out,* which I wrote with my daughter's preschool director, Mary Hartzell, lead to many parents asking when we'd teach them to meditate. Mary and I had taken the scientific findings of *The Developing Mind* and translated those principles of interpersonal neurobiology into stories and summaries of science for parents to make their own and use to make sense of their lives. Not meditators, Mary and I were puzzled by the queries. For me, having left full-time academics, I was already pushing the envelope to say

that relationships shaped the brain, and what we did with our mind could also change the brain's structure. Even though my wife, Caroline, had been meditating quietly each morning for decades, at that time meditation seemed like something "too out there" to take on in my professional life. Mary and I had used the term *mindfulness* to mean parents should be conscientious and intentional in their ways of being with their children.

Soon after we published the book and were offering workshops, I was asked to be on a panel with Jon Kabat-Zinn—an international expert in the field of bringing mindfulness to the medical world with his stress reduction program. Preparing for that meeting, hosted by *Psychotherapy Networker* magazine, I read as much of the limited research literature on the science of mindfulness as I could find. What struck me was that the outcome measures for mindfulness training seemed to overlap almost perfectly with the outcome measures from my own field of research training, attachment. Secure attachment and mindfulness seemed to be parallel processes. I became fascinated with the possible overlaps between mindfulness and secure attachment relationships that were based on empathy and compassion, kindness and care.

What might the mindful awareness cultivated within meditation and the empathic communication of attachment relationships share in common?

On the panel I offered questions about the odd overlaps among mindfulness training, secure attachment, and an area of the brain—the prefrontal cortex—that linked widely separated areas to each other. As we discussed in the last entry, this region links cortex, limbic area, brainstem, body-proper, and social signals into one coherent whole. Five sources of differentiated energy and information flow become linked so they can be coordinated and balanced by the prefrontal region's integrative fibers.

Could these three aspects of human life be reflecting some common ground in integration? Could mindful awareness be the mind's integrated state, secure attachment the relational state of integration, and the prefrontal region representing the embodied brain's integration?

Unfamiliar with the practice of meditation and trained as a researcher in attachment, I was urged by Kabat-Zinn to seek direct experience with mindfulness meditation. Over the next eighteen months, I went to a series of trainings in mindfulness which ultimately led to my writing a book about a newcomer (me) exploring mindfulness called *The Mindful Brain*. I also taught a conference with Jon and two other colleagues who had been with us at that Networker meeting, Diane Ackerman and John O'Donohue, which we called "Mind and Moment," and found myself filled with more and more questions. When I sometimes watch the video recordings we made of that event with my students at the Mindsight Institute, I am filled with a sense of immediacy and energy. Could that really have happened over ten years ago now? It still feels as if it were happening right now, the presence we experienced there filling the room as we watch, a presence somehow eternal. And maybe that is exactly the focus of this entry on the meaning of time in the mind. What do the past and our subjective experience of time really mean?

Our subtitle for that mind and moment meeting was "Neuroscience, Mindfulness, and the Poetry of Transformation in Everyday Life," and the event continues to fill me with a deep feeling of gratitude and awe for the insights of Diane as a naturalist and poet, of Jon with his translating of Buddhist mindfulness practice into a universally accessible application in his stress reduction work, and of O'John (our nickname for John O'Donohue to distinguish him from Jon) as a philosopher, poet, Catholic priest, and, in his own words, a mystic—someone who deeply respected the invisible world around us. We each watched the video of that meeting and decided to release it to the public hoping it might connect people to these ways of exploring life and awakening the mind. But before these formal plans could be concretized in writing, O'John died unexpectedly. As I am writing these words to you now, I turned to his last book, *To Bless the Space Between Us*, published just before his death, and re-read some of the book's passages, including his descriptions of his father and a friend in the acknowledgements. Of his father, Paddy, he expressed beautifully what I feel about O'John himself: "His quiet facility for presence altered space, his gentle eyes in love with the invisible world." Well, O'John wasn't exactly quiet, but his eyes were peaceful, beckoning, and questioning all at once. And how he wrote of his relationship with a friend who also died too young, O'John articulates what I also feel right now about him: "Never expecting death to come so soon, I am lonesome for all the conversations we never had."

How our relationships with one another so deeply shape and transform us. With all this love and life, it felt then as something new and deeply moving was arising. As I reflect on all this now, I can see that as my professional passions expanded in the new millennium, my mind opened to new vistas I never had imagined would be a part of my life. The personal was not separable from the professional; the subjective was interwoven with the objective. The whole of me felt open to whatever was emerging, an emergence beyond my understanding, beyond my control. I had a feeling that the best I could do was to let things happen, put myself into the world with a general stance, an open mind, and not try to make life go in one narrowly-predetermined direction or another.

That first decade of the millennium saw the birth of new friendships with these and other wonderful individuals, and my mind was invited to open to new ways of seeing who we were and ways of exploring and expressing the mystery of our existence.

During that decade, I also became connected to Jack Kornfield, one of the first individuals to bring mindfulness practices from Southeast Asia to the West and make the insight meditation practice of the Theravada Buddhist tradition available to a wide audience in the United States back in the 1970s. Jack had been one of Jon Kabat-Zinn's first insight meditation teachers. I was fascinated, in learning from Jack and in our teaching together, by how many overlaps seemed to exist between the teachings of this 2600-year-old tradition with its cultural values and meditative practices and the independent findings of interpersonal neurobiology. The connections Jack and I have made continue to invite me to explore the overlaps between spiritual practices and science in new and sometimes startling ways as we'll soon discover.

Back then during that first decade of the new millennium, I was primarily practicing as a clinical psychiatrist, working with children, adolescents, adults, couples, and families. Yet I was also obsessed with these ideas, possessed by these questions about time and what we could see and not see, wondering what our relationships and embodied brains might be doing with mindfulness practice. What was the overlap between our relationships and meditation?

As a research-trained psychiatrist, too, I attended a range of scientific meetings. One was an intensive day on the neuroscience of autism. Empirical findings were being reported, and the issues of how the challenges to social communication, emotion regulation, and sensory processing were being discussed. Just before I was heading out for a lunch meeting, a research project was described in which individuals with autism were found to have significantly diminished gamma waves on a magnetoencephalogram (MEG) study. MEG enables us to peer into the functioning of the brain, and this low level of gamma waves suggested that integration was low. I had heard whispers, ones that later became published data, that other studies had begun to show similar functional and anatomic evidence of impaired integration in the brains of not only individuals with autism, but also those with other non-experientially caused challenges like schizophrenia and bipolar disorder.

Later on at the end of that first decade of the new millennium, while working with 15 interns to revise the first edition of *The Developing Mind*, we reviewed some of the thousands of research articles that had been published in the last dozen years. I requested that the interns take on the project to disprove the first edition's proposals, such as the mind is embodied and relational, not just enskulled in a single head, and that it might be a self-organizational process cultivating integration as health. They were surprised about the request to find empirical data to disqualify, nullify, and negate these ideas; but I reassured them it was the only way to be sure we weren't simply selectively finding random but supportive data to bolster the initial claims. We needed to find data, even if a minority view, that disproved the ideas. Let's just write a new book, I suggested, and discard the old one.

What we found, the interns and I, was that the vast majority of proposals that had not yet been empirically supported but were derived from scientific reasoning ended up having independent research labs' findings consistent with these hypotheses of mind and integration. A few ideas were discredited, revealed to be invalid by new research data. The main finding was the earlier statement about emotional intensity being higher in the right hemisphere. We found it was perhaps more directly related to the body, but couldn't find a reliable source for intensity being less in the left.

What the interns and I largely found scouring the literature were supportive findings, results predicted by the hypotheses, but not proof of their validity. It was very exciting to see independent labs coming up with empirical findings that were predicted ahead of time by the basic framework of interpersonal neurobiology, about the mind's self-organizational nature and the central role of integration in well-being. Science progresses by these small steps, finding support study by study, and it was exhilarating to have the companionship of these young minds in pursuit of knowledge. It was our intellectual adventure to seek and deepen consilience across a wide range of disciplines. We had a lot of fun and laughs, and I remain grateful for our collaboration. It was thrilling to see the picture we've been exploring together, you and I in this journey, continue to come together with these basic fundamentals having such wide support in the science.

When I heard that MEG studies showed diminished gamma waves and asked what the researchers thought might be going on, they said they weren't sure. Other studies suggested that, for some unknown reason, the brain of some individuals with autism spectrum issues stops differentiating its parts during the first two or three years of life. I wondered, could lowered gamma waves be revealing impaired functional integration as a result of this impaired anatomic integration? Could these neural challenges to integration lead to some of the challenges to interpersonal integration in these individuals?

Interpersonal neurobiology offered a framework, here and now, across a range of situations and science.

Back at that autism meeting in D.C., I had to leave, so I stepped out of the auditorium and into the hallway of the lower level of the hotel. I walked toward the elevators but was stopped by a dark suited man with an earpiece hanging from the left side of his face. I looked beyond his imposing physique and saw other federal security agents lining the hallway. A smaller man dressed in orange and red robes came out of a nearby room, escorted by agents and other colorfully dressed Tibetan Buddhist monks. As the group approached the elevators I was waiting to ride, suddenly two janitors came out of the adjacent restrooms. There was a flurry of agents running toward them as they were just a few feet from the first robed monk, and just a few yards from me. The monk walked past the agents and headed straight to the janitors, two men who appeared to be in their forties or fifties who looked both puzzled and pleased. The monk greeted each with a clasping of both their hands, peering deeply into their faces, first one, then the other. Then he held onto the left hand of one, the right of the other, so that they formed a tight three-person circle as the security agents were in a tizzy. The three men talked for what seemed an eternity—three minutes, perhaps four—and when they seemed to be done, to have felt filled up by one another, they released their clasped hands, the monk waved to the sweating security agents, and moved toward the elevators.

As the monk turned to leave and I could finally see his face, I realized this was the Dalai Lama. He had taken the time, all the time in the world it felt, to connect with the janitors who had somehow slipped beneath the radar of security protecting this head of Tibet in exile, this spiritual leader of Tibetan Buddhism.

In the present moment, it seemed to me, the Dalai Lama was creating interpersonal integration. He was honoring the differentiated beings of these two janitors, and linking with them with clasped hands, a locked eye gaze, and a loving mindset. (I am not sure the security forces would agree with this interpretation when it came to their experience of fear and failure.) I stood back, soaking in the sight of such a moment suspended in time. But that's exactly the point of this entry: There is a possibility that there is nothing to time but now, and this moment was a moment of meeting, a moment of integration that illuminates the timeless reality of our lives.

That week I attended two meetings where the Dalai Lama was speaking in Washington, D.C. At the Mind and Life Institute's meeting there was an opportunity for me to hear numerous articulate scientists present their research findings on meditation and its impact on the brain to "His Holiness," as he is called (HHDL is the short way of putting his name together). Later that week I attended the huge annual neuroscience meetings, with over thirty thousand in attendance, and watched on as ten thousand neuroscientists waited over two hours to file into a gigantic conference center room to be in the presence of The Dalai Lama, HHDL, and hear his keynote address to the prestigious group. What struck me was that there were numerous satellite rooms where the live video feed would be projected, allowing you to sit near the screen and see the whole scene, up close and personal as they say, if only two-dimensional. That's where I chose to sit, having been in the room with him for three days prior in the relatively smaller audience, and not wanting to miss a moment of the other amazing presentations happening at that groundbreaking meeting. For some reason, those researchers really wanted to be in the same room with him. It was striking to watch these academics simply stand in line to be in his physical presence. What was going on here? This observable phenomenon in the meeting hallways, and even the fact that the Dalai Lama as a symbol of compassion and contemplation was invited to be their speaker, seemed to reflect a shift in our cultural mind. Or perhaps it was simply curiosity about a celebrity. But maybe it was something more than stargazing. Could it reflect a deeper longing that HHDL embodied and articulated?

Something is collectively arising beyond celebrity fascination. The growing interest in social and emotional learning in schools, the emergence of a modern focus on mindfulness in education and our larger organizational and societal worlds, and the fact that *Time Magazine*—what a fitting name—published a cover story in 2014 titled, "Mindful Revolution," all support this impression. Some interest in finding a way to strengthen and perhaps connect our solitary minds seems to be afoot in a collective mental process, a joining across cultures and arenas of human endeavors, from corporations to curricula, boardrooms to living rooms.

During that meeting I went to the formal presentations that included Sara Lazar reporting on her new discoveries of how mindfulness meditation could change the structure of the brain. Someone had whispered in my ear at the Ether Dome where I had been giving a lecture earlier that year, as we discussed before, that Lazar and her colleagues would be coming out soon with a publication revealing that integrative regions were thicker in those who did decades of mindfulness meditation. And those neural areas that were changed? They were areas that linked widely separated regions to each other, such as the hippocampus and areas of the prefrontal cortex, including the insula. Later studies, too, would reveal that the corpus callosum linking left and right hemispheres to each other grew with mindfulness meditation. In a nutshell, these studies can be summarized this way: Mindfulness training integrates the brain.

I asked my interns to find one form of self-regulation that did not depend upon integration in the brain. They could not. Regulation of attention, affect or emotion, thought, impulses, behavior, and relationships depend upon the linkage of differentiated areas in the brain. We can now say that mindfulness training likely enhances self-regulation by promoting neural integration.

At the same neuroscience meeting in D.C., I also attended the poster sessions where research data are put up on panels for members to simply read over and then ask the researcher directly about the findings. At one such poster on mindfulness meditation, I asked the young researcher about his data and we began discussing our work. When I told him I was curious about the possible connection between mindfulness meditation and attachment, he told me I had no idea what I was talking about. I agreed with him, acknowledging I had never meditated and was going to attend a research-based week-long silent mindfulness meditation retreat soon. No, he said, it was the concepts I was talking about—they were all wrong. In Buddhist philosophy and practice, he urgently told me, you try to get rid of attachment. What? Yes, get rid of attachment he repeated.

I took him and one of his students to lunch, and we explored what that all meant, from his training and research. I had no scholarly background in Buddhist thought. In broad terms, Buddhist philosophy, he said, suggested that the source of suffering was the effort to avoid or cling to things. When you let go of such aversions or attachment, he said, suffering subsides. That was a goal of mindfulness meditation, to find a deep acceptance of what is, including the reality of the transience of all that unfolds.

Wonderful, I thought, having no deeper knowledge of Buddhist thought or meditation. But, I told him, the attachment I was referring to was not about clinging, it was about love. A loving relationship involves the attachment system in the brain that we as mammals all have, and that we, as humans, all share. He wasn't familiar with the field of attachment from a relationship research point of view, so you can imagine how lively our lunchtime talk was. His graduate student seemed amused by the meeting of these two minds, of these two ways of knowing. There ultimately was no resolution, but for me, there was a deep sense of excitement about bridging these various fields of knowing, of science, of meditation, and of everyday life, into one coherent whole.

I went on to do that weeklong silent mindfulness meditation retreat, surrounded by about a hundred scientists I would have loved to have spoken with, but we were all in silence. It was "noble silence," meaning yes, no talking, but no, not even eye contact or other forms of non-verbal communication. I thought I was going to lose my mind. But then I found my mind. Beneath all the busyness of attending to the external world, including the needs of others, after a few days of the noble silence there was a deep sense of peace and clarity that emerged. I was surprised by this inner sanctuary, and sad when it seemed to slip away as we left our silence and re-entered the world of scientific discussion and social engagement.

After training in Jon Kabat-Zinn's mindfulness based stress reduction (MBSR) work, and then teaching our conference called *Mind and Moment* with Kabat-Zinn, Diane Ackerman, and the late John O'Donohue, I was filled with ideas and questions. Later that year I was asked to be on the faculty of the Mind and Life Summer Research Institute, and wandered around the poster sessions with Sara Lazar, pondering with her and others how mindfulness meditation might influence the brain, mind, and relationships.

All of that wandering and wondering led to the writing of a book about my journey into this new world of contemplative neuroscience, *The Mindful Brain*. In that exploration as a novice, I hypothesized that some form of attunement—internal for mindfulness, and interpersonal for secure attachment—might be their common core.

Research was revealing that health emerged from supportive relationships early in life. As the adverse childhood experiences studies (Felitti, et al., 1998) have revealed, difficult times in our lives early on can lead to significant challenges to our mental as well as medical well-being. For example, developmental trauma, early experiences of severe abuse or of neglect, lead to impediments to the growth of integrative fibers in the brain, including the hippocampus, corpus callosum, and prefrontal cortex (Teicher, 2002). Most of those nine prefrontal functions we discussed earlier, from emotional balance to morality, are the empirically proven outcomes of secure attachment relationships. Those same functions are a research-proven outcome of mindfulness training as well. Secure attachment—relational love, and mindfulness—befriending and loving yourself, seem to both emerge from and cultivate integration within, and between.

New research was affirming what wisdom traditions had taught for centuries—being aware of the present moment without being lost in judgments, being mindful, leads to well-being. It seems being present creates well-being. It is important to clarify that not being swept up by pre-established views of how things should be is what is intended by the term, "non-judgmental." The mind is always making top-down filtering in the form of evaluations, appraisal, and judgments. And in many ways, as Jon Kabat-Zinn said to me recently when we were teaching together, what he means by that term is simply realizing that our judgments are mental activities we don't need to be imprisoned by. This is how having mindsight—seeing the mind of self and other and promoting integration—includes this way of viewing what it means to be mindful. Also, we can have discernment in which we use a reflective mind to make skillful assessments of a situation. Some would use the word "judging," such as a structural engineer judging whether a bridge can withstand a certain number of cars safely. Discerning judgments are an important part of living a healthy life. Presence is a term that embraces this way of being open to things as they are, and not getting overtaken by pre-existing beliefs and expectations, our judgmental imprisonments that separate us from being with what is actually present. That's what we mean when we say "non-judgmental," even though that term itself can seem quite, well, judgmental. When we are present for ourselves, we have mindful awareness. When we are present for our children, we create secure attachment. Both secure attachment and mindfulness promote well-being—physically and mentally.

A range of studies now show that mindfulness meditation can help improve the medical condition of those with psoriasis, fibromyalgia, multiple sclerosis, and hypertension. Mindfulness has now been shown to improve immune function and even raise the level of the enzyme telomerase, which maintains and repairs the ends of chromosomes. Mindfulness training also has been shown to have psychological benefits as well, with a decrease in the symptoms of anxiety and depression, binge eating, attention deficit, obsessive-compulsive disorder, and substance abuse.

But how might presence, internal or interpersonal, cultivate well-being?

My bottom-up mind as a conduit enabled the flow of all that had been transpiring, soaking in ancient views and modern discoveries. My top-down mind was generating and interpreting, processing information and seeking meaning in all that was unfolding.

The Dalai Lama's connection with the janitors in those moments was a form of integration. The areas of the brain that seemed to be activated and growing with mindfulness meditation were integrative. When I was immersed for the first time in that silent retreat, I could experience the differentiation of many streams of awareness, for example, distinguishing a sensing stream from an observing stream. For me, being mindful seemed to be a way to integrate my mind. At a minimum, it seemed to distinguish what we can now call the aspect of the mind's conduit from the constructor, to differentiate the experience of conduition from construction.

Was my top-down constructor mind simply projecting a previous belief on this new world into which I was now being immersed? Or could integration really be what was at the heart of both mindfulness and attachment?

The psychiatrist Eric Kandel was awarded the Nobel Prize in 2000 for discovering the basic molecular mechanisms of how learning changes neural connectivity. Research with those with psychiatric disorders, as we discussed earlier, is now revealing what anatomical and functional differences seem to underlie these conditions. Preliminary work in psychotherapy suggested that even without medications you could change the function and structure of the brain.

Perhaps one of the most compelling and relevant discoveries from research was that what you focus your attention on shapes the physical structure of your brain. With the leadership of luminaries like Kabat-Zinn and his colleague, Richie Davidson, careful research had been carried out on participants in the by then well-established Mindfulness Based Stress Reduction (MBSR) program, as well as on long-term meditators. The results were unambiguous: Training the mind to focus attention on present moment experience without being swept up by judgments could not only help you feel better, it could change your physiology and improve your physical health. As this new millennium unfolded, even preliminary studies began to emerge that suggested doing a daily practice of mindfulness can optimize how the non-DNA molecules that regulate gene expression, those epigenetic regulators we discussed earlier, are configured in areas of your genome that help prevent inflammation, conditions that may be related to certain kinds of diabetes and cancer. Mindfulness seems to optimize epigenetic regulatory changes in your genome.

Kabat-Zinn views mindfulness as a way of paying attention, on purpose, non-judgmentally to the present moment (Kabat-Zinn, 2005). Others, such as Shauna Shapiro and her colleagues, would see mindful practice as how we pay attention to the present moment in "an open, kind and discerning way" (Shapiro & Carlson, 2013, p. 1). In each of these views, mindfulness is imbued with a sense of what the COAL acronym embodies, curiosity, openness, acceptance, and love. Mindful awareness is that open state of presence. Yet these leaders in the field focus on attention. Let's review what attention is.

Attention is simply the channeling of energy and information flow. So what could this attention be doing in the brain? Energy and information flow—the essential feature of mind—leads directly to the activation of neuronal firing. As we discussed before, we can recall the connection between attention, neuronal firing, and neuroplasticity with this sequence in mind. Where attention goes, neural firing flows. Neuronal firing, in turn, activates genes, enabling any of the four possible neuroplastic changes to be initiated. Recall these include growth of new neurons, shaping of synapses, laying down of myelin, and modification of epigenetic regulators. In other words, the mental process of energy and information flow shapes the physical processes and properties of the brain at the molecular and anatomic levels. Mind changes the physical nature of brain—both its function and structure.

I once said this at a public talk and one of the audience members politely and firmly suggested I had made an error when I spoke. Specifically, I said that the "mind uses the brain to create itself." He said I must have meant that the "brain creates the mind."

This is the heart of the issue: If one aspect of mind is defined as that higher order, regulatory, self-organizing, emergent property of a complex system, then this regulation of energy and information flow manifests itself in our bodies—including our brains. It manifests itself in our relationships—even the one happening now between you and me.

So perhaps it is more complete to state that the mind uses the body and its brain along with our relationships with each other, and with the planet, to create itself. By "create itself" I mean that the self-organizing aspect of mind emerges moment by moment to create our experience of mind. At a minimum, energy and information flow is emerging in the present moment. As each moment unfolds, energy and information self-organizes. This is the *when* of mind—the self-organizing movement toward linking differentiated elements of our system, a system that rests in these bodies we inhabit, as our mindscape, and within the mindsphere in which we live.

The *when* of mind is this emergent property of *now.*

The nows that are then, as we've discussed back in the beginning nows of our journey, have the quality of being fixed. We can't go back and change then, but it is still a now—or, in physics terms, what is called an "event," something that has occurred in the past. And the nows that are the future, these are not yet emergent, but are open. Fixed as "past," open as "future," and emergent as "present"—these are the qualities of the ever-present now events of the *when* of mind.

When we realize this state of mind of now can be with kind regard, acceptance and non-judgmental care, we can sense that mindfulness creates the conditions of being fully open to the present moment. We differentiate the stream of observation from the stream of sensation, and we are open to whatever arises with a state of presence to all there is—the present moment. With the love of secure attachment, we can feel how being present with others enables us to join, honoring one another's subjective experience and linking in compassionate communication.

Being present is the portal to integration in each emerging moment of our lives.

Attunement, Integration, and Time

We connect with one another in the moment. One way to illuminate how secure attachment between a child and a caregiver leads to well-being is to focus on the nature of integration and how it unfolds, moment by moment. Secure attachment is based on attuning to the inner world of a child, and when this attunement is not possible, the rupture can be repaired and a reconnection established. Attunement can be simply defined as the focus of attention on the internal world. Interpersonal attunement is the focusing of kind attention on the internal subjective experience of another. Internal attunement is the

Photo by Madeleine Siegel

Photo by Lee Freedman

focusing of attention on the internal subjective experience of a personal self. This all happens in the present, and it happens with the parental presence at the heart of secure attachment. With others, or with ourselves, we can enter states of disconnection and disregard. Repair of these inevitable ruptures is the norm; perfection in parenting and our relationships with ourselves, as with anything else in life, is impossible. Expecting the impossible makes life filled with strife.

In other words, it turns out that for much of our experience relating to others, and perhaps to our interior selves, we are not in alignment. But at moments in need of connection, when emotions run high, for example, we need to be seen by another. When those moments of need are not met with the presence of the other that rupture hurts. It is the repair of ruptures, the realignment of minds, that is the key to security. Even realigning with how we connect with our own interior mindscape after being disconnected can lead to a deep sense of coherence and wholeness. Presence permits repair by being the gateway that enables integration to naturally unfold.

Interpersonal attunement allows two differentiated individuals to become linked in that moment as a *we*. Differentiation remains, but linkage transforms an isolated self into both a *me* and a *we*. The subjective sense of that joining is what that patient of mine expressed long ago when I was first in training to be a therapist: "feeling felt."

When we feel felt by another person, we feel our inner life is seen and respected by the other. When we are present, when we attune by focusing attention on the inner world of the other, when we resonate and become changed by this interaction, we develop the state of trust. This, you may recall from our prior nows, now fixed, is the presence, attunement, resonance, and trust that reveal the PART we play with secure attachments. This is interpersonal integration at the heart of an integrated relationship. Each person is honored for their differences and then linked by compassionate communication.

How might this parallel mindful awareness? Is integration a part of what being mindful is all about? In my own experience of mindfulness training, I could sense that separate streams of awareness were each differentiated and could become linked within the spaciousness of being mindful. We've explored two, one of sensation, one of observation. This is but one way to imagine the integration of our interior life.

In my experience, these flows of energy and information into awareness include a sensory stream with which we feel sensations arising from direct experience—what we see, hear, smell, taste, and touch. This extends to what we feel from inside the body, even the sensations of our thoughts, emotions, and memories, and the sensations of our connectedness to others. That's direct sensation, the conduit of the mind, offering an experience as bottom-up and direct as it can be.

There is also an observing stream. This is a more distant sense of awareness, one that can give rise, as we've discussed, to a witness that allows us to be bit more distant from a sensation or impulse but present as the one who bears witness. We may not be constructing concepts yet, but we are not fully in the flow of sensation, of conduition. Such observation gives space within the mind to not identify with ongoing experience as the totality of who we are. We can observe and witness, then perhaps put our experience into words, pictures, or music as we narrate our lives from this observing stream. Remember the acronym, OWN, for observing, witnessing, and narrating? We've travelled from the conduit function of a sensory stream to a bridge of observation—not only conduit, not only construction. When we continue on that bridge, we've moved to witnessing, still closer to construction as it is embedded with a sense of one who bears witness. Further along that bridge, we've moved to narrating, as we dance, sing, recite poetry, tell stories, or offer lectures. All of this is symbolic construction. Not bad, just different from the conduit function of sensation, and even of the first step of the bridge of observation.

In my own experience at least, as my doubting OWN circuit urges me to write to you, there are perhaps two other streams, if not more. One is of *conceptualization*. Having ideas or concepts about things is not really sensing them or observing them. We've gone beyond witnessing, even beyond *descriptions,* with words that construct the symbolic but directly telling aspects of the narrative of our stories. This descriptive aspect of narrative is the way we can "show and not tell" within direct, emotionally evocative story-telling. With the conceptualization stream, we are fully into construction. You may have a concept of a dog, and that concept is a mental model or schema that shapes how you see the dog. Your previous learning shapes your concepts, and gives you expectations and judgments about the world, so this is where the top-down, deeply constructor element reveals itself. With words this is more like an *explanation* than a description. Conceptualization is akin to the telling of an expository essay rather than the description of a fictional dramatic scene. This conceptual stream is important in the functioning of the mind in order to streamline our responses. We don't need to see every dog as a new organism, we can know that this animal is a dog and is or is not dangerous. Again, each is a differentiated stream, important, unique, in need of care, cultivation, and connection.

One more stream beyond sensation, observation, and conceptualization seems apparent as well. Some deeper form of knowing, a sense of coherence of the larger picture, a feeling of wholeness, a sense of truth, seems also to be present. There is something about this wise sense that is not the same as the term *knowing* when it comes to simple awareness, in that you know you are reading these words right now. This is a knowing of wisdom, a knowing of truth, not simply sensing, observing, or conceptualizing within the simple knowing of awareness itself.

In my experience, this *s*ensation, *o*bserving, *c*onceptualizing, and *k*nowing are the SOCK of mindful awareness. (Sorry about another acronym if these are not your thing—it's just how this particular mind works.) In my own inner experience of mindfulness training, each of these streams was differentiated and then linked. For me, internal attunement is a form of internal integration, differentiating at least these four streams of awareness and then linking them to each other.

As I write this now, I recall the sensations of these differing streams as quite distinct. Perhaps they were streams of attention, distinct ways of sending energy and information flow into my singular awareness. Perhaps my constructing mind took these distinct sensations of the conduit of experience and labeled them as "streams of awareness," put them together as an acronym, and now I relate this to you. Whatever linguistic symbols we choose, be it different streams of attention with one awareness, or distinct streams of awareness, we come to this: differentiated flows become linked into a whole.

One way in which we can lose touch with the present moment is by fretting over the future or becoming preoccupied with the past. In many ways, these can be seen as the constructive mind building representations that pull us away from the conduit flow of now. Fearful fretting about the future or preoccupied pondering about the past pull us from the present. Yes, we are presently fretting or preoccupied—but we become lost in these constructed images, these re-presentations of the fixed or open nows that we then manufacture in our mind as something to worry about, to attempt to control, to strive for a different outcome. In many ways, these are the aversions we try to avoid or the attachments we cling onto that are highlighted in Buddhist practice as the source of suffering (see Jack Kornfield, 2008). From our view of mind, these are the outcomes of mental construction.

Conduition is the first step to regaining presence. But it is also possible to open to both conduition and construction in the present moment. And the key is cultivating consciousness to discern, within awareness, the nature of all that is emerging, moment-by-moment. What is here now? Everything that arises is now, because that is all there is. The sense of "time" is simply an awareness that things change. What is here as a sensation, what is here as an observation? What is here as a concept, what is here as a knowing? All is now, but discerning these layers of the stream of awareness offers liberation from being lost in preoccupations and fretting that move us from presence. Presence is what permits integration to arise naturally. But how?

Another way of viewing internal integration is through the notion of the integration of consciousness. In a practice I developed during the Decade of the Brain called the Wheel of Awareness, the approach aimed to differentiate elements of consciousness from each other then link them with attention. If consciousness were needed for change, as a review of various forms of psychotherapy and education revealed, then what would happen if one could integrate consciousness?

If consciousness at its most basic conceptualization could be seen as a basic sense of knowing and that which is known, could knowing and known be differentiated from one another? With the focus of attention systematically distinguishing the knowing from the known, and the various knowns from each other, perhaps we could then integrate consciousness?

When I brought my patients around a table in my office, the central glass hub of the table could represent the knowing, the outer rim represented the knowns, and the spoke of attention would be represented by the table's stand. Instead of a "table of awareness," I called this the Wheel of Awareness (see figure on page 91).

This reflective practice entails differentiating the knowing of the hub of this metaphoric wheel from the rim of the known. Here "knowing" may be different from the knowing stream of the four flows of SOCK as we discussed, as it is more about the two differentiable aspects of consciousness itself—knowing from the known.

The rim contains the four segments representing the known of 1) outer sensations (hearing, seeing, smelling, tasting, touching); 2) internal sensations (interoception of the body's interior signals from the muscles, bones, and internal organs) called the sixth sense; 3) mental activities (emotions, thoughts, memories, beliefs, attitudes, longings, desires, intentions) we can call the seventh sense; and, 4) even an eighth sense—our interconnections to others and the larger world in which we live, our relational sense.

The Wheel of Awareness practice could promote integration by differentiating the knowing from the known of awareness, and also differentiate the various knowns from one another by the systematic movement of a metaphoric spoke of attention around the rim. With all these words it may seem abstract. If you haven't had a chance yet to try it, dive into the wheel practice and see what it is like for you (you can access this from our website on the resources tab at drdansiegel.com).

The positive responses in my therapy practice, and later from my students of psychotherapy, and then later still in workshops with various individuals around the globe, suggest the power of integrating consciousness for transforming chaos and rigidity into harmony and well-being.

In many ways, the Wheel practice reveals that integration of consciousness may be an important way we bring healing and health into our lives. In the next entry, we'll explore one possible view that could perhaps explain why integration is so helpful and illuminate possible mechanisms beneath consciousness itself.

Internal attunement is a way of defining the integration of consciousness that may be at the heart of a range of ancient practices that cultivate mindful awareness. Energy and information become integrated with such practices. Interpersonal attunement, inherent in secure attachment, enables the experience of feeling felt to be a reliable (not ever present, but consistently available) experience of being differentiated and linked to a caring other.

The proposal that could be made with this new view of the mind as an embodied and relational, self-organizing process led to a workable hypothesis explaining the potential overlap between secure attachment research and the outcomes of mindful awareness training, supported by Kabat-Zinn recently stating in a lecture, "Mindfulness is Relationality." This view overlaps with our notion of energy and information flow that happens within and between. We have a relationship with our inner flow—our mindscape—and we have a relationship with our inter flow—our mindsphere. Being present with kindness and compassion is being mindful.

What emerged with the review of the literature and exciting new findings from a range of research were foundational discoveries, which made the interns working on the second edition of *The Developing Mind* jump around the room with excitement. I can't even believe, with my doubting mind, that the following simple statements could have so much support in the science. Here are reminders of some simple views, discussed briefly earlier, that we could not disprove, but only found supportive empirical findings to point us in this direction: *Interpersonal integration promotes internal neural integration.*

What this means is that when we are in the PART with others—when we respect differences and cultivate compassionate linkages, we stimulate the activity and growth of fibers in the brain that link widely separated areas to each other.

Here is another simple truth for which we found only support: *Neural integration is the basis of healthy self-regulation.*

When we looked for the neural mechanisms beneath forms of self-regulation, sometimes called executive function as we've briefly explored earlier, we found that the regulation of attention, affect or emotion, thought, self-awareness, social relatedness, impulse-control, and behavior were dependent upon fibers in the brain that connected widely separated areas. That connection of separate regions could be seen within new technologies, such as diffusion tensor imaging, which now could reveal the connectome, or interconnected anatomy and function of the brain. This interconnection of differentiated regions is what we can simply call internal or neural integration.

Later, as studies continued beyond the first decade of the new millennium, we'd discover again and again that mindfulness meditation leads to changes in the brain that can be broadly summarized as enhanced neural integration. For instance, growth in the hippocampus, corpus callosum, and insula are examples of integrative areas that grow. Amazingly, as we've seen, these very areas are those that are compromised with the developmental trauma of abuse and neglect. Studies of what we've discussed earlier called the default mode network, the set of interconnected regions that mediate our awareness of others and the self we've dubbed the OATS circuitry, reveal that this area plays a major role in wellness or unwellness (Zhang & Raichle, 2010). Mindfulness meditation increases the integration of the default mode network (Creswell, et al., Doll, et al., 2015).

What an amazing time to be alive on this journey. These science-inspired ideas were finding more and more support as research progressed. Opening to the present moment catalyzes integration—internally and interpersonally. The interns and I were struck by the simplicity of these statements, and it all seemed to boil down to this: Our focus of attention with presence in awareness and our attuned relationships with ourselves and each other, promote neural integration, the basis of healthy self-regulation.

Reflections and Invitations: Awareness and Time

As that second half of the first decade of the new millennium continued to unfold, I was asked to be on a number of panels with the Dalai Lama, talking about a range of topics from compassion in the brain to mindsight in education. On one of those occasions, on a panel in

Seattle, he gave the three other scientists and me a homework assignment: Could we come up with a secular approach to how to make a more compassionate world? I thought about that question incessantly over the ensuing year. I reflected on the paper I had written for the Pope on the biology of compassion, thought about all we've been exploring on this journey about the mind as an embodied and relational process, and wondered about religion and science in general. When I was on a panel with HHDL in Vancouver the following year, I offered him my homework assignment and suggested that perhaps integration was the common route toward a more compassionate world. If health arose from integration and each person had a right to live a healthy life, then if we promoted integration we'd be promoting well-being. Since it just might be true that kindness and compassion are integration made visible, that integration is the underlying mechanism of a compassionate life, then promoting integration would be cultivating the seeds of a more compassionate world as well as the roots of resilience and well-being.

A few years later I was asked again to be with the Dalai Lama, this time in Rotterdam, facilitating a day of discussion with Dutch business leaders and students from kindergarten through twelfth grade. Could we promote more emotional and social intelligence in our organizations and schools? With integration as a linking concept, we dove into a daylong conversation, the school kids presenting their efforts to make a more compassionate world, group by group, as the business leaders listened and then joined in the discussion.

In all this traveling, I have felt the power our individual minds muster when wrestling with these ideas. I have also felt patterns arising in our larger cultural dialogue, not just in science, as at the neuroscience meeting in Washington, D.C., but as a human family throughout the globe. The relevance of mindsight and these notions of integration have led to invitations to speak at conferences focusing on climate change involving corporations and governments. One of those gatherings was with a group of about 150 scientists, mostly physicists. Our topic: the connection of science and spirituality.

I was fascinated by the theme. A few years earlier, I had spoken, along with HHDL, in Germany at a conference for the 550th birthday of the University of Freiburg. The topic was spirituality and education. Not having any background in the former, I asked my workshop participants to teach me what the term spirituality meant to them.

One by one, the participants offered their view. Each said something related to two features: Spirituality, for them, meant 1) being a part of something larger than a personal self, being connected to something larger; and 2) having a deeper meaning than the details of everyday life, something beyond survival alone. With that notion of connection and meaning, we dove into a fascinating discussion of how to cultivate those ways of living in education.

So I was primed, in a sense, to attend that next meeting in Italy with those physicists back in 2009. The elements that I felt compelled to explore beyond science and spirituality, and perhaps underlying them, were energy and time. Energy and time are the specialty of physics, and the opportunity to live, dine, walk, and talk with people devoting their lives to the study of our physical world was captivating.

It might not surprise you, now that you are getting to know the patterns of this mind, of this body, in this life, getting to know "me," that I was possessed with a drive to ask questions. The conversations never ceased. Rather than relate them one by one I'll share some take-home points. The first general caveat is to say that, as is the case with everyone, scientists have many differing viewpoints. The second is that what is published in the acceptable professional literature doesn't always fully convey either the uncertainties or the passion of the people doing the writing and the research. As a therapist by training, and a human fascinated with things of the mind, I was naturally curious about how these folks saw the world, not just how they publically made statements that fit their profession's perspective.

All this is background. Here are the two major points that arose, one about energy, the other about time. When we think of the *when* of mind, these issues become central to deepening *when* the mind might be.

Energy is a word we use for a broad phenomenon that manifests in different forms. We have light as photons, sound as airwaves riding along the motion of molecules, electricity as electrical charge flows. But when you ask what each of these forms of energy have in common so that they are each called "energy," a first response from scientists is often that we don't know. "Oh, come on," I'd push back. "You are a physicist. You are a specialist in energy. What is it?" "Well," they'd say, "it's a potential. A potential to do stuff."

How do we observe this potential? The potential of energy manifests as degrees of certainty, many but not all physicists I met with would say. These degrees of certainty are like a probability curve, one that moves between near-zero certainty on one end, to 100 percent certainty on the other.

So, I asked, if we say that energy flows across time, is it fair to say that this refers to the movement of the position on the probability curve, between near-zero to 100 percent certainty? Yes. And is the position along that curve toward zero percent certainty, when the predictive probability of the position, say, of a photon is unknown, is that what we could call a "sea of infinite potential" or an "open plane of possibility?" Yes, that made sense to them (see figure on page 59 or 254).

So energy was the movement of a potential between openness to certainty as the position on an energy probability curve moved.

I know this may be challenging to get a feeling for on first pass, or even on second pass as we've discussed it a bit earlier along our journey. But sticking with it can be quite useful. We'll dive more deeply into this aspect of reality in the next entries, but here let me outline the basics of what we can apply to our wrestling with not only the *when* of mind, but also to our broader journey into the essence of what the mind is.

In my own wondering mind, I felt driven to try to grasp what these physicists were saying. Now the question in my mind was this: If movement along an energy probability curve is what *energy flow* literally means—a movement between openness and certainty that transverses across a range of probabilities—does this happen in space, time, across potentials without time or space, or what? What does energy flow really mean?

Why would I be so obsessed with this question? With our definition of at least one aspect of mind as being the emergent, self-organizing process that arises from and regulates energy and information flow, knowing what energy flow really means is essential. Information for some physicists is a pattern of energy with symbolic value, as we've seen. Even Einstein's equation, energy is mass times speed of light squared (E=mc2), reminds us that the universe is filled with energy, even mass as condensed energy.

So in this view, energy is the manifestation of potential flowing between infinite and finite, the movement between uncertainty and certainty. Flow means something changes. Energy flow signifies the transformation of a range of probabilities, from wide open if not infinite, to increased probability or certainty, and into an actuality of the realization of potential. This position on the curve can also move back down as it shifts its position along the energy probability curve. That curve is a way of simply labeling a mathematical image of how this potentiality-probability-certainty might be represented as a graph. That's the probability curve—a visual image of the range of values of probability, spanning a curve that has at one extreme openness, and the other finiteness.

This attempt to define the word *flow* in relation to energy shifts brings us deeply into the issue of time. Time, as we've addressed on our journey, is a word we use to describe a sense of something that flows that might be a mental construction rather than an actual reality. We do have the feeling of time, as something that passes, for example, but that, as some physicists tell us, may literally be a creation of our minds, a label we use to denote positions within our four dimensional world, what is called "spacetime." Time is a word we use to denote our position in that world; and the passage of time, this movement along the dimension of time, is something we measure using various clocks. A clock is something that has a repeating and consistent recurrent event, a signal, a pattern. We have heartbeats and other biological clocks within our neural and endocrine systems that measure out time; and outside of our bodies, we sense the passage of the sun to map out a day, the moon for a month, the seasons for a year. In other words, *time* as we mentally construct it and name it may not necessarily be anything like what our minds imagine it to be: Time may *not* be something that flows, or something you have too little of or run out of (for some accessible examples, see Barbour, 2000 and 2008; as well as Feng & Crooks, 2008; also for a range of overviews, see the brief essays of the FQXi 2008 contest on the nature of time at www.fqxi.org such as Ellis, 2008; and Weinstein, 2008; and for historical perspectives, see Hawking & Ellis, 1973; Dorling, 1970; Prigogine, 1996). For some physicists, for those that were interested in this area and willing to discuss it, and from all the reading of publications I've been able to devour, one view of time is simply that, as we think we know it as its own entity, it actually does not exist.

Let me offer a brief overview of some of these mind-bending notions, drawing on fascinating views from physics as expressed by one of its leading theoreticians, Sean Carroll (2010). Time can be considered a concept we have that describes a feature of the universe that refers to how events unfold in spacetime. In other words, we have height, width, depth, and time—our four dimensions. In physicists' string theory, there are many more dimensions of reality, but for now, let's stick simply with these four dimensions. Some use the term "block universe" for a notion that these four dimensions exist as a huge block of reality. Time, in this view, is simply a designation of where in that block we are in the moment, and where in that block we are referring to as something in what we call past or future—where we are not right now, but where we've been or where we might go within that block universe. Time is simply one of the four coordinates used to locate ourselves in the four dimensional block universe.

Time itself is not something that flows. Time is simply a label we use, as humans, to refer to the universe we live in. We change location in the block of spacetime, and time is simply a reference for that change in location. Yet since we have clocks in this universe of ours—inside our bodies, with respiration and circadian rhythms, or outside our bodies, like the quartz crystals of modern watches and now our interconnected smartphones—we can use their repeated signals to map out the "passage of time"—the movement along the fourth dimension of spacetime. Albert Einstein made the proposal with General Relativity Theory that spacetime was not flat and uniform—that it might have curves and could be bent so that the seemingly constant intervals of time could change *relative* to the speed and the gravity of an event. The faster you go toward the limit of the speed of light, and the stronger the gravitational field, the "slower" time moves—meaning, more accurately, the wider the time intervals are relative to the place you started at slower speed or less gravity. So speed and gravity alter the shape of the spacetime block. There is nothing flowing, simply a four-dimensional world that stretches and bends. The role of gravity in all this is complex and quite a mystery, and the connection of all of this to quantum physics is still quite an active area of research and theoretical exploration.

One area this physics background hasn't really addressed is what the relationships are among past, present and future. With the fundamental laws of physics that govern the behavior of small particles, and with quantum physics itself, there is no distinction made between past and future. Even a deep exploration of the ways observation "collapses the probability wave function" of quantum physics, which appears to have a directionality to it, has been demonstrated to be simply a matter of a classical Newtonian imposition on a quantum state function—

and so it leaves quantum theory by itself still without a directionality of time. The fundamental particles and energy of the universe create what are called "microstates." The physics of microstates is symmetric, meaning it is reversible and without an inherent directionality to it. This is what we mean when we state that the fundamental physics equations that predict microstate behaviors suggest they can go in one direction or another, they are symmetric.

But the world of large collections of microstates as configured into macrostates, like your body as a collection of many molecules arranged in a wide array of configurations, actually has a directionality to how it unfolds. It is not reversible; it is not symmetric. With macrostates, there is a directionality connecting past, present, and future, just as there are causes and effects. This directionality of time for macrostates is revealed, for example, if we consider the event of mixing blueberry juice with strawberry juice and how it ultimately yields a purple mixture. The purple concoction does not spontaneously revert back into one section of blue and one of red. This direction of unfolding of macrostates is called the *Arrow of Time*. This arrow refers to the asymmetry of unfolding, the irreversibility of change, the directionality of time. We cannot go back and change the past; we can anticipate and influence the future. This is what we mean when we've stated that the past is fixed yet the future is open. This seeming irreversibility of time is because we live in a macrostate world.

Even our mental lives reveal the Arrow of Time in what is called epistemic unfolding—the way we can know the past and do not know the future—even if we can anticipate and plan for what happens next. Some people may not agree with that in their own experience, I know, but this is the perspective of modern physics at the macrostate level I am describing, not simply a view of our mental limitations.

But why do macrostates have an Arrow of Time? Especially when the fundamental laws of physics, the empirically established rules of energy and the basic particles of the universe that govern our microstates, do not have such directionality to the unfoldings of change? A first place to address this question is to state that the directionality of unfolding of macrostates in the universe, the Arrow of Time, is actually not a property of time itself. Odd as this sounds, time's directionality—the fact that we experience past as different from future—is a feature of the universe, the universe of macrostates, and not of the time dimension of spacetime itself. One of the ways of understanding the directionality of the Arrow is the "Second Law of Thermodynamics," an accepted physics law stating that a contained or closed system, like the universe, moves toward increasing entropy.

Entropy on first pass is often described as states of disorder or randomness. But this view is not always the case. A more complete definition of entropy is the number of *microstate configurations* that look the same at a given *macrostate* level. A microstate is made of the smallest components of a system—depending on the level of analysis, for example, this could mean energy, parts of an atom, or sometimes the atoms or interconnected atoms we call molecules—that comprise the macrostate assembly—the atoms, molecules, and larger configurations of all these microstates clustered together. To get a feeling for this, let's go back to that weird fruit concoction we mixed together earlier in time—in the past, back at that prior location in that block of spacetime. Picture the blueberry and the strawberry juice in separate containers. Now we pour some of each juice into a glass bowl, filling it to the top. What happens? Let's say we had placed a divider in the bowl that initially made it so that the two juices were on separate sides—red on the right, blue on the left. We remove the divider, and then what do you imagine happens? Yes, in our universe with its Second Law, the molecules, at first in a certain state of entropy, would now move to increase the entropy. How? They would do this by beginning to diffuse throughout the bowl. Why? Because a mixed juice configuration has many more possibilities than both juices staying in separate places, the probability is simply very high that they'd take on the many more options to become a mixed bowl rather than remain separate. Is there a chance they'd simply stay put, or instead simply change sides? Sure, a very, very, very small chance. But the sheer number of options for microstates that configure into a macrostate of a blended purple mixture is simply much, much higher, making that the likely process you'd observe. Over time—the unfolding changes in the universe—you'd see the entropy increase as the macrostates changed with movements of the microstates of the individual juice molecules, their positions in the bowl. In color terms, the purple mixture would be more probable than the separate red and blue macrostate assemblies.

Because our sense of past, present, and future are so fundamental to our mental lives, let me beg your indulgence (especially if you don't particularly like physics) and spend just a few moments with you here explaining the gist of this Second Law and why it is important for understanding our minds.

Physicists have an accepted notion, a fact, about the universe we live in that they call the "Past Hypothesis," a notion which basically states that what happened *before* had a lower state of entropy than right now. This fits with the Second Law that states that what happens *next* in the universe overall has more entropy. Not all individual macrostates need to have more entropy in and of themselves—as living beings, for example, we can lower our personal entropy when we clean our desk—or even clear out our minds, perhaps. But the Second Law does state that the sum total of change for the universe as a whole goes up as time unfolds. So even when you've cleared up your desk and lowered your personal entropy, the heat you gave off from the work will increase the entropy of the world. Okay, so big deal, you may be thinking. Why care? If we care about how our minds live with preoccupations with the past and frets over the future, this directionality of time makes a huge impact on mental experience. So how does this explain the connections among past-present-future, and what does this tell us about the Arrow of Time? And if increasing entropy is what drives this Arrow, this direction from past to future, we can see that in many ways, this directionality of time is directly related to probability as entropy is the higher probability that the many microstates comprising a given macrostate will be emerging.

You may be wondering what this has to do with living beings, especially with us. Here is the answer: No one really knows why spacetime has this direction, except for the implications of the Second Law of Thermodynamics and its companion Past Hypothesis view of the movement of entropy from low to high as we "move through time." Consider this: There is no arrow for space by itself. We can move back and forth in space, and we don't have a set of rules governing where we are likely to go. But in the time dimension, at the macrostate level, we do have this Arrow, this directionality to where we go within the spacetime block universe.

The empirical findings that this viewpoint gave rise to, you might imagine, is the cosmological view from physics and astronomy of the Big Bang Theory—the notion that long ago entropy was at a low point that couldn't go any lower. Scientists have calculated that this was 13.7 billion years ago, a moment when time began and our universe was very tightly configured—a few centimeters—and had very low entropy, very few variations that could create that particularly dense state. After this beginning explosion, the universe soon expanded and became filled with all sorts of stars and their galaxies and the property of the universe called gravity.

This tiny initial state of the universe was quite simple; it was not very complex. As the universe began to expand, the Second Law of Thermodynamics states things would gain more entropy. This movement ultimately would lead to an endpoint, some views suggest, in what is called thermal equilibrium—like the juice mixture reaching maximal entropy by equilibrating the blue and red to make purple. A blended world without differentiated features. But even those end-states of equilibrium are actually very simple, not complex. So where do we get the complexity of our biological entities here on Earth, or where do we get the wide variety of 100 billion galaxies each with an estimated 100 billion stars and their accompanying planets? That's also a fair amount of complexity. Well, it turns out that high states of complexity arise on the journey from extremely low entropy to high states of entropy in the universe as a whole.

In other words, as Carroll proposes, it may just be that the closed universe's movement toward maximizing entropy—for now at least in this mid-range state we're in at year 13.7 billion—is what enables us to exist as life forms. Life is not so easily defined, but one thing we do in our open systems' way is fight against thermal equilibrium. But calculations of the overall effect of life on earth as we take in the energy from the sun as photons, grow here as plants and animals, and then release energy back out into our atmosphere as infrared heat, is to increase the entropy of the universe as a whole. So while a single organism, like you or me, is fighting thermal equilibrium, we actually are contributing to increasing entropy in the universe.

While it may not seem fully satisfying, the state of the science of time itself at this, well, time, is to say that the directionality of time from past to present to future has an asymmetric macrostate feature, the Arrow of Time, because of the Second Law of Thermodynamics and the Past Hypothesis. As living creatures, our lives are filled with the Arrow of Time reality as we live in a closed universe with this Second Law and we live at the level of macrostates. But interestingly, quantum theory does not reveal this Arrow of Time, as the Second Law is actually a part of a classical Newtonian view that governs macrostates.

One fascinating implication that arises from this disparity between quantum microstates having no Arrow of Time and classical macrostates having an Arrow of Time is the following proposal we can make: Could the mind experience both a microstate arrow-free emergence and *also* a macrostate arrow-bound movement of the probability flow? Let's keep this notion in the back, side, front, and time-axis of our spacetime minds, a proposal we'll explore in more depth in the entries ahead.

Time is simply a term for how macrostates unfold from past to present to future, in that direction. Time is a location within the block universe we've named spacetime. With clocks, we can measure intervals of time, just as a ruler measures intervals of space. Events happen, change occurs across what we've named time, and time has an arrow, it happens in a direction. And now we have at least an initial understanding of what the term "time" actually means and why there is a directionality to the unfoldings of the macrostate aspects of our lives.

So unfoldings happen within what we are calling space and across what we are calling time—they unfold in spacetime. But these changes of microstate configurations that make up macrostates are comprised of energy and of particles which themselves are made of energy. And so while we can imagine a position of these compact energy forms, these particles, moving along spacetime, they can also change in other ways. Change without space? How can that be? One way of conceiving this is the movement of the position along the energy probability curve. That transition in position along the energy probability curve can occur without a change in space. That's what flow of the mind may mean, a shift in probability state. And we don't really know the full implications of what these changes in microstates really mean. Some may relate to the Arrow, some may have other qualities to them, as we'll soon see. Especially when it comes to the mind, this may be the essence of what "energy flow" may entail, movements along an energy probability distribution curve. Our awareness of such change may be one source of the mentally constructed experience of mental time. For example, when we see something new, or when we are given more details about something, our mental perception of time's passage feels as if time itself is slowing down. Imagine being in a new city and wandering around its novel streets. An afternoon of strolling there can feel like a whole week; the "same time interval" of an afternoon back in your familiar home turf may feel quite fast. Some equate this with the density of information bits we soak in somehow shaping our experience of the passage of time. But beyond such basic subjective senses of time, the mind's experience of change, of the unfoldings across the Arrow of Time, may also have elements from the nature of time that relate to our journey into mind.

What follows are a number of insightful reflections from a range of academicians on the nature of mind and time I share with you here to give a feeling for the kinds of issues being discussed in the professional literature. We've been diving deeply into mind and emergence and time, and these citations offer a glimpse into the background academic discussions underpinning our journey. Philosopher Craig Callender states,

> At a time when researchers in quantum gravity regularly propose speculative theories with no time at all, a better understanding of time in physics is all the more important—even if only to see what is lost by its absence...[T]he temporal direction is that direction on the manifold of events in which our best theories can tell the strongest, most informative "stories." Put another way, time is that direction in which our theories can obtain as much determinism as possible. Time is not only the "great amplifier" (Misner, C., Thorne, K. and Wheeler, J.A. (1973) *Gravitation*. New York: W.H. Freeman.), but it is also the great informer. (2008, p. 1)

And so from this view, time shapes our experience, but is not really something flowing—it is a process based on probability we use to inform our lives.

The physicists Rodolfo Gambini and Jorge Pullin go on to state:

> In quantum gravity there is no notion of absolute time. Like all other quantities in the theory, the notion of time has to be introduced "relationally," by studying the behavior of some physical quantities in terms of others chosen as a "clock." In this and other aspects of quantum mechanics, the standard view of time is challenged in empirical and mathematical ways. (2008, p.1)

The philosopher George Ellis brings us a bit closer to what we may be accustomed to:

> The most important property of time is that it unfolds. The present is different from both the past and future, which in turn are completely different from each other, the past being fixed and the future changeable. The present is in the instant of transition between these two states. The time that is the present at this instant will be in the past at the next instant. This process of coming into being rolls on in an immutable way: while we can influence what happens in time, we cannot influence the way that time itself progresses on. As stated by Omar Khayyam (E. Fitzgerald, 1989, The Rubaiyat of Omar Khayyam, New York: Penguin) "The Moving Finger writes: and, having writ, Moves on: nor all thy Piety nor Wit, Shall lure it back to cancel half a Line, Nor all thy Tears wash out a Word of it." (2008, p. 1)

One aspect of our sense of time is that it moves forward and cannot be reversed. But even this "time-asymmetry" may not always be present at microscopic, quantum levels. As the mechanical engineering professor Seth Lloyd explains, "Time and reality are large concepts and hard to grapple with. Our work shows that there are counterintuitive subtleties in the physical nature of time and is an example of how small quantum information processes can be used to explore big questions about nature" (Lloyd, interviewed in Six Degrees to the Emergence of Reality by Carinne Piekema for www.Fqxi.org., January 1, 2015). On the microscopic level, time appears to be symmetric—"time reversal symmetry"—but on the macroscopic level, time is unidirectional—that is, it is asymmetric, moving only toward what we call future.

Lloyd's physicist colleague, Jacob Biamonte, addresses their new theory of quantum complex networks by exploring the analogy of how the quantum aspect of reality is like the trees, but the whole forest is the classical (Newtonian) level of reality. In this way, this higher level of complexity—the classical level—is actually an emergent phenomenon of the lower level components, the trees or quanta. "One of the oldest examples of emergence, and arguably the most important, is the question of why the world around us seems too often well described using classical physics, while the world we live in is, in actual fact quantum."

The philosopher Jagdish Hattiangadi in a chapter entitled "The Emergence of Minds in Space and Time," explores some related issues to our whole journey. Here are some excerpts that highlight some of these relevant perspectives:

> [A] contemporary and indeterministic account of the physical world, a part of what is often called the orthodox interpretation of quantum phenomena, would suggest that emergence is not only quite coherent with it, but the best account available to us...[I]t will be shown to be relevant, and even crucial, to the development of an adequate theory of minds as they emerge in space and in time. (2005, p. 79)

Further, Hattiangadi states,

> In any system in which the arrangement of physical conditions in the state of the system in not fully determined by the laws, together with the previous states of the system, emergence is possible. Any theory of emergence will need to invoke a chance event or events that lead to the formation of a particular configuration of lower-level things. This configuration has a stability that is very different from the usual expectations of the lower-level things. Any stable structure will now have properties that the substrate fails to exhibit. We need only note that every emergent entity is in this sense a stable whole whose origin needs a special explanation. Once it has arisen, its stable existence can be understood as self-perpetuating. (pp. 83-84)

In exploring the relationship of quantum mechanics and the contributions of Niels Bohr to issues regarding the mind, he states, "Quantum mechanics is not being invoked because it is authoritative. This is not an argument from Bohr's authority. It is relevant to study because it lies at the lowest level that physics itself addresses" (2005, p. 86).

And in going on to explore the points of view of reductionism, the author goes on to state:

> So it is important to note that it is in micro-physics, at the level below the atomic, that the doubt about reductionism is clearest. The analysis carries this antireductive feature all the way up to the highest of levels whenever entities are emergent and irreducibly so. What this analysis shows is that though a whole is always com-

> posed of its parts, sometimes the types of things that constitute the parts cannot be fully described in all causally relevant respects without describing how they interact with the types of things that are the whole *as wholes* that are composed out of them. (p. 87)

Our journey into mind is diving deeply into the notion that the whole is greater than the sum of its parts, the emergent self-organizing phenomena that is based upon integration.

Hattiangadi goes on to discuss the implications of Niels Bohr's views:

> His message is that we do not need to accept the doctrine of reduction. The reason is that even physics is not reducible to mechanics. The most basic laws of the world cannot be understood without studying their interaction with the higher level of entities. There is no reason why we should expect life or mind to reduce to mechanics, or even to an intermediate level...The causal efficacy of the emergent whole upon the entities at the level that constitute its parts (downward causation) is a feature that can be illustrated at every level of emergence. (2005, p.91)

At this moment, there is quite an intense discussion about physics, the nature of time, and their relationship to our experience of mind. But one thing for certain at this moment is that we are not certain what the experience we call "time" is really all about. This uncertainty invites us to keep an open mind to experiences as they "unfold" across, well, across what we can simply call time as a linguistic placeholder for the inevitable changes that occur as life unfolds across events, as experience emerges, moment-by-moment. Emergence from levels of complex systems seems to be a solid place to rest along our journey into time and mind. And this emergence happens in the present moment. Being open to the unusual unfoldings of life events as they emerge is what having an open mind may be all about. Here is one I'll never forget, or possibly fully understand.

While lecturing in Seattle one year, I went far from my hotel where I was speaking to meet some friends for dinner. Two of us got up from the table before we ate to go find a restroom. Unfortunately the restaurant's bathroom was broken, so we were directed to go across the street to use the facilities in the neighboring hotel. We passed the rainy courtyard and entered the lobby of the hotel, looking for the restroom. A man in a trench coat came up to me and said, "Dan?!" Startled, I looked at him through the raindrops dripping off the hood covering his head, and said, "Yes, that's my name." He took off his hood and there in front of me was one of the physicists who'd been at the meeting in Italy on science and spirituality. We greeted each other with an embrace, and I asked him what he was doing in Seattle, since he was from Massachusetts. "I've come here to hear you speak." "But I'm across town," I said. Well, he just happened to pick this hotel.

That physicist, Arthur Zajonc, was also the President of the Mind and Life Institute, the organization that was co-created by Francisco Varela, a neuroscientist, and headed by the Dalai Lama. The aim of Mind and Life is to focus on the science of meditation and other ways of training the mind toward compassion and well-being. Arthur came and joined us at our dinner, and having eaten already we just shared dessert. But while we ate, I asked him if he could share with us his view of what physicists were thinking lately about time.

Time, he said to us, is not like what we sense it as. Time may be more like a distribution of probabilities than something that flows. So what we call *now* has a higher probability of being predicted than what we think of as the future. In certain ways, this fits with Callender's discussion of time as an informer, a phenomenon that helps us tell the best story—the one with highest probability of being true. The further from the present moment, the less certainty there is.

In reading more into the nature of time, as these discussions reveal, this view of the central role of probabilities keeps emerging, moment by moment. As we've stated earlier, one general consensus, with some dissent, is that events of now that have occurred in what we think of as a "past time" are actually present moments, nows, that are fixed and cannot be changed. I know that sounds odd, and some physicists don't even buy into this view. But the sense that something is flowing that is independent of change—something we've named "time"—is what the issue is really all about. For some scientists, there simply isn't any evidence supporting a notion that something is actually flowing. Depending upon the level of reality—macro- or microstates—the directionality of time may become more or less apparent as change emerges. But time may not be "something" at all, but rather a reference perspective we create in mind and in social communication for marking change. Nothing in reality may be flowing even if events unfold in what we call time. In this very specific sense, the statement "time, as we know it, does not actually exist" is what we are highlighting.

How we view the past, the present, and the future does, in fact, shape our lives. For example, in the field of developmental psychology, we've learned that how a parent makes sense of his or her own past childhood experiences is the best predictor of how the child of that adult will become attached to them (Siegel, 2012a). And children as well as adults who focus on the future in what is called "prospective memory" actually have better outcomes than those that do not anticipate and prepare for what may happen next (see Schacter, Addis & Buckner, 2007; Spreng, Mar, & Kim, 2009; Miles, Nind & McCrae, 2010). And in the field of anthropology, David Scott reveals that the way a culture tells the collective story of what happened in the past as romance or tragedy can directly shape how the future is created (Scott, 2004, 2014).

All of this tells us that while there may be nothing that is a "time that flows," how we reflect on and make sense of the past moments of now and also anticipate and plan for the future moments of now have important impacts on our well-being. In many ways, these can be seen as forms of temporal integration—how we link our differentiated nows across what we'll call "time," which really means the changing unfolding moments of life.

This view has huge implications. All we have is now. Yet so much of our mental lives are consumed with what we are now proposing is a mentally constructed concept, based on our perceptions of change, and of our consciousness of these perceptions of change, this "time" that is limited and limiting. We become preoccupied with the past, fret about the future, and come to feel that, as we've said, this time stuff rushes by, escapes us, runs out on us, or is something we can ever hold on to.

Instead, change does happen—life flows, not time. When this flow of change is locked into place, when it is fixed, it has achieved a high degree of probability. That's what we can generically call a "past event." That's a lot of certainty. It is unchangeable. In contrast, events that are happening in this present now are emergent, as they arise with some amount of uncertainty. That is a moderate degree of probability. And events of what we call the future are open, having the highest degrees of uncertainty. We really don't know what is going to unfold in the nows we call future.

For the questions we've been asking about mind, and our proposal that mind is an emergent property of energy and information flow, this discussion of time and the nature of energy is hugely important. If it is true that the subjective experience aspect of mind, the felt texture of lived life, is also emerging from energy and information flow as a prime—something not reducible to something else—then what does *time* mean here? If time is the unfolding of potential into actuality across a range of degrees of probability, then what this means is that mind emerges as energy flows— as it makes its way across the curve of probability's spectrum of values, from certainty to uncertainty, potential to actuality.

Now we come to a slightly different sense of ranges of probability. First, we explored the nature of time and have seen this spectrum of fixed, emergent, and open. This spectrum corresponds to what we subjectively experience as change and what we often call time, labeling it past, present, and future. Second, we are now looking deeply into the present moment's emergence. So this is not exactly about fixed, emergent, and open as across change. This is a focus on this here-and-now present moment emergence. Let's dive into this now in a deep way. Hold on to your hat.

In any given moment of now, I am suggesting we consider, energy is in a particular location along a probability curve. No matter where it is, as part of this present moment, it is emerging constantly. Some moments we can be up at the highest point, a peak of certainty. An example of this might be a thought. For illustration, imagine this as 100% certainty. You know you are thinking about the Golden Gate Bridge. Then, in another moment, we can be at a bit lower certainty at a sub-peak value, say of 80%. Now you are not with a singular thought (or image or memory or emotion). You are in the process of thinking, in this case, about all the bridges you know of or have seen in your life. Next, you drop that energy probability curve down to 50% and you are simply resting in a wider expanse of thoughts, imagining all the architectural structures you have ever seen. Then you drop the curve even further down, say to 20%, and you are simply letting arise whatever arises in your thinking (or imaging or remembering or feeling). As we'll discuss in detail in the next entry, you can even experience when the energy curve moves toward a near-zero probability (we can simply label this 0% for convenience, acknowledging that it may actually never reach absolute zero). As we'll explore, this may be the origin of consciousness itself. More later on that possibility!

This view of the facets of mind as information-processor and even of consciousness and its subjective experience suggests that these aspects of mind may be understood deeply as emergent properties of energy flow's movement along the probability distribution curve. A few years later when I'd present these ideas to a number of physicists, including Arthur Zajonc, there was a great deal of support and excitement expressed by this framework relating this quantum view of energy and the nature of mind.

And the other facet of mind as a self-organizing emergent property of the complex system of energy and information flow that is both embodied and relational may also be illuminated in a different way with this discussion of energy and time. Self-organization, as we've been seeing, enables maximal complexity to arise as it links differentiated areas to one another. We have called this linkage of differentiated parts integration, fully acknowledging that this is not a term mathematicians and physicists would choose to use because for them, that term means addition. We name the linkage of differentiated parts *integration* because it's a commonly used term that reveals how the whole is greater than the addition of its parts. Integration is a natural outcome of self-organization.

Probability shifts as potential moves to actuality, the uncertainty of open possibility transforming to the certainty of a potential manifesting itself as actuality. That's the feeling many use to describe subjective experience. We are proposing that while this subjective experience may ultimately be an internal process, meaning we feel what we feel from our embodied mind, within our mindscape, perhaps we still have a relational and embodied aspect of self-organization. But perhaps we also feel something in the subjective sense of what happens between us as well. Subjectivity and self-organization may each be aspects of an embodied and relational energy and information flow—sensations and regulations that arise as shifts from potential to actual and back into potential.

This is what *energy flow* really means. This shift may also be what self-organization harnesses in its ever-present push toward integration to maximize complexity, to create harmony. When this flow is thwarted, when probabilities cannot intertwine differentiated elements into a coherently linked larger whole, when we are not integrated, we move toward chaos or rigidity.

Chaos and rigidity can be seen as particular patterns of the movement of the energy probability curve. Rigidity would be one set of fixed certainties or increased probabilities happening in the moment, over and over again with little change. We could be in a prolonged depressed mood of increased probability, and we could have repeated thoughts of being worthless or guilty of things we never did as incessant actualities of our mental life. Chaos would be highly diverse probability states co-occurring simultaneously in a given moment. In contrast to this chaos and rigidity, integrative harmony would involve the diverse fluid movement across a dynamic probability curve that moved freely from an open plane to plateaus of probability to peaks of actualized certainty.

What might this look like to the eye?

I drew out such a figure for some students (see below) on the train leaving that conference on science and spirituality. Now we can use that view of flow as shifts in probability to explore the deep nature of mind—in both its subjective experience and self-organization.

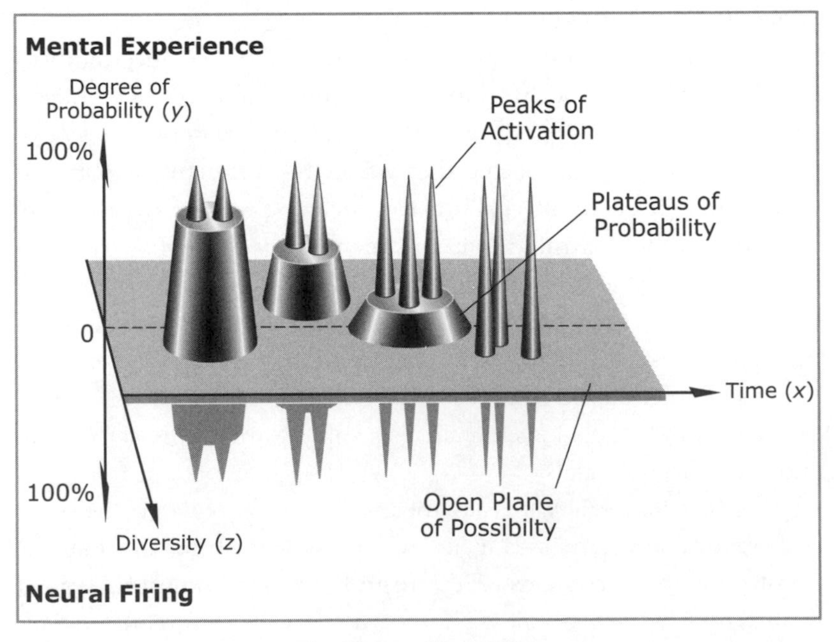

The Plane of Possibility

Since the Decade of the Brain, I had been using the Wheel of Awareness reflective exercise we mentioned earlier at first with patients, then with colleagues, and eventually workshop participants. This view of energy and time helps deepen our questions about the nature of mind and illuminates what the Wheel practice might be revealing about the deep nature of consciousness, and of the mind as a whole.

We'll dive deeply into the Wheel and these issues of the mind's flow in the next entry. I know this has been a lot, but the *when* of mind question invites us to take a step back from what appears real on the surface, like the subjective sense of time, and dive more deeply into the flow of change and the nature of mind.

In your own life, can you sense how changes in the likelihood of feelings or thoughts or memories may arise, moment by moment? This subjective experience may reflect internal shifts of energy flowing within you—your body and its head—or within your interpersonal relationships, and even your relationships with the larger environment around you. Here, now, we are exploring what the sensation of your mindscape and your mindsphere *feels like*. And here, now, we are considering what it would be like for you to try on the conceptual constructive stream that these shifts may actually be shifts in probability functions, movements along the universal energy probability distribution curve.

Now why would this matter, how you conceptualize these shifts? Here is the notion: If you develop your capacities to distinguish conduit from constructor within your moment-by-moment experience, you'll be taking a step toward integration. How? You'll be differentiating the most basic experiences of mind.

Next, imagine expanding that framework a bit to include the SOCK of sensation, observation, conceptualization, and knowing. As we've discussed, sensation is the core of conduition, observation is the bridge between conduition and construction, and conceptualization is construction. The knowing of truth, the sense of wholeness and "yesness" of things may be some weaving of conduition and construction. That is the integration of seeing clearly. That is the integration of well-being.

If we then add in the reality that if we take on a new way of conceiving what the mind is, we actually take a step towards relaxing some of the top-down filters that may be unknowingly keeping us from living fully in the present. For example, the notion of time as something we run out of, that thought, can make us anxious and feel existentially bereft. As I mentioned earlier in our journey, O'John, my friend John O'Donohue, described this as the sensation that "time was like fine sand in his hand." One of the conversations I wished we now could have had since his passing in 2008 would be about time as not being something that can slip out of reach. As we'll explore in the coming entries, embracing these notions of the awareness of change being the origin of our subjective sense of time-as-something-that-flows can actually liberate us from that existential angst.

The invitation is to consider that time is actually change we experience in mind as shifts in probability. Your mind emerges as probabilities shift, not as any object or noun-like thing is shifting as time is flowing. This lets us rest in the present moment, be open to whatever arises across all streams of awareness.

I invite you to take in these concepts and reflect on them, observe your experiences as you consider them in your life, sense the present moment with this framework as a background, and open your knowing stream to simply soak in all this sensation, observation, and conceptualization fully. The idea is that with this SOCK woven together, you may open your life more fully to the power and possibility of presence.

This presence is a portal to integration.

If you haven't done the Wheel practice yet, I again invite you to consider trying that experience out from my website, drdansiegel.com.

In the next entry, we'll see what might arise in our continuing journey into mind. Imagine mind itself as a journey, a process of unfolding and discovery. This journey does not end, it simply is. Now is all we have, as the nows unfold and bend possibility into the birthing of actuality, moment by moment.

CHAPTER 9

A Continuum Connecting Consciousness, Cognition, and Community?

IN THIS SECOND TO LAST EPOCH ENTRY, WE'LL EXPLORE MORE deeply the place where we've arrived in sensing energy as a distribution of probability that ranges from certainty to uncertainty, high probability to open possibility. With this framework, we've suggested that consciousness may arise from a sea of potential, a plane of infinite possibility. Mental processes, such as intention and mood, arise as this energy curve moves toward higher degrees of certainty that we are calling plateaus of probability. Mental activities, such as emoting and emotion, thinking and thought, remembering and memory, are viewed as elevated positions on the curve, sub-peak values preceding the emergence of the peak value of an actualized possibility. This is how the mind can be viewed as an ever emergent unfolding of potential into actual. We've also proposed that beyond consciousness and the subjective feeling of lived life, and perhaps beyond information processing, one aspect of mind can be defined as an emergent, self-organizing, embodied and relational process that arises from and regulates energy and information flow, within and between. Offering this definition of one facet of the mind as a self-organizing process empowers us to define a healthy mind and ways to cultivate mental well-being. In this entry we'll explore possible answers to some of our questions and continue the natural flow of inquiry that itself gives rise to fascinating and unanticipated new vistas into the essence of who we are.

Integrating Consciousness, Illuminating Mind (2010-2015)

I awake from a brief rest. It is 4 a.m. and the stars remain in the early morning sky. I know this book is coming to a close, our journey, for now, nearing some kind of resting moment, some kind of pause, in this penultimate entry. Below these cliffs overlooking the Pacific, the pleading wails of sea lions bellow amidst the chorus of crashing waves. My 25-year-old son is here, asleep in the loft upstairs, traveling with me on this teaching week in Big Sur, at the Esalen Institute, a 50-year-old center that played a central role in the history of the human potential movement in the United States. One of my son's songs is playing in the conduit of my mind, its lines with me as I wake with ideas about this entry: "Half of my life, is gone, gone, gone . . . I need a good leg to stand on; I had too many questions . . . for all the answers I was told." (Good Leg, music and lyrics by Alex Siegel). The words echo inside me, the surf echoes below, stars glimmer above, and these words reach from within me to between us to within you now.

Photo by Caroline Welch

All one continuous movement.

The answers I was told in my own education never felt quite right. This journey into mind over these decades, and shared in these entries, was initiated by that restless feeling. What has driven this exploration forward is a longing to see clearly and share fully what seems real, what seems true. The responses of patients and colleagues, students and readers, has sustained this drive to question the nature of our minds and inquire more and more into what this heart of being human may truly be.

In a few hours I'll take the stage in a large room where the 150 participants will gather. Our conference on the science of compassion, gratitude, forgiveness, and mindfulness has filled this entire sanctuary by the sea; I've been listening to the faculty's beautiful work on these themes. I will present the Wheel of Awareness to them first, diving deeply into the experience of integrating and exploring consciousness, then explore what the science-based notions of the experience of this Wheel practice might be. It feels, here from this cliff now, stars in the sky, waves below, that this whole journey is some flowing process, and I am here for some service, to see or say something that might be of help. I'd rather be sleeping right now, but my mind is abuzz, not so much with ideas as a sense of something, set of images, feeling in my body, a cotton ball in my head that needs expression somehow. I know that makes no sense—a cotton ball in need of expression—but that is simply what it feels like.

The shore here feels ancient, the tides coming and going for eons. But in our epoch, we humans have changed the face of the planet. The modern world we have molded is molding the inner landscape of the mind, our mindscape, and the inter mindsphere that shapes the culture in which we are immersed. This inner and inter mind, this mindscape and mindsphere, are at the core of who we are, and this mind of ours in these hectic and harrowing days needs tending. This may be why we are all here. It feels as if this is why I am up so early this morning.

Being human is more than being a brain at the helm of an isolated ship lost at sea. We are fully embedded in our social world and fully

embodied beyond the solo skull. This embodied and embedded reality means we are truly open systems. There is no boundary within which we can feel we are completely in charge. There is no programmer to whom we can turn for solace that all will be well. Even when we fully embrace the most rigorous of scientific beliefs and most cherished of religious beliefs, this view urges us to embrace our humanity filled with humility.

Yet we are so often told, by society, school, even by science, that we are solo players in a competitive, dog-eat-dog world. We live for a limited time, so make the most of it. We are told to accumulate stuff for the sake of our individual happiness, achieve for our individual sense of accomplishment. But what feels so misguided by these efforts is the implicit assumption that the self lives only in the body, or in the brain. For all the answers I was told, I had too many questions. My son's sentiment is exactly what creates this restless feeling inside me, driving us forward on this journey of discovery filled with questions. The self is not limited in time, as time as a unitary something that flows may not even exist. The self is not limited by the skull, or boundaries of the skin. The self is the system in which we live, our bodies the nodes of a larger interconnected whole in which we are inextricably embedded.

I hear the call of the sea lions below. The stars appear less bright, and as dawn approaches, the bats swoop over my head, seeking out the bugs that are now buzzing about. We are all awake, all here, all a part of this wholeness of life, this wholeness of mind and our interconnected reality.

I had a dear friend I mentioned earlier named John O'Donohue, who died suddenly in the last decade. John and I would teach together, he from his background as an Irish Catholic priest, poet, philosopher, and mystic; I from the interpersonal neurobiology perspective we've been exploring. We first met on a rocky shore just like this one, along the Oregon Coast. The last time we saw each other, the last time we taught together, was on the rocky shores of the West of Ireland where he grew up. John and I shared so much, loved to be with each other, and were forever in a state of unfolding. John used to say that he'd like to live like a river flows, surprised by the nature of its own becoming. Being and teaching together, we never knew exactly how we'd be becoming. John and I were the same age, in our early fifties back then, and he died just months after we taught together near his childhood home. We were created in our connection—not only our relationship of we, but each of us individually. I am who I am even now, years since his passing, because of who we were.

Perhaps instead of writing, "had a friend" in the paragraph above, I could have, or should have, written, "have a friend." John lives in me still.

Our essence, our mind, is truly relational.

People often ask me if the mind needs a brain to exist. Do we need our bodies to be alive for the mind to survive? I can imagine that John's experience of subjective reality may have depended fully on his embodied brain being alive—now that his body no longer lives, this aspect of his mind no longer unfolds. Others might believe that a soul continues on, an essence of a person that survives after the body dies. I am open to that, and John not only may have held that view, he may be living it now, too. I hope so.

But the self-organizing aspect of John's mind may certainly still be alive even though his body is gone. What I mean by this is what perhaps you may have experienced yourself, how after the loss of someone you love, someone who has shaped you, changed you, altered the course of your development from the inside out, they are still with you though their body is gone. In that way, our deep connections to others live on. That's the relational aspect of mind.

The self-organizational aspect of mind may also continue on, after our body is gone, in how we influence others whom may never have known us personally. Since John's passing, numerous individuals have approached me after I've spoken about John or mentioned his books and audio programs. After reading those works or listening to recordings of John's mellifluous voice, they tell me what many said when John was still alive: they feel deeply transformed by their connection with him. For me, this is a gift John gave us all, a reminder of the magnificence of our days and the magic, mystery, and majesty of our lives.

I used to tell John that we may have an "eternal imprint" that persists in reality even though we seem to have moved forward in time. What I said to him was that if we imagine an ant crawling along a ruler, walking from inch two to three, three to four, and then four to five, the ant would believe that the only thing that exists at this moment is inch five. Yet the ruler did not disappear. Inches two, three, and four continue to exist even though the ant's ongoing perception is only of inch five. John loved that idea and image, and loved the phrase of an eternal imprint. Maybe it is wishful thinking, wishing his presence could be here now with me overlooking this ocean, on this particular inch on the ruler of life. But if time truly does not exist as something that flows, then in fact the time of John's body being alive is still "here" in the eternal imprint of life, just not at this inch now along the Pacific. I hope that is true, and it feels good, to imagine that what we have had in life can never be taken away. It is an eternal imprint, an ever-existing location in spacetime within our four-dimensional block universe. It is a reminder, too, to embrace the awesome privilege of living with gratitude for each day of our lives, and for each other.

Nows have passed through these flows of energy and information as they transform, calendar pages turning alongside the chapters of this journey as summer edges toward autumn, winter moves toward spring.

This morning, too, I awaken with the immediate memories of a weekend's events from earlier this year. The stars fading in the dawn sky, bats still flitting about, sea lions' calls dancing with the surging surf, am I in a dream? I've been listening to the manuscript for this book transformed electronically into a robot's voice while I drive the dozens of miles to and from my teaching engagements. Is this journey of discovery really unfolding? Are we shaking up the surface world of mind and revealing its inner layers? Did the weekend gathering at All Saints Church in Pasadena a few months ago really happen?

Rodolfo Llinás once said in a lecture on consciousness I attended long ago that consciousness was essentially a waking dream. We don't really have a distinction in the neural events, he said, of sleep narratives and the unfolding of everyday life's awareness. Being alive is a dream. In his work (Llinás, 2014), he describes a neural correlate of consciousness as involving a 40-cycle–per-second (40 Hz or hertz) sweep between the thalamus and the cortex. Anything swept up in that oscillating neural wave we then experience within consciousness. That view is consilient with other discussions of consciousness, such as that of Tononi and Koch (Tononi & Koch, 2015) who propose essentially that the degree of complexity achieved with the linking of different parts of the brain and perhaps other systems—the degree of integration—somehow yields the mental experience of consciousness. These authors, and the many others who've proposed a range of potential NCC, or neural correlates of consciousness, have offered intriguing suggestions about how the subjective experience of being aware might go along with the brain's activity (see Damasio, 2000 and 2005; Edelman, 1993; Edelman & Tononi, 2000; Graziano, 2014),.

As I awaken this morning, I think that my generating constructor mode of mind is fully engaged, waking me from my slumber. With my constructor reflecting upon itself, and now the constructive function being activated, I generate the information, knowing the top-down date of our human created calendar, and know that this is months since All Saints. My constructor, too, interprets that knowledge and finds meaning in it, the ABCDE's of meaning in the brain: the *associations* of that weekend event earlier this year; the *beliefs* that life is not just on the everyday surface, the *cognitions* of information flow that unfold about that weekend, my communicating with you, and our discussion of consciousness itself; the *developmental* period that this experience of epochal entries evokes relevant in time and space; and the *emotions* that arise, those shifts in integration that emerge not only as feelings but changes in my state of mind. This is how we interpret and generate meaning in the mind.

I can sense that the gathering at the Church was not a dream in that it was not manufactured by the generating mind within this now awakening body. Perhaps it was a dreamed of event, something created in the minds of the Episcopal priest who envisioned its happening, and now it has happened. Somehow all this unfolding seems related to the ongoing life of John's self-organizing mind.

Ed Bacon, an Episcopal priest at All Saints, first heard of John O'Donohue years ago. Inspired by John's writings and recordings and after hearing of John's passing, eight years ago now, eight annual inches on the eternal imprint ruler of life, Ed went on a pilgrimage to be in the land where John had lived that had so inspired John and his writings. Ed found his way to John's family, and upon meeting with his brothers who heard he was from Southern California, they told him about me and my relationship with John.

I soon received a call from Ed, and we met in Los Angeles to discuss our mutual interests. After being a member of our monthly community mindsight meetings in town, Ed came up with the idea of doing a weekend retreat at his creative and socially active church to explore the connection between science and spirituality. We set up a "Soul and Synapse" event, the name of workshops I had been offering that explore consciousness and the nature of mind with experiential immersions in the Wheel of Awareness practice, along with scientific discussions on human life. I retrieve my journal entries from this computer, and share some of those passages written just a day after our event with you here.

That weekend at All Saints was a powerful experience for me, joining with 300 people and diving into the Wheel practice, sharing direct first-person subjective realities of the experience, doing the Wheel again, sharing more, and then beginning to reflect on what might be going on. As the facilitator of the retreat, I emphasized the importance of realizing that subjective reality was real, and that we needed to honor that data as people attempted to translate into words the felt experience of the practice. I could write a whole book on what this and other retreats focusing on the Wheel reveal, and maybe that's a good idea for some day, but here I will simply share some highlights for our journey together.

For some, the Wheel practice is disorienting. They describe a sense that the way they see the world changes after the practice. Some descriptions from the first day suggested that this new way of perceiving the world within was uncomfortable for them, and at times they became filled with a hostile voice saying that they were doing things wrong, or that they should be worrying about one thing or another and not following along. For others, the shift in perspective is more like a new way of seeing, of noticing differences in the temperature of air, for example, as they focus on the breath coming in and out of the nostrils. Some felt excited by this difference in perception, some felt calm and relaxed. These are perhaps common findings not just for the Wheel practice, but when people begin reflective exercises of all sorts that invite them to look inward, not just outward.

On the second pass through the Wheel, I add the bending of the spoke component, so that at this point the participants are now not only differentiating hub from rim, knowing from known, by way of the movement of the spoke around the rim, but are also experiencing a direct awareness of awareness. The spoke of attention is directed right into the hub. For some, this is hard to experience, and their minds simply "go off" and they get lost in one rim activity or another—be it thought, memory, or perceptual streams. But for others, there is a powerful and for me now familiar reporting of a sense of expansiveness and peace. One person said, "I found myself in a place in my mind I've never experienced before. There is nothing to do. There is nothing to grasp, nothing to get rid of, just being, just being right here. It was incredibly peaceful." Another person offered that "the boundaries of where my body defines who I am went away and I have this amazing feeling of being not just connected, but being a part of everyone, and everything." Someone else said, "I had the experience that who I am is the universe, and it was magnificent. Maybe that's who I really am." And one other person offered that she had "never felt so much peace before. This was God, this was love, and I don't think I'll ever lose touch with that."

By the second day, we were exploring other experiences, with people reflecting on how the movement away from a closed, bodily-based identity seemed to give them a deep sense of change, a "transformation" as many individuals would name it. This change in identity, this shift within awareness of a sense of who we are, seems to have a powerful impact on a person's sense of well-being. One person even came up to me at a break and urged I invite people to write down their responses, as she didn't feel comfortable sharing with the larger group. I asked her what her experience was like, and she said, "Well, I didn't want to be seen as a bragging person, so I didn't want to say. But it was the most amazing experience I have ever had. I felt so whole, so big, so infinite, and I felt so at peace. Thank you. I don't think I'll ever be the same."

It does not always work out so well. On the first day, for example, a few individuals with a traumatic childhood history offered that they felt a sense of panic with the practice, and with the focus on the hub they initially felt a sense of dissociation—disconnecting from the experience and being flooded with uncomfortable images, emotions, or bodily sensations. They "could handle" these in this setting, they said, but it was not a comfortable experience. By the second wheel practice on the first day for one person, and on the second day's immersions for several others, this discomfort transformed, and as one person articulated, "I feel free from the prison." When I asked her to say more, she said, "I am not just those memories, not just those feelings in my body. Somehow that hub is now my friend, and the wheel is not my prison. And I feel deeply freed."

A marvelous reality we seem to have encountered is that we are not passive in all this activity of mind and awareness. With consciousness comes the possibility of choice and change. But what might this really mean with regard to our new understandings of mind, energy, and of time, and of the nature of consciousness?

After teaching the Wheel of Awareness to thousands of people in other workshops, and receiving feedback from some of the more than 750 thousand people who've downloaded the practice from our website, a fascinating pattern of experience has been revealed. No matter the cultural or national background, whatever the extent of education or the age, whether they have had no prior meditation experience and are newcomers to reflective practice or have had extensive meditation training in a range of disciplines, including those who are the directors of meditation centers and monasteries, the responses are remarkably consistent.

Not all people by any means find the practice compelling, but when they do, no matter their background, the following pattern has been expressed about the Wheel practice: When discussing the rim, a wide range of sensory descriptions of the outer world or bodily sensations is offered. When reflecting on the aspect of the rim that symbolizes mental activities, many state that inviting feelings, thoughts, or memories into awareness somehow calms the mind and they simply feel a clarity and stability, sometimes for the first time in their lives. When they have a "bring it on" stance, the mind becomes clear. Nothing need be repelled, nor sought. Being open invites one to simply be. When we get to the fourth segment that represents our relational or eighth sense, people often feel a deep sense of connectedness. Experiencing the interconnections we have with other people and the planet gives many on this fourth segment of the rim a deep sense of gratitude and belonging.

When we discuss the experience of bending the spoke of attention 180 degrees so there is the direct experience of awareness of awareness, some common terms, like the ones articulated that weekend, have been used to describe what happens: openness, being as wide as the sky, as deep as the ocean, God, Pure Love, Home, safe, clear, spirit, the infinite, boundless.

Imagine what might be going on. With a simple metaphor like the Wheel of Awareness we assign a visual image of the wheel with the mental aspects of the known on the rim and the knowing of awareness on the hub; we systematically move a spoke of attention to explore each in turn, to differentiate each of these from one another and then link them in the whole practice to create an integrative experience. This brief reflective practice can result in profound and commonly expressed experiences. Even though I have done this with over ten thousand people in person, it still amazes me each time we dive into the practice and then hear the offerings of descriptions of what transpired.

The next question we asked in the workshop is this. If the Wheel of Awareness is the metaphoric practice, could it be that the probability curve of the plane of possibility, discussed in the last entry and explored further in the pages ahead, might actually be the underlying mechanism of mind? Let's dive more deeply into that possibility.

One view of quantum physics is that energy moves along a probability distribution curve, as we've discussed earlier. Let's review the important components of this view and how they may relate to the first-person accounts of the Wheel practice. At one end of the probability curve is 100 percent certainty; at the other is zero or near-zero certainty. The figure of the plane of possibility represents a graph of this energy curve, from near-zero certainty up to the certainties of actualities. When we map possible mental correlates onto this visual mathematical depiction of the probability curve, it includes a potential notion manifesting as an actual, realized thought that would have sub-peak and peak values at the 100 percent point, the peak. When we have an intention or mood, this would be a plateau of elevated probability, one that sets the stage for and limits the direction of which possibilities might subsequently arise from that plateau.

But what would the plane be? I ask the participants. Could these practices of the Wheel be indeed revealing some aspect of consciousness, some way that energy flow in all its mystery, even to physicists, might be the source of awareness? Could the plane of possibility be the source of consciousness? This suggestion comes from taking in the empirical data of thousands of first-person reports of the Wheel practice. At the time of the one-eighty turnaround of the spoke, the repeated reports of being open, infinite, expansive, connected to everything, and the fact that these statements were so similar, no matter the culture, education, age, or experience of the individual, suggest this: the metaphor of the hub represents the mechanism of the plane.

For nearly 100 years, quantum studies revealed that the act of observing, of human awareness some would propose, directly collapses the wave function of a photon into a particle property. A wave has a wide range of values, or locations; a particle has one value, or one location. The wild finding is that taking an impression, taking a photograph for example, of where the photon is shifts it from a wave to a particle. For many of those physicists who subscribe to what is called the orthodox Copenhagen interpretation of quantum mechanics, the view of these empirical findings is that this observation of a camera implies the consciousness of the person taking the picture. For others, this is not about consciousness—it's some yet-to-be explained issue about a camera and assessing the specific value of a probability wave that might involve other issues, such as multiple universes. The implications of this debate are huge, the controversies unresolved. But for some physicists, the issue is clear: Consciousness seems an inherent part of the universe (see Stapp, 2011; Kafatos & Siegel, 2015).

In simple terms, observation moves energy from uncertainty to certainty along its probability distribution curve. This quantum reality, that observation somehow shifts the nature of energy, is not disputed, even though interpretations of the equations and their meaning within this branch of physics are highly debated.

If the specific interpretation of the role of consciousness is indeed true, that awareness itself has something to do with altering energy's probability distribution function, then the Wheel practice might be revealing a continuity among mental experience from awareness in the plane, moods and intentions in the plateaus of elevated probability, and mental processes just below the fixed peaks of mental activities. I've made this proposal to quantum physicists who say this is not stated by quantum views, but is completely consistent with the accepted laws of quantum mechanics.

Such movement of energy, this flow across degrees of probability, might be a way to explain how thinking becomes thought, emoting becomes emotion, and remembering becomes memory. This would be depicted as sub-peak values moving toward a peak of certainty. The open plane of possibility may be how awareness arises, how we can experience a "beginner's mind" from which to release ourselves from the at times automatic top-down filters that can imprison us from a more open way of knowing. This would be a mechanism of the space between impulse and action, between stimulus and response. Top-down filters would be the plateaus and peaks of prior experience that create our constructed experience of the world, from ideas and concepts to our narrative sense of self. When we develop access to the plane, we drop beneath these filters and enter this spaciousness of mind, a stillness, a pause before action, which enables us to see clearly, to be fully present for life in the moment, rather than lost in preoccupations with the past or frets for the future.

When we consider our discussions of time in our last entry, we can imagine that there might be some distinction between our experience of mind emerging from mind as microstates which are arrow-free, and those that might be emerging from macrostates that are arrow-bound. Recall that quantum views explore microstates that have no Arrow of Time; classical (Newtonian) physics includes the Second Law of Thermodynamics and the Past Hypothesis which help us understand why our macrostates unfold with a directionality we call the Arrow of Time. If energy can be experienced as microstates where quantum properties dominate, or macrostates where classical properties dominate, then the mind—as an emergent property of energy flow itself—may have two ways of being experienced: One with an arrow, and one arrow-free. Change or flow would be in both forms, but the quality of mental life, of subjective experience, might feel quite distinct. Here is the notion: Could our experience of consciousness as illuminated with clarity with the Wheel practice, be revealing a plane of possibility that is more of a microstate with an open, spacious quality? This is how the hub metaphor and the plane mechanism would reveal a microstate flow of mind. In contrast, we feel more directly in the movement within the spacetime block universe as we live on the rim of the Wheel, those above-plane values that might actually be macrostates of a range of configurations of energy patterns. Let's keep this proposal in mind as we continue our journey into the experience of being aware with the possible contrast of the "timeless" knowing and the "time-bound" known of consciousness.

With this view of a deep underlying mechanism of mind seen as a distribution of unfolding probabilities that we can actually influence, influence with our minds, we would now have one way of describing the knowing of consciousness. The knowing of consciousness arises from the energy curve at near-zero certainty, the equivalent of near-infinite possibility. Even if it is not quite infinite, it is wide open, and we can immerse ourselves in this the open plane of possibility. This is a sea of potential from which arises probabilities and then actualities.

This view of the plane has been a powerful framework supported, not proven by any means, by the direct first person experiences offered in reflections before the plane is ever introduced in these workshops. It is consistent with quantum physics, though not a part of quantum physics. It can both fit with and predict reports of individuals experiencing their mental lives, including the explorations during the Wheel practice.

As a working hypothesis, the plane of possibility gives us a frame of reference to begin to make sense of experience. Yet as a framework, we need to keep an open mind for the top-down ways it can limit us as a linguistic model and a visual metaphor. Perhaps it is true, perhaps it is partially true, or perhaps it is wrong. Let's be open to that wide range of possibilities.

One intriguing aspect of the plane is that when offered this perspective, many participants often intensify the work and expand their sense of "what is going on" in terms of feeling clarity about shifts in their field of awareness; they feel empowered to bring more equanimity and connectivity to their lives. Some—and this happens at nearly every workshop I have taught—even come up to say that the chronic pain they've had, in their knee or neck, back or hip, has disappeared. Emails I've received suggest that the pain stays abated. What could be going on? If pain is on the metaphoric rim, then strengthening the access to the hub gives relief. From a plane mechanism framework, if chronic pain were a persistent plateau giving rise to intense peaks of pain, then accessing the plane literally drops the subjective experience of the mind down, down into the plane of possibility and away from the persistent prison of the pain's peak.

And what is the sensation of "being in" the plane? For many, accessing the plane is associated with the feelings we've discussed: openness, peace, infinite potential, love, connection, wholeness, and serenity. If I were reading these words from afar, not familiar with the experience personally, I wouldn't believe them myself if I hadn't heard them reported directly from participants, over and over again.

Knowing about the plane itself is unnecessary for the Wheel to work, but this understanding seems to not only allow things to fall into place conceptually and practically, but also facilitate the empowerment people experience. This is an intriguing observation, but in no way evidence of the validity of the hypothesis.

At the end of our weekend retreat, we had a summary session in which Ed and I shared the stage and focused on a range of topics, including an initiative they were working on called "The New Narrative Project," an effort to find a cross-discipline, inter-faith way of linking science and spirituality to bring people together in many walks of life. I looked out on the audience, seeing people of all ages from diverse backgrounds, and felt that Ed's passion to take his training in religion and blend it with a scientific immersion in mind was fueling a potential shift in all who shared our experience that weekend.

Ed asked me about how we might connect some of the reflections that had arisen with the Wheel practice with people's experience of God and science. I listened to his question and paused, breathless from the clarity of the query, and hesitant to know how to respond. I felt an awareness of John O'Donohue, recalled in my mind our many discussions of science, society and spirituality, of his experiences as a Catholic priest, views of Celtic mysticism, work in philosophy, and how I experienced him, as not just a poet, but a walking, breathing poem. John and I would teach about "awakening the mind" through these two lenses of spirituality and science. Somehow the self-organizing soul of John's mind was fully there, in Ed, me, and the room.

I took a deep breath. When I think about that moment now, I imagine that my own Wheel of Awareness had a four-layer circle around the hub that streamed in whatever arose as energy flowed—as actualities transfigured, a term John might have used, from a sea of potential, a plane of possibility. This was the SOCK we've spoken about on our journey: a *sensory* stream filling me with the bottom-up conduit flow of sensation; an *observational* stream, giving me a bit of distance still as a conduit but also a witness to the unfoldings of now, not just a participant in sensation; a *conceptual* stream, a top-down set of constructed categories, ideas, and language, perhaps the source of a narrative, that both illuminated and limited a sense of understanding and organized my perspective; and a stream of *knowing*, something beyond simply being aware, beyond conduition and construction perhaps, something that had a deep sense of purpose and truth beyond and before words or concepts.

I began to respond to Ed by suggesting that a truly scientific view would embrace the reality that the entities asking and responding to the questions were emerging, in part, from within a body. The body's apparatus, including its brain, was limited in its pathways of perception, and perhaps even of conception. A scientific view ought to embrace the possibility that things that are real may not now, or perhaps ever, be fully knowable from a scientist's lens. That being said, I went on to offer that the weekend's immersion enabled us to gain insight into the possible nature of the essence of mind, and the essence of Ed's question.

If by "God," Ed meant that there were structured forces at work in the universe that were invisible to the eye and perhaps unknowable to the rational mind, then the principles of science ought to be able to consider the likelihood of that invisible reality. Not everything that is real can be seen, just as not everything important can be measured. If Ed meant that we as human beings might be best served by exploring how those in spiritual practice and scientific pursuits might find common ground, then I had a notion of how we could do that. With our view of spirituality as a life of connectedness and meaning, and with a broad sense of science as illuminating the deep nature of reality, of shaking up the surface appearance of things and revealing underlying mechanisms, then there ought to be a way where the two find not only common ground, but also mutual inspiration.

What I then said, I can't easily reconstruct here. This is what I was thinking before I spoke aloud, keeping in mind all that you and I have been exploring in our journey together here, and all the connections I felt with Ed within his question—if the mind has subjective textures, subjective reality, and self-organization as two fundamental components, then in many ways, as we've been exploring, the mind has a mind of its own. Whether we are being primarily influenced in this moment by a bottom-up conduit, having senses stream into awareness without our control, or by a top-down constructor's conceptual filter forming and shaping how we see and what we do without our conscious control, awareness can feel as if something "other than ourselves" is in charge. Yet if we get out of our own way, if we sometimes let that flow of sensation and concept simply arise, when we let our observing and knowing streams take part too, without trying to be the driver of this unfolding process, sometimes helpful experiences emerge.

That's what I thought, and that's what I felt, before I spoke.

Then I felt my mouth move, heard the words come out, sensed John's self-organizing presence, saw Ed's caring eyes, gazed out at the 300 faces sitting still, focused on this conversation, and began to speak. I think what was said was something like this: If we imagine that energy is the fundamental essence of the universe, that even matter is condensed energy, then patterns of energy are the essence of reality. If we imagine that energy manifests itself as potentials, traversing a range from open possibility to probability to actuality, back and forth, then we can begin to sense how we may experience our mental lives. If we consider that there is a natural push for an open system to move toward linking differentiated parts, then this integration may be an invisible force that drives our self-organizing minds.

When we do the Wheel practice, I went on, we are differentiating distinct aspects along the spectrum of energy's probability distribution curve. This spectrum is the experience of mind. Various rim elements are experienced as emergent mental activities, arising and dissolving as actualities melt back into probability and then into possibility. When we move the spoke of attention that directs that energy and information flow around the rim, we are innately differentiating not only rim elements from each other, but also distinguishing the hub from the spoke arising from the rim, the energy patterns above the plane, the manifestations of probability and actuality arising from possibility.

This movement of the spoke strengthens the hub in that it harnesses the ability of the individual to access the plane of possibility. When we next turn the spoke of attention around, bending it 180 degrees right into the metaphoric hub, we gain a direct experience of awareness of awareness.

What the hub may represent is the plane of possibility, I said. And from this open plane, there may be a feeling of the infinite. There may be an emerging awareness that your hub and my hub, that all of our hubs, are essentially the same. Infinity is infinity—each of our planes of possibility represents a sea of potential, a pool of the infinite, or nearly infinite. These are identical, or nearly identical. I looked out over the group, at all of us assembled that weekend to dive into this experience, to explore the mind. Though we live in different bodies, I said, we share the same essential plane. Even if the plane is near-zero certainty, even though we live in a body, it may be possible to get as close as possible to this open sense of infinity through this exercise. When I looked out at the beautiful and handsome faces in the group, it felt as if each person was of the same essence, each individual a part of a collective being, each of them were me, each of us connected. We were connected through the hub, connected in our planes of possibility.

Putting this together, I said to Ed and the group, nervous about a science view that might offend a religious sensibility, it may be that one view of God might be considered as a mystical sense of the generator of diversity, the open plane from which all arises into being. And somehow, it may be possible that the very experience of being aware, of being conscious, is a sacred experience. Arising from the open plane, our experience of awareness comes into being.

Ed grasped my hand, a beautiful smile on his face relieving the anxiety within me that I might have offered something unwelcome in that house of worship. The group seemed electrified, too, as we brought our immersion together to a close for that now of all the nows that had unfolded, together.

I also thought of you, back then when I spoke in front of the group, and now, as I write these words connecting us to each other. I know here in this penultimate epoch of our present journey we have not answered some fundamental questions. Maybe we've had fun focusing on the mental, but the questions still remain wide open, the doubt respected, the uncertainty embraced. Even if there are neural correlates, of which regions are activated simultaneously in a 40-cycle sweep, or an integrative linkage reaching certain levels of complexity in the brain itself, what may ultimately be found to be the brain-basis of consciousness, how does this knowing of awareness actually happen? Well, we just don't know. No one knows. If it does not ultimately depend solely on a brain, how would a spectrum of energy probability values give rise to the subjective experience of mental life? How might a plane of possibility give rise to the knowing of consciousness, within the brain, within the body as a whole, or even perhaps beyond the limitations of the body? We just don't know. But we are moving into deeper questions along this journey, questions that can give rise to helpful approaches, ones that might provide relief for suffering in the individual, and a transformation of identity that might serve the well-being of the individual, others, and the planet.

I feel sad, but also kind of exhilarated, to bring up these unanswered questions with you. A part of my mind wishes for the certainty of answers. I drop down into my plane, and plateaus of worry and peaks of petrified thoughts melt down into that sanctuary of equanimity, that sea of potential. It feels like home. It feels like a place you and I can share, a place where we can deeply find each other.

We began our journey stating that it may end up that we come to understand that it is the questioning that is the journey's value, not the coming to a final destination, a final answer. Perhaps that is the beauty and the bounty, the power and the potential, of this plane-of-possiblity view of mind.

So for now, we can take a deep breath and simply say something like consciousness may be a prime of the plane of possibility. The plane reveals a state of near-zero positioning on the energy probability curve. What the neural correlates of this energy probability position along the curve might be is open to empirical investigation. Would such a state involve markedly decreased neural firing, some neural equivalent to a subjective state of openness to what might arise? Might this be the neural correlate of the pause, the stillness, the space between impulse and action? Or would it involve some higher state of neural coordination and firing, some higher state of integration as suggested by some views of consciousness? How the mind may use the brain with mental practice to create more access to such a near-zero probability state in its embodied experience of consciousness also remains an open and researchable question.

This energy state of open possibility, this sea of potential, may not only be the origin of consciousness, but also the mind's portal for integration, enabling the innate arising of self-organization's natural push to differentiate and link. Especially when blockages to integration are present, the open awareness of the plane may be needed to drop down beneath the restrictive or chaotic plateaus and peaks that are shutting down the innate drive toward self-organizational integration. Whatever the associated neural correlates of this mind state may be, perhaps they involve an integrative complexity within the brain. This naturally gives rise to the important, perplexing, and often-controversial questions of mind and body. Could consciousness arise without a brain? Does consciousness use the brain to create itself? Does consciousness have influences beyond the brain? If orthodox quantum interpretations of the pivotal role conscious observation plays in collapsing the wave function of a photon into a particle property are accurate, then turning a range of possibilities into one particular actuality at that moment is influenced by consciousness. If this is true, as the empirical data suggest it might indeed be, then we could suggest that yes, consciousness seems to have influences way beyond the skull and skin of the observer. But having influences that extend beyond the skull, and beyond the skin, ways of impacting the world beyond the body, does not necessarily mean that consciousness could exist without a body. This is a hugely debated issue, one we won't solve here.

For those who feel uncomfortable with the notion that consciousness extends beyond the brain or doesn't need a brain, we could propose the following line of reasoning that at least can open our minds to the questions at hand. Integration across a range of neural regions seems to be associated with consciousness. These are called the NCC, the neural correlates of consciousness. But how that neural integration, even in part, could give rise to consciousness is basically unknown. We just don't know, even with various intriguing statements regarding microtubules, sweeping neural firing patterns, representational processes related to our social brains, and integrative neural complexity, how the subjective experience of being aware might be caused by neuronal activity patterns. And so in ways still—and perhaps forever—unknown, in ways that for now are mysterious and magical—meaning we don't understand them, and perhaps never will—we'll need to rest in some kind of ease and intrigue with the wonder of it all. It is amazing we are here, and amazing, too, that we are aware we are aware of being here. Again, our species name is not homo sapiens—the one who knows. It is actually homo sapiens sapiens, the ones who know we know. That is a prime of being human, and that is wild and wondrous.

Consciousness, Non-Consciousness, and Presence

Consciousness is one process that the Wheel of Awareness illuminates.

But we've seen that the mind has other dimensions beyond the experience of knowing in being aware. We also have the knowns of awareness. Whether these are the energy flows of the conduit of sensation, bringing in the outside world through the first five senses, or the signals from the interior world of the body with the sixth sense, we can have a subjective experience of these forms of sensation, the knowns of the physical world derived from outside and within the body. We can also have the knowns of construction. These knowns can be our emotions and thoughts, memories and beliefs, what we've named our seventh sense of mental activities. These can also be the known of our connectedness to other people, and to the planet, what we're calling our eighth sense. Still within awareness, these are flows of energy transformed into information by the constructor of our minds.

We are proposing that the fundamental element of the system of mind is the flow of energy, some of which is the symbolic form we call information. This information processing is driven by energy flow and can arise within awareness, the knowing of these knowns. But much, if not most, of what happens in the mind is outside awareness. This can be envisioned as flows of energy and information that are not experienced within consciousness, and so they are not "known" by the experience of "knowing." Sigmund Freud (1955) would likely have called this the Unconscious. Modern scientists affirm that these non-conscious information processes are real and impact our everyday lives (Hassin, Uleman, & Baragh, 2005; Sato & Aoki, 2006). Named the unconscious, subconscious, and preconscious, I try to avoid the confusion often associated with the use of these terms by a range of disciplines and simply use the term *non-conscious*. This term simply means energy flow not in awareness.

Non-conscious energy flow does not involve subjective experience, as it is not in awareness, but it is still real. These non-conscious ebbs and flows of energy, sometimes with symbolic value, sometimes just pure energy, happen all the time. Whatever label we call these aspects of mental life, we have offered a simple definition of what they are: energy and information flow. Since a sense of control may arise from the impression that being aware means you are in charge, these non-conscious flows can make a person sometimes feel discomfort in that they cannot control or even know what impact these flows are having on their lives. In other words, for some, the notion that stuff is happening outside awareness can be terrifying. But the story isn't as simple as consciousness means control; non-consciousness means willy-nilly wild abandon.

Self-organization does not depend upon awareness. If there is a natural push toward integration, this may occur without the conscious experience of knowing. A take-home lesson about this finding is that sometimes we need to get out of our own way to let the many layers of energy flow simply arise and let self-organization happen. We can call this state of openness *presence*.

Presence is the portal toward integration.

If our conscious minds try to control all that arises, fixed plateaus and peaks may be created that limit the natural drive toward integration. The result of such clinging to a sense of control would be propensities toward chaos and rigidity. These states reveal the movement of a system outside the harmony of the river of integration, the mind's state now moving to either bank outside that river, the banks of being stuck in rigidity or overwhelmed in chaos.

We've gone even further in our journey to actually offer a definition of what this flow may be all about: the *who, what, where, why, how,* and *when* of mind. We've offered a suggestion that you know of, perhaps only in your non-conscious mental life and triggered with our communication, now, in your conscious mind: an embodied and relational, emergent self-organizing process that regulates energy and information flow. The flow of energy and information is not limited by skull or skin and neither is the mind that emerges from it. The mind is within and between. This flow is not limited to consciousness. It happens within and without consciousness.

As a regulatory process, the mind relies on monitoring and modifying that which it will regulate. And in these ways, the Wheel practice helps strengthen the mind as well as reveal the mind's underlying architecture. The wheel guides us beneath the surface and lets us see the miraculous mind's mechanisms.

Strengthening monitoring means sensing energy and information flow with more stability so that what one senses is with more depth, focus, and detail. So often the mindsight lens through which we sense energy and information flow, within and between, is unsteady. The result is a blurry picture, and modification of that flow is difficult. One way of seeing this stabilization of mindsight's lens is with the three O's of objectivity, observation, and openness. We sense the knowns of the rim as objects of attention, not the totality of who we are. We have a sense of an observing function, one a bit distanced from the thing being observed, a sense, perhaps, of the observer, the knower of experience. Openness means we are receptive to what arises as it arises, letting go of the expectations that filter our experience or get us to avoid or cling to what emerges. These three O's form a kind of tripod that stabilizes the lens through which we sense the mind. One way of summarizing this openness, observation, and objectivity is with the term *presence*.

Strengthening modification means setting an intention to enable the flow of energy and information to move toward integration. Sometimes this means detecting chaos and rigidity so that the flow can be directed toward differentiation and linkage. How do we set that intention? Presence. Since integration is a natural drive of a complex system, sometimes what we need to do is simply "get junk out of the way"—or simply get out of our own way—and let integration naturally emerge. This is how presence is the portal for integration.

With some understanding of how a complex system functions, we've taken our journey another step further into questions and potential answers. As an emergent property of the complex system of energy flow—some of which has symbolic value we call information, some not—we can explore aspects of the science of energy and flow. We've seen that one possible view is that energy manifests itself as a spectrum of values along a probability distribution curve. The changes in probability, between certainty and uncertainty, are how this energy flows. Even if time is not something that flows, or even exists as some physicists are telling us, these alterations in probability can be seen to change in the unfoldings of now. When we see the term *time*, we can always replace it with the word *change*, and we'll be on solid scientific ground. Life and reality are filled with change. That change we experience happens in what we've come to call time and space. But if in reality there is no flowing thing like time, probability curve distributions, where we are on that spectrum, still can change. As energy and information flow, they shift and transfigure their probability functions. These views from science and subjective reflection, combined together on our journey, deepen our understanding of mind.

Here is the wild proposal we've come to explore in our journey: The *experience* of *knowing* of awareness arises from an energy probability of the infinite plane. *What* we are aware of, the *knowns* of mood and intention, thinking and thought, emoting and emotion, remembering and memory, arise as energy moves from possibility to probability to actuality in a flow of nows. We can know these unfoldings, letting the conduit of our sensory experience flow them toward awareness and then, amazingly, we can shape them as we interpret their meaning

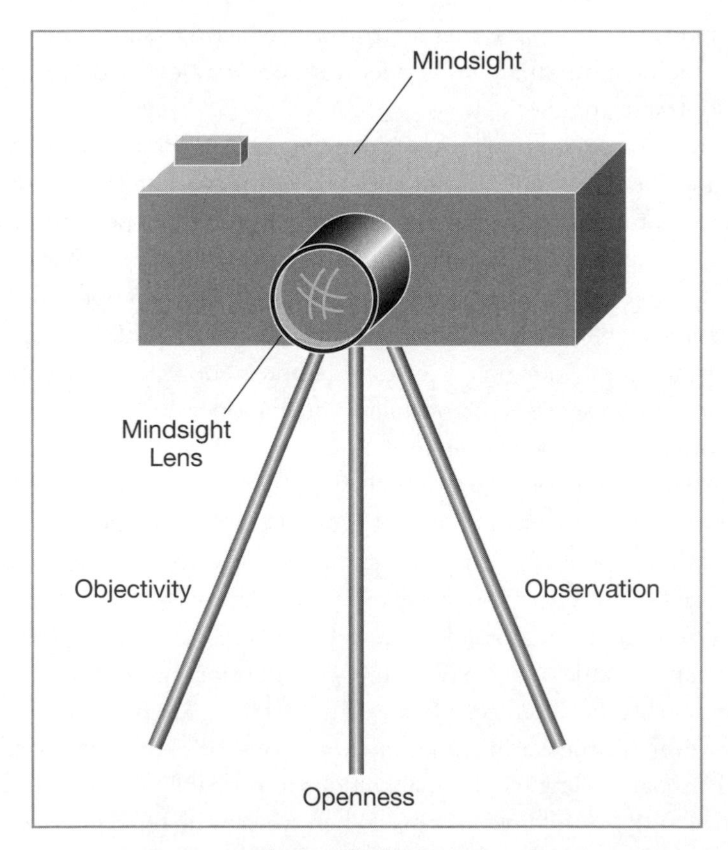

The Mindsight Tripod

and construct narratives and action. These flows can happen without awareness, as the non-conscious flow of energy arises and falls without much dipping into the source of knowing, the plane of possibility.

Open awareness, *mindful awareness*, the receptive state of presence, offers the possibility to embrace the reality of now, a state of accepting that now is all there is. It is from this state that automatic patterns can be intentionally altered if we have become stuck in chaos or rigidity. What this suggests is this role of consciousness in our lives: Consciousness catalyzes choice and change. More than an important gateway to subjective experience, being receptively aware allows us to choose a different path. In this state of presence, this presence of mind, we come to this startling insight, a possible view into the true nature of mind and reality.

The plane's presence moves us toward health by releasing the innate potential to cultivate integration.

Presence is the portal for integration to arise.

This may simply be the prime of how the mind works, a fundamental force of our mental lives. When we are stuck, presence can open the way to freeing us from the prisons of unyielding plateaus and unhelpful peaks. Depressed, fearful, or anxious moods can be shifted; negative thoughts, frightening images, or worrisome preoccupations can be shifted. As we've seen, even chronic somatic pain can be shifted with the development of the presence of the plane of possibility, of accessing the hub of the Wheel of Awareness.

How can this happen? Our awareness of energy can shape the energy itself by altering its position along the probability curve in the present moment. Just as some quantum physicists assert that the act of conscious observation of a photon transforms its probability curve from wave to particle, it may also be true, perhaps, that awareness alters the flow of energy and therefore information by way of shifts in the probability curve. This is the *how* of the question, how awareness could transform mental activities. These activities are the knowns of energy and information flow. And beyond awareness, some suggest that intention has a direct impact on energy flow (Stapp, 2011), and information processing both within and between individuals (Maleeh, 2015). Intention may be some way that energy probability flow patterns are also shaped, with or without awareness. Intention might be represented on our plane of possibility figure as a low plateau, an energy probability position only slightly elevated above the plane that sets a direction for other mental activities that arise. As we've discussed earlier, mood and state of mind may also be plateaus, perhaps a bit more elevated but still lower than those sub-peak processes like thinking, emoting, and remembering. Peaks would be the actualization of those sub-peak values into their certainty forms of a thought, an emotion, and a memory (see Plane of Possibility figure in Chapters 2 and 8, pages 59 and 254).

Attention—that process that governs the direction of energy and information flow—may be some mechanism that shapes how probability patterns unfold, how they change and transform plateaus, sub-peak processes, and peaks.

Self-organization may differentiate and link the complex system of energy and information flow of this mindscape world of energy distributions utilizing directed awareness called focal attention, or non-conscious mechanisms such as intentional shifts in mental processes, in energy and information flow. As we've discussed earlier, non-integrated flow patterns can be depicted such that periods of rigidity are extended plateaus and peaks without much change across the x-axis, which we've labeled as "time"; chaos would be represented as great diversity along the z-axis at one particular moment.

Consciousness, as we've discussed, may be what we need in order to be aware of subjective experience, of the lived texture of life. This would be depicted in our graph as the tapping into the various above-plane peaks and plateaus. Beyond merely enabling us to be aware of the subjective experience of these knowns, awareness may shift the knowns themselves. This is the power of consciousness to change the flow of energy and information. Cultivating intentional states within awareness, developing a state of mind with purpose and direction, as in various forms of mental training such as mindfulness practice, or mindsight skill trainings as in the Wheel of Awareness, may also be important in shaping our mental lives—both our mindscape and our mindsphere, our within and between. Then, even without awareness the persistence of those intentional states—those elevated probability states, those low plateaus we can call mental vectors not needing consciousness—may also influence the pattern of energy flow shifts. Intention is a mental vector or stance, which in very real physical ways shapes the quantum dynamics of energy as it moves along the probability distribution curve. Sometimes we are aware of these processes;

sometimes, perhaps most often, we are not aware of all the mental factors that shape our minds.

This is all wild and wonderful, and we're taking it one step further in our hypothesis. Accessing the plane opens us to the natural push toward integration. We don't need physical space and we don't need time to be something that flows or even real to reveal the nature of mind. What the mind is all about is now, along with the transformation of the presence of possibility into the knowns of probability and actuality, the emergence of degrees of certainty in the forever unfoldings of now.

While this is our working proposal for what mental activities may be, the movement from possibility to probability to actuality and back again, this view also gives us insight into how consciousness might fit into this picture.

This, we are suggesting from numerous in-depth first-person reflections from the Wheel of Awareness practice, may be the origin of awareness. How exactly this happens, we don't know. But the notion that energy passes through a near-zero probability level along the energy probability curve's spectrum of values enables us to see how this could be experienced as an open plane that is wide, deep, expansive, and has a sense of being infinite. This is our proposed origin of consciousness. In many ways this can be described as an energy state filled with potential. As we discussed earlier, perhaps this near-zero probability state, this plane of possibility, has the quantum properties of microstates that do not have an Arrow of Time, and perhaps this is why some describe this as "timeless." This is a sea of potential, a plane of possibility that exists when the probability distribution curve is at the near-zero value. Mental activities are on a continuum with consciousness, as probabilities and actualities bubble up from this sea of potential, as they arise from this plane of possibility.

Within the plane we can feel an expansive sense of freedom. The idea is not to try to be in one aspect of the energy curve or another, but to gain flexible access to all, freely moving along the full spectrum of experience. This linkage of differentiated positions along the curve is the mechanism of harmony, bounded by chaotic or rigid patterns, that yields a sense of coherence in life. That is what integration entails—the linkage of differentiated positions along the energy probability curve.

It may just be that the closest you and I can get to each other, the place we can find deep resonance, is from this plane of possibility. This is how our individual mindscape access to the plane can shape our mindsphere experience of openness and connection. But such interconnection is not limited to the plane. My plane and your plane are likely quite similar, if not identical. Infinity is infinity in each of our bodies, each of our mindscapes. This gives us a starting place to connect, link and identify with one another. Integration is more than linkage, it also involves differentiation. The plateaus and peaks that form the core of our personal identity are important ways we differentiate from each other. Integration is the honoring of differences and the cultivation of compassionate linkages. Differentiation naturally unfolds as plateaus and peaks, within and beneath awareness, arise and form our mental life. But accessing the plane allows us to move beyond a sense of separated, isolated selves. Yes, our peaks and plateaus are unique in each of us; but honoring these and accessing the plane enables us to find common ground. We come to sense our connections, our commonalities, not just our differences. The plane is the portal for the linkage of integration to arise. It is through this plane of possibility, too, that we realize we are not only connected to other people, but also deeply connected to the planet, our common home, this place we call Earth.

Through the plane we can find a pathway to deeply receiving each other's plateaus of probability and peaks of certainty. This is how we honor each other's differences, how we respect and cherish, even nourish, our differentiated ways of being. Through the mental presence of the plane, we can be open to the particular propensities and predispositions of our personalities. Your personal identity, history, predilections, and vulnerabilities, each of these would be patterns of plateaus and peaks. These we can explore within our autobiographical reflections and the stories we tell of our embodied, individual mindscape lives. These are not "the bad guys" or something to rid our selves of, they are merely aspects of the full spectrum of mental life to embrace, but in whose rigid or chaotic patterns not become imprisoned.

We can learn, with intention and practice, to move more freely from particular peaks and plateaus that may be restricting us, limiting our identity, imprisoning us in habits and beliefs that keep us in rigid states of depletion and stagnation, or chaotic states of overwhelm and flooding. How can we do this? By intentionally training the mind to access the open plane of possibility, we can learn new ways that energy can emerge from its space of open possibility and uncertainty toward manifestations as the actual, toward certainty and realization. And then we can move back again. In this way, developing access to the plane opens us to connect both with others through our common ground, and also to a vast array of possibilities within our own mindscape's plateaus and peaks. Awakening the mind is freeing the movement of energy along these new patterns of probability, shifting from paths that have become stuck or chaotic in our lives by gaining the freedom to experience new unfoldings, a liberation that often requires we let go of a need to control and instead relax into the embrace of the unknown and empower the natural drive toward integration and harmony to be realized.

The fear of uncertainty for some is the push away from resting comfortably in what may be an unfamiliar plane of possibility. Needing to know, a fear of not being in charge, can limit our access to the plane. Yet when we move beyond these needs and through these fears, we can transform our lives as we free our minds. This would be how the integration of consciousness facilitates the honoring of differentiated places along the energy probability curve, while linking those differences within and between us. I can respect your particular plateaus and peaks, and I can find union with you from our planes. In Wheel terms, we are different on our rim but we find commonality in our hub. This is how we awaken our lives and liberate our minds.

Reflections and Invitations: Cultivating Presence

Being aware of something by itself does not mean we are in a state of receptive presence. We can live with an unfocused attention, with things entering a blurry awareness, and we can live with presence and a sharp focus of what we experience in awareness. When the lens with which we sense the mind is stabilized, with openness, objectivity, and observation, we can see with more depth and clarity. From this open state of mind, from this presence, we can enable changes to arise with intention and self-organization—we can best move toward integration.

Here, we'll reflect on these experiences of being aware and explore what may help illuminate the ways we can build access to this plane of possibility and empower the presence that research has affirmed can cultivate well-being. As we move through these reflections, you may find it helpful to imagine how these layers of being aware arise in your day-to-day life, how your mind may wrestle with your brain, and how ultimately you can use your mind to create more integration in your day-to-day experiences.

How, you might ask, can we be aware without being fully present? Our plane of possibility model may help us explain such a finding

of how the experience of consciousness may work. Let's assume for the sake of these reflections that the knowing of awareness is somehow an emergent property of energy flow that arises when this flow is at the near-zero plane of possibility. This assumption is consistent with our proposal that mind—including consciousness—is an emergent property, a prime, of energy flow. Given that energy flows along this curve, we've been discussing the observation that the knowing of awareness may overlap with this position along that curve.

It may be that open awareness, or what some might call pure consciousness, is an arrow-free quantum property at the microstate level. When energy probability values arise above the plane, perhaps they then have the classical macrostate features in which time is arrow-bound—meaning that our feelings, thoughts and memories, for example, have a sense of both movement and a unitary direction "over time," from past to present to future. In contrast, consciousness would have a timeless quality. The tension of our mental lives, if this is true, is that with regard to a sense of the passage of time, we might have a quantum consciousness with a classical array of mental activities. In addition, we live in a macrostate classical arrow-bound world of physical objects yet we are aware of this classical environment with a quantum mind's awareness. This tension between the classical and the quantum may indeed be a fundamental conundrum of our human experience.

In addition, we have the wide range of states of consciousness we can be in at a given moment. It may be that some blending of the positions of the energy probability distribution curve can help us envision what may be going on. If we assume that the knowing of awareness emerges from the plane, and that mental activities like thoughts, emotions, and memories emerge from the curve above the plane, then perhaps non-conscious and conscious mental life can be seen in the following way. Several neuroscientists suggest that some linking processes are found when a subject is conscious of something (Tononi & Koch, 2015; Edelman & Tononi, 2000). For example, in the formulations mentioned earlier, if a 40-cycle-per-second sweeping event in the brain is happening as Rodolfo Llinás suggests, and integrative linkages arise as Tononi and colleagues have proposed, then consciousness might depend, in part, on a range of neural areas contributing to the brain side of awareness. These are some of the leading notions of the neural correlates of consciousness.

But what would pure consciousness—the knowing of awareness without the known—actually correspond with in the brain's activity? Would it be a remarkably low level of firing? Would there be a hint of how some uniform microstate may be predominant—the quantum source of the knowing of awareness—without the macrostate features of above-plane aspects of the known? And when we are aware of something, how might this blend these two potential properties of the classical and the quantum? We just don't have the empirical data to know the answers to these foundational questions about the brain's activity and the knowing of being aware.

On the mind side of consciousness, we would examine energy flow, not merely the changes in physical location in the brain or anywhere else, but shifts in various states of probability. What might that flow of energy be like?

Let's examine the plane of possibility figure and focus on the upper and the lower halves. As with any drawing, this figure of course is simply a map, a visual metaphor, and not the "real thing." But metaphors, as a kind of map as we've discussed, can be quite helpful if their limitations are acknowledged. In this mapping of energy probability distributions, the upper part of the graph we've been exploring represents mental life, some of it experienced as "subjective experience" within consciousness. At the lower half of the graph are the corresponding shifts in probability that might occur at the same moment within the brain. These are the neural activities that would be the neural correlates of consciousness. One implication of this display is that sometimes the brain may lead the way, with patterns of neural firing holding sway over mental life; at other times, the mind may lead the way, pulling neural firing in a direction it might not ordinarily move. This is how the mind can shape the brain's firing patterns, a now established capacity.

For now, we'll be referring to the position of the energy probability as in-the-plane or not-in-the-plane. For the mind, this would be activity represented above the plane of possibility; for the brain this would be activity below the plane. How energy flow in mind and in brain correlate with each other would be reflected within the mirroring of what is above and below the plane. However, our proposal that subjective experience and neural firing are not the same phenomena is reinforced by this graph's depiction of the neural and the mental *correlated* but not identical. In other words as we've discussed throughout our journey, they are not the same even if they directly influence each other. Now let's look into how consciousness might involve both neural and mental processes through the lens of these probability shifts.

If we imagine that being aware, the knowing of consciousness, requires some combination of energy flow to have a plane component, some movement of the energy probability distribution curve into near-zero certainty, then that, as we are proposing, could be the basis, energy wise, of the knowing of consciousness.

The quality of consciousness, whether it is blurry awareness with only vague details or receptive presence with clarity and openness, might be determined by different ratios of how much of the 40-cycle-per-second sweep or some other such sampling process would include the plane (knowing of awareness) and the peaks or plateaus (the knowns of awareness). What this suggestion means is that the known and the knowing would be on a continuum, each representing a different location along the curve, that is, each a different state of probability. If we then take this suggestion and refine it, we may be able to see how different amounts of the plane (knowing) and of the non-plane plateaus and peaks (knowns) may combine to reveal different ways we can be aware.

If the ratio of non-plane curve activity of the knowns is slightly dominant, perhaps we are more aware of things happening without them being fully experienced in a mindful receptive state, a place of openness with choice and change. We are wonderfully lost in the experience without a wide expanse of awareness that might include self-reflective observation. Could this be an example of when we bask in the conduit of the mind's streaming of sensory energy? Perhaps this is simply the state of what Csikszentmihalyi calls flow, as we are aware with a dominance of a sensory stream but without much, if any, self-observation (Csikszentmihalyi, 2008). Perhaps the energy distribution curve profile would have the knowns of flow arise directly from the plane without much filtering by restricting plateaus that alter perception. In this way, energy in the plane would have deep awareness while energy beyond the plane, as rising peaks, would be in as pure a state as possible, not constrained by prior expectations. Perhaps this bottom-up energy profile, balancing peaks and plane, is what we might find with flow. The key is the origins of peaks and the ratio of plane to non-plane positions.

Flow can feel wonderful, but it is simply not the same as a wide-open awareness that would include self-observation. With flow, the rising peaks may fill our sense of the present moment, and we become lost in the activity, fused with it, inseparable from it. When we add more of a ratio of the plane to the experience, perhaps the broader sense of a consciousness that might arise would then include the sense of knowing itself, and even a sense of the knower, beyond the known of the flow experience. This wide-open state seems to access conduition as well as construction, and offers a window for us to sense the infinite expansiveness that is perhaps the heart of receptive stillness, the core of consciousness itself. This is how presence can choose to engage deeply in flow, but is not the same as flow. That presence is a state open to all, a "bring-it-on" stance that invites the minutia and magnificent all into one huge, accepting embrace.

With states other than the clarity that comes with flow or open awareness, we may be in a massively dominant above-the-plane state, but are simply not very aware of the details of an experience at all. At such a time we might be vaguely aware of what is happening with blurry details, having a cloudy focus with a fleeting sense, unstable image, or foggy notion. We are easily distracted in attention, our awareness simply going for a ride here and there without stability, without bathing in the majesty of flow, or resting in the stillness of the wide-open expanse of receptive awareness, of presence. In such a foggy state, we know perhaps in some cloudy way that these events are there, happening, but we are not filled with the richness and detail of flow, nor the expansively embracing, open experience of presence. This blurry state is how we might have awareness without flow, and awareness without presence. Here we would imagine that the energy curve sampling as it sweeps by is predominantly above the plane, with just a small access to the plane giving a touch of awareness, but only in a vague way. In other words, to be aware, we would need some energy curve elements to dip into the plane—the less dip, the less the clarity.

If the non-plane curve activity as plateaus and peaks is exclusive, if the blending of the 40Hz sweep in the moment, or any other integrative process yet to be discovered, does not include the plane but only distributions beyond it, then this would be a possible profile revealing how non-conscious mental activities could occur. Peaks and plateaus would be unfolding without much, if any, dipping into the plane. This is a proposal for what the non-conscious mind's profile in this framework would look like, and what the neural correlates might be. This would be the mind's information flow that would be beneath or before consciousness and neural activity that does not yield awareness. The thoughts, feelings, or memories are there, but we are simply not aware of them—not even in a blurry way. We don't even have a direct subjective experience of them, even if ultimately we can sense their shadows on other aspects of our mental life that do enter awareness.

Here is the exciting thing about where we've come. We can now take the fundamental proposal of mind as an emergent property of energy flow and see how, by looking deeply into what that flow means as probability shifts, we can understand the notion of conscious and non-conscious mental activities.

The greater fraction of the plane included in the linking sweep, the more receptive awareness we have. No plane yields no awareness, and in this situation these non-plane distributions are non-conscious mental activities. A small amount of plane, and awareness is blurry or vague. A medium amount of plane, and perhaps we are in sensory flow. The greater fraction of the plane swept up in the integrative sweep, the greater the receptivity. The specific defining ratios distinguishing the experience of flow from a broader mindful presence may not be fixed or perhaps ever clearly demarcated, varying from moment to moment, individual to individual. This of course would make empirical studies challenging in their effort to measure such processes. But an important issue is that the integrative sweep's ratio can change in any experience; we can be in flow and then need to access more self-reflection, so we dip more into the plane, increasing the ratio as needed. In other words, we can use the mind to alter what we can call the *sweep ratio*, changing the quality of consciousness in that moment. If we practice the Wheel, or are simply naturally in the plane as a feature of who we are, we may be more capable of enhanced plane activity and therefore have presence as a trait of our lives. This would mean you can increase the ratio of in-plane to non-plane positions along the probability curve during the integrative linking, during the sweep.

Can you have presence and flow simultaneously? Perhaps the issue is about timing, even if time as something that flows isn't really a part of reality. Then this question becomes one of an issue of probability distribution patterns that arise, moment-by-moment, the change in probability that exists without time-as-something-that-flows, and even a change that would exist without space. In other words, even without time and space, change can happen in the probability curve position of energy, moment-by-moment. What do these shifts have to do with open awareness? It may simply be that with presence, you have the capacity to *choose* all sorts of ratios that serve different purposes. In this way, with presence you can choose flow. But perhaps not all those who experience flow are able to tap easily into presence. That is a researchable question that so far we don't have an answer for. Here is an example of what we might consider: If we become absorbed in anger and strike out at someone in a violent action, we can be in the flow of our rage but not be accessing the flexibility and morality of a mindful state of presence. Presence can include flow as a choice, but flow can exist without presence.

As we've stated, presence is the portal for integration. Presence is what may be needed for choice and change, releasing ourselves from impediments to the natural push toward integration. In such a state of presence, peaks and plateaus may be readily accessible as smaller components of the sweep ratios compared to the high degree of plane represented in that integrative sweep in the moment. In other words, a sweep takes place and in some sense defines what "now" may be experienced as, what this moment is in neural and mental terms. This would be a probability profile we can suggest that depicts the core elements of mindful awareness, of being aware with acceptance and love in this present moment. A more equal distribution of the sweep into both plane and non-plane values would allow for greater fullness in the awareness of the flow of experience.

The more plane components that fill a moment's sweep ratio, the more a sense of wide-open awareness we may feel. But there is a component to this that can be potentially troubling. I met a colleague recently for dinner and we discussed her long-term meditation practice. She asked me how I distinguished the "nothingness of non-dual reality" that she had become immersed in over the past decades of meditating from the experience of openness and fullness others had described earlier that day during a workshop with the Wheel practice as their experience of the hub—the plane of possibility. For her, she said, accessing the deep sense of pure awareness in her own practice over these years made her realize nothing existed. When I asked her how that felt, she said, "Completely empty." Her face was blank. She then added that everything else, to her, was simply delusion. Nothing, in reality, existed. This was her reality: The only thing that was real was nothingness.

If we look to our discussions about energy flow, we can perhaps understand what this thoughtful person was wrestling with. She is not alone. Some consider that a "non-dual" view means that anything that seems to separate the self from the larger world is a delusion. In essence, there is no real separation of anyone from anything. Yet this non-dual reality might mean that life is full, not empty. If the real non-dual world is full, how did this woman, in her words later on, become so "trapped" in a void? The probability curve framework may help us understand her experience. The realization of a nearly infinite set of possibilities into probabilities and actualities is, we are saying, the source of our mental life. Okay. From a non-dual perspective, what and who and how and when and where and why we are, our ways of being, emerge as transformations of energy patterns, ceaseless movements along the probability distribution curve. That's it. And that is full of endless potential, limitless emergence. Awareness, we are proposing, the knowing of our mental lives, emerges when the probability position in the moment is in the near-zero plane. Now one way of understanding the experience of being in that plane by itself, of learning to be only in that plane and not in the non-plane plateaus and peaks, is that you can subjectively feel this "zero percent probability" state as the experience of nothing. Literally, there is "no thing" there, just potential. The no "thing" is the lack of the "things" of plateaus and peaks, of the knowns of the mind. And if you have come to distrust the emergence of things—such as intentions and moods as the plateaus of increased probabilities, or thoughts, emotions, memories, perceptions, sensations, or whatever as peaks of actualities—then you'll think everything outside of that plane of pure awareness is delusion, is false, is not real. If that line of, well, believing or thinking gets you to only rest in the plane as a place to avoid what you consider to be the delusion of all else, a mistrust of the knowns of the mind, then in some ways this plane is experienced as completely empty. You don't trust anything that might arise from it. The plane becomes a haven at first, and then, as in this person's case, a prison of emptiness.

In contrast to my colleague's experience, the equivalent of zero or near-zero probability can be experienced as infinite or filled with open possibility. Instead of being empty, the plane is actually full. Full of what? Full of potential. And for many, including me and the thousands of others who've described it, this plane is a source of peace, openness, the infinite, joy, clarity, and connection. In this view, what arises from the plane is not a delusion, but rather an opportunity. Life emerges full of freedom, not imprisoned in the plane.

Emptiness or fullness. The plane can be experienced in both ways. Some may experience it as a prison; some may experience it as a palace. How you sense the plane as emptiness or fullness and whether you subjectively experience it as prison or palace may depend on many factors in your life. For example, some find the experience of uncertainty to be terrifying and dropping into the plane is an uncomfortable experience, at least at first, as we've discussed. Others, in contrast, find the vastness of possibilities to be soothing, filled with a sense of the infinite that brings a deep sense of connection not only to a wide open path, but also to a wide open world as their sense of self broadens and deepens.

How does this probability distribution curve and consciousness arising from the plane relate to self-organization? We can describe their connection from the view of integration.

An integrated consciousness allows full and free access to the entirety of the probability distribution curve. Embracing all of these emergences, resting in the plane is simply one aspect of a fullness of being. There is nothing to do, but everything to experience. In this perspective, the knowing, the knower, and the known are part of one continuum. This way of sensing a non-dual view that has many aspects lets us see that things in the world, the knowns and the knower, are not delusion. They are not only real, but they are really important as differentiated aspects of reality that become linked within the harmony of integration. It may even be that the knowing has an arrow-free sense of timelessness whereas the known and perhaps even the knower—if it is a mental construction of "I" and not just pure awareness—may have a more arrow-bound macrostate quality to them, as we've proposed earlier. In this way, my colleague may have become highly focused on the arrow-free state and eschewed an arrow-bound state of above-plane mental life that she only saw as unreal, untrustworthy, and basically a lie. She seemed to have become trapped in a view that all that was real was the plane, which for her, as she described it, was a prison of emptiness in which she was now rigidly stuck. She felt helpless to change her life and was looking for relief. How, she wondered, could she use the process of integration to achieve the self-organizational flow that could free her by embracing the differentiated positions along the probability distribution curve and literally open her life and free her mind?

Let's imagine that if the ratio is plane-dominant, then we feel a sense of expansive fullness filling us as we rest in the magnificence of this moment. Some might call this mindful awareness, we might simply call this presence. At times we may drop completely down into the plane and feel a sense of dissolution of a separate self, experienced in many different ways, as a *we*, as the divine, as spirit, as our fundamental interconnected, interdependent, non-dual nature. As we've seen, if we get stuck in the plane, as my colleague may have done inadvertently with her belief that everything else was delusion, then we can make the proposal that she was no longer integrated. The plane, for her, had become a prison of rigidity. For others, not having access to the plane at all may create a life of automatic pilot that could result in rigid or chaotic experiences, too. But life does not have to be like that. Integration of the plane, plateaus, and peaks, linking the arrow-free knowing and arrow-bound knowns, honoring each of their differences and cultivating their linkages, reveals how we can transform such potential rigidity or chaos into the harmony of integration. I mentioned all of this to my colleague and she seemed open and relieved, if not also challenged and terrified all at once.

Coming back now to this writing after our meeting today along the Pacific coast, here were some of the new verbatim statements made about the Wheel practice, especially when the spoke was turned 180 degrees back into the hub: "Wow." "Awesome." "We are all connected, there is no separate me." "You found the golden lock, and now we have the key." "I was bored until the focus on the hub, then I entered an altered state and stayed there." "I had a sense of hugeness, of eternity, joy, comfort, ease, bliss." "An expanding cocoon of awareness."

Increasing access to the plane of possibility can make awareness enter a state of presence. Presence does not need to be a prison or something to trap oneself in—it offers the potential to live fully. It is the portal for integration.

These are simply ways to consider as we build on where we have come, imagining how our mental lives, conscious and not, common and unique, isolated or connected, each may arise from the same fundamental mechanisms of mind that emerge from energy and information flow.

These reflections give rise to invitations to sense what our own awareness is like.

If you have done the Wheel practice, what was it like for you? When individuals expect to have the same experience as descriptions they have heard, sometimes it can inhibit their own natural emergence within the practice. I share these descriptions so you can appreciate the first-person data supporting the notion that something profound happens, whether it is at a church gathering or here along the cliffs of Big Sur, that are dramatically similar, no matter the background or experience level of the participant.

How has accessing this hub of the wheel, this plane of possibility, influenced your own life? What did it feel like for you? What were the various parts of the rim like? As you reflect on these experiences, imagine what might happen if you could readily access an inner place of clarity and choice, the hub of your mind. What are ways you can imagine using this model of the mind to inform how you approach your day-to-day living?

You have the possibility to use your mind to focus your attention and open your awareness—the top of the Plane of Possibility figure—in ways that can make your brain—the bottom of the figure—move in new and more integrated ways. Even if your automatic mode is to be non-integrated, to be prone to chaos or rigidity, you actually can choose to strengthen your mind and get it—the regulation of energy and information flow—to intentionally drive energy flow through your brain in new ways. Now that you know about the centrality of integration, we can say that you are empowered to use your mind to integrate your brain and enhance your sense of well-being. As you experience this way of living, you can become an active part of the community of individuals that promote integration. This is the way your own mindscape work becomes woven with the mindsphere around you. This is the continuity of consciousness, cognition, and community.

At the meeting this week, some were complaining about the severe traffic. Today I suggested that while peaks of angry thoughts and plateaus of irritated mood may arise while driving, with the practice of gaining ready access to the plane of possibility one can have an inner sanctuary easily available, even in traffic. We all are a part of a larger world, including the flow of other human beings in traffic. This is a physical manifestation of the mindsphere. And how we connect our inner mindscape work to that mindsphere environment is the key to living fully in an embodied and relational way. This is bringing integration into the world, from the inside out.

These experiences for me create an inner sense that something important rests within our journey, something that can make a difference in people's lives, something that, if made clear, might help us see a larger, more comprehensive, and therefore more accurate view of mind. This clearer view of mind might just provide a more effective approach for anyone helping others, whether within the connections of our family homes, the learning in our classrooms, the transformation in our clinical centers, or the communication within our local and extended communities.

Studies reveal that we are deeply interconnected to each other in our communities and culture (Christakis & Fowler, 2009). We influence people who are at least three degrees separated from us, even without intention. In other words, how we live and how we behave can inspire individuals we may never meet. Can you imagine what life would be like in your community if you and others developed more access to the plane of possibility? If we all live mostly outside the plane, without presence, we can live an isolated life, being swept up by non-conscious cognition that puts us on automatic pilot. But consider what developing access to the plane might mean. For your individual life, you gain more self-awareness, freedom, and flexibility. For your immediate relationships, you develop more empathy, compassion, and connection. In your community, and your larger culture, you can have new conversations about the nature of our lives as you inspire others to find that same source of clarity, flexibility, and connection. You and I, we, are empowered from the inside out to make such changes a reality. From the sea of potential, we can make that sea accessible to ourselves, creating cultural waves of positive influence for others.

In these reflections, I have invited you to explore your own direct experience. If you SIFT your mind, how is the subjective texture of your sensations, images, feelings, and thoughts as you reflect on what mind is? If you begin with the simple task of asking how and where these four aspects of your mental life arise, can you sense, in the plane of possibility, the embodied and relational origins of your mental experience? Can you feel the potential of this personal and shared mental sea?

In this moment, I feel the excitement of this journey and also a sense of sadness about its end as that moment approaches. I am filled, right now, with a sense of gratitude that you and I have been on this journey and can even ask these fundamental questions. In our next epoch, we'll explore how we can embrace our connections with one another and cultivate awe for being alive, for simply traveling on this unfolding journey we call being human.

CHAPTER 10

Humankind: Can We Be Both?

HERE WE ARE, OUR LAST ENTRY. THIS EPOCH BEGINS NOW, AND takes us into the open nows to come. We'll explore further some of the central themes that have emerged on our journey together, from the inner and inter nature of mind to the centrality of subjective experience and necessity of integration in cultivating a healthy mind throughout life. One simple view is this: Kindness is a natural outcome of integration. Integration of the self enables the differentiation of an individual "me" with an interconnected "we." Being kind to one another, respecting differences and cultivating compassionate connections, is living an integrated life. It's a privilege to be on this exploration with you, and I thank you for your companionship along this journey of discovery into the heart of being human.

Being, Doing, and Integrating Mind (2015-eternal present)

Sunrise, New Year's Day. The oranges, blues, and greens of daybreak along the shore at the edge of North America fill the sky with luminescence. The sound of waves gently unfolding now, as they have for

infinite nows, in patterns beyond imagination, creates a gentle soundscape enveloping my mind in a lullaby beckoning me back to bed. This body needs more rest after last night's New Year's Eve festivities with friends and family. But I am up, here with you, wanting to express something of this journey in words we can share, together, in these nows that forever wrap us in existence, life, and the journey of these lived moments we've come to know as mind.

Are we the sunrise? Are we the lapping waves? Are we the creation of time, the denotation of a passing of something marked as a day, month, year, a demi-decade like these epochs that have organized our journey? The hooting and hollering of celebration for this mind-created edge of a year across the world, the display of fireworks in the skies across Earth, the screens shared among billions of humans across the planet: are each of these some shared construction of our collective mind?

We create meaning from an infinite set of energy patterns and make information come alive. We are the sensory conduits enabling bottom-up to flow freely in our awareness; we are the interpretative constructors, making sense of and narrating our lives as they unfold. There is in reality no "new year" anywhere beyond our mind. As we've seen in a prior now, when we leave the reality of now-here, we move the hyphen and become no-where. That is the risk of believing the mind's illusion of time existing as some unitary something that flows rather than a facet of our four-dimensional spacetime reality. We can become preoccupied with the past and fret about the future. Yes, there is the passing of the planet around the sun in some periodicity that shapes our relationship to that glowing orb, that reminder of the source of all energy since the universe began. Yes, the turning of Earth on its axis each period of what we call time and have named a "day" demarcates the boundaries of a pattern we call diurnal, a way to sense how now unfolds in some consistent way. Our body's wide array of circadian rhythms, too, reveal shifts in our physiology based on the relationships of Earth to sun. The sun rises from the edge of the horizon, floats seamlessly upward, and then sets. Each of these changes in spatial relations we interpret as something passing, as time unfolding, is a clock we use to measure out time. We have a pattern of nows that we call upon daily. The movement of the white circle reflecting the sun's light we call moon; that glowing reflection of the sun shifts in repeatable patterns as well in what we call a lunar cycle, something we denote roughly as a month. The angle of the sun with respect to that horizon changes over what we call time, too, in a repeating fashion of nows we call the seasons that follow a pattern we call annual. If we place a stick in the ground at just the right angle, we'll have reinvented the sundial to map out the time of day across the ever-recurring seasons of our circling-the-sun years. All of this is about changes in the relationships among objects in the world. With a little shenanigan we change the number of those days, put on a few meaningful names, and we've got twelve of them, those months, marking our calendar of nows.

By every outward appearance, time is a singular something that seems to flow.

Yet in reality all of this is a way that patterns of energy we perceive with our minds give meaning to the world and let us share those perceptions with one another. Our minds create a sense of time flowing. To do all that we need, to communicate with one another, we draw upon the interpretative role of our top-down constructor. We generate the perceptions of the world around us, including the sense of time. We construct out of all this energy change in the world around us the information of a symbol, concept, an idea we call time, and set off fireworks to celebrate the seeming flow of our constructed creation.

What of the world within? Our minds can also see themselves, perceive energy from inside the body, sense patterns within. As with dreams, reflections on our inner mental landscape, our mindscape, can feel timeless—perhaps because we are immersed in the experience of our plane of possibility with its arrow-free quality of emergence. This mindsight perceptual stream can be used to see others' patterns as well, to create what our minds call *empathy*, to glean a possible view of another's inner mental life—the life of subjective experience internally, feelings and thoughts, memories and beliefs. Insight and empathy let us know the inner world of self and others.

When another is filled with suffering, we build on that empathy to sense their pain, imagine how to help, and we carry out acts of compassion to alleviate that suffering. Those energy patterns we sense from others are the signals of their inner life, sent by the body, sculpted by the body's brain. The brain's structure governs its patterns, its development dependent upon experiences and the epigenetic and genetic factors that alter its unfolding structure. This synaptic architecture directly influences how energy flows and is transformed into patterns of information.

This is the startling reality of living in these bodies. We did not choose to enter this bodily world, but here we are as quantum-level mental awareness living in a classical world of macrostate configurations. Information from our top-down constructors creates a story of the life of the body, the life of the personal self, across the passage of time, these unfoldings of our macrostates and mental states. Yet now may be all our minds experience directly as we live in the moment, remember the past, and imagine the future. The now that has already happened, as fixed as it may be as a moment, has created a synaptic shadow that is cast upon the neural circuitry that plays a central role in how the now of now unfolds, not only in our subjective experience, but in energy flow patterns shaped by these neural pathways. Since the brain is an anticipation machine, we live in a body carving out the horizon of now before it even happens. Trying to live in the moment, truly in now, challenges us to move before and beneath this anticipatory molding of how we live life. The more experience we have, as we've seen, ironically the more our expertise may keep us from seeing clearly and living fully.

To awaken from this slumber of expertise, from the cloud of construction, we need to reimagine the very nature of mind and existence. It may be as simple and sensible as being present. That may be the art, and the science, of living well.

Preoccupation with the apparent past, fretting about the seeming future, we become distant from the presence that awakens the mind to reality, to immersing ourselves in the life that is within and between us.

Living in the shadows of memory and the horizons of anticipation, focusing on past and future, the mind emerges from neural patterns of firing. The mind is not the same as those firings, but intimately interwoven with them. The story is not the wires nor the screen, but is shaped by those mechanisms that mold energy's flow.

And somehow, in ways wonderful and mysterious, we can even know all of this stuff of life. We can have this miraculous way of being aware. And in this awareness, the magic continues. Shaped by our inner and inter realities, energy flow streams in many layers. We can sense as directly as these bodily-based nervous systems allow, the breadth and depth of a world our minds desperately try to organize. This conduit facilitates the flow as purely as somatically possible while the constructor interprets input and narrates to try to make sense of our lives, including the life of the mind and our sense of time. The mind emerges from these energy flows, conduit and constructor, within and between.

Our proposal of at least one aspect of the mind as the self-organizer of that flow, detecting patterns and making meaning, points to the notion that this emergent aspect of mind is both within and between us. Our stories are shared; our understanding of one another links us. We, our collective mental lives, are distributed across individuals to create something larger than a self-alone. We have an inner mindscape, yes, and we have an inter mindsphere. We are both.

In the now, in living here, we collectively and individually assemble those patterns within and between in ways that differentiate and link. As this integration unfolds, the mind creates, and the mind subjectively experiences, the emergence of harmony. Your mind and mine have been exploring along our journey the fundamental notion that well-being emerges from this internal and interconnected integration. We do not become each other; we maintain differentiated identities and become deeply linked. This is how integration creates the reality that we are a part of something larger than merely the sum of our parts. We is more than simply you plus me.

The sun is now higher on the horizon. The brilliant reds and oranges are folding into muted greys and faded yellows. Now is day, the fixed now is daybreak, and ahead, the open nows of what will come for you and me.

Integration may also involve our sense of time. We can embrace the reality of the Arrow of Time moving us ever forward, from future openness to present emergence and past fixed macrostate assemblies of life, while simultaneously experiencing a timeless sense of the arrow-free emergence of life arising within the plane of possibility. Integration involves embracing the tension of apparent opposites in our lives.

I stand here, typing to you, aware we are in our present epoch on this final entry of our journey together. I feel a need now to summarize where we've come, reflect on where we've been, consider ways we've imagined the mind might be seen, might be shared, and do an overview as an inner-view and inter-view with each other. As we conceive of mind as both embodied and relational, I come back to the question of whether the sun and sky, the cold wind and scattering sand, here, now, are, in their essence, my essence, the essence of me. As I share these words, can't they become a part of you, a facet of the you that is us? The illusion, belief, top-down, accepted, interpretatively-constructed perspective that I am fully separate from you and separate from this larger whole seems, as Einstein said years ago, to be an "optical delusion" of our consciousness (Einstein, 1972).

He did not say an illusion, but a delusion. A delusion is a psychotic belief, something not consistent with reality that may trap us in dysfunction and distress.

Just as time may be a construction of the mind, so, too, may be our sense of a fully separate identity. In helping address the grief of a father who had lost a child, these were Einstein's words (Einsten, 1972):

> *A human being is a part of the whole, called by us, "Universe," a part limited in time and space. He experiences himself, his thoughts and feelings as something separated from the rest—a kind of optical delusion of his consciousness. This delusion is a kind of prison for us, restricting us to our personal desires and to affection for a few persons nearest to us. Our task must be to free ourselves from this prison by widening our circle of compassion to embrace all living creatures and the whole of nature in its beauty. Nobody is able to achieve this completely, but the striving for such achievement is in itself a part of the liberation and a foundation for inner security.*

The ability of our minds to take in these sensory experiences, consider this knowledge, and then take on this perspective empowers us to create integration in our lives. This integration is the essence of a healthy body, mental life, and relationships. When we think deeply about all this, we come to realize the scientific underpinnings of perhaps an ancient truth from many wisdom traditions. This consilient view is that the outcome and process of integration, from insight and empathy to emotional balance and morality, may be the basis of a life well lived.

Seeing the macro view of our relationships with each other enables us to understand what a wide range of studies on happiness, longevity, medical well-being, and mental health all share in common: Supportive relationships are one of the most robust causal factors to create these elements of well-being in our lives.

And on the micro level, science has revealed the powerful ways in which developing presence, being aware of what is arising as it arises, optimizes epigenetic controls to prevent certain diseases, raises the enzyme telomerase to repair and maintain the telomeres at the ends of our chromosomes to support the health of our cells, improves immune function, and enhances our overall physiological well-being. Mindful awareness training has been shown, too, to change the structure of the brain toward integration. Developing presence, learning to strengthen our mind to cultivate an open awareness to whatever arises without being swept up by top-down automatic judgments, can enhance our health. That's now a proven fact. Presence cultivates well-being. What we do with our mind matters.

From this somatic view, we can also examine how our vertebrate nervous system is regulatory—helping balance our internal organs and interactions with the environment. As we evolved from fish to amphibians, then reptiles on to mammals, this nervous system became more intricate. As mammals, our ancestors evolved into primates and we, us humans, came on the scene in our earliest incarnation in the last few million years and found something close to our modern form in the last hundred thousand years or so. For at least thirty thousand years, some scientists affirm, we've been a story-telling family, sharing our experiences through drawings and likely spoken language with each other to help make sense of the experiences of living (Cook, 2013; Lewis-Williams, 2002).

As social primates, we need each other to survive in the groups in which we live. As human beings, we've developed even more social complexity, beginning with the unusual practice Sarah Hrdy writes so eloquently about in *Mothers and Others* (2009), where she discusses alloparenting, in which we share our child-rearing responsibilities with trusted others besides the mother. Imagine your dog or cat passing their puppies or kittens on to the neighbor's pets. Doesn't happen, does it? Even most other primates don't share this most precious of responsibilities, the care of our young and dependent offspring.

This alloparenting sets up a fascinating and forceful social environment: We survive by relying on others we can trust. We need a mindsphere to connect us to one another. The profoundly social nature of our lives has had powerful influences on the development of our social brains and the nature of our conscious minds (Dunbar, Gamble, & Gowlett, 2010; Graziano, 2014).

Trust has mechanisms in our relationships (see Gottman, 2011) and in the circuitry in the brain, in which we activate what Steven Porges calls the "social engagement system" (Porges, 2011). When I teach workshops, I often do an experiential learning immersion during which I say "no" harshly several times, pause, and then say "yes" soothingly several times. The result, no matter the culture or background, is quite similar. "No" evokes a harsh sensation, a feeling of withdrawal, an urge to flee, tightening of the muscles, an impulse to attack back. All just from some guy simply saying "no."

And "yes"? "Yes" often brings up (unless a person is stuck in the hostility, fear, or paralysis of the no) a feeling of calm, openness, engagement, relaxation.

What I think this exercise reveals is our two fundamental states: "No" evokes the reactive state, "yes" evokes the receptive state.

Reactivity has its roots in our ancient reptilian past, the 300 million-year-old brainstem being activated with a state of threat that gets you ready for the four f's of fight, flee, freeze, or faint. In contrast, a younger 200 million-year-old mammalian circuit soothes the brainstem's alarm bells, turning on the social engagement system that opens us up by making us receptive. Our muscles relax, we can hear a wider range of sounds, see a wider range of things in front of us. This is the neural correlate of our open, receptive state, ready to connect and learn.

My guess is that on the mindscape side, a receptive state emerges as we gain access to the plane of possibility. We become more aware, awake, and ready to engage.

Even studies of other mammals reveal that when certain physiological states are created, an individual is more likely to engage in pro-social behaviors. What is that physiology? Piloerection. This is the standing up of the hair follicles, and is what we experience when we get the goose bumps, in this case, with the feeling of awe. Studies of awe by Dacher Keltner and his colleagues, as well as others, (Shiota, Keltner, & Mossman, 2007), have revealed that experiencing awe opens our minds and subordinates one's self-interest for the benefits of the larger group. With awe we become more oriented to focusing on our community and quieting self-focused involvement. These ingenious studies have individuals look at an awe-inspiring scene, such as a magnificent grove of trees on the University of California—Berkeley campus, rather than an imposing modern building nearby (see the photos of the campus that follow in this entry—including the one of a pedestrian missing the awe of the trees in the captivation of a cell phone). When looking at the trees, subjects are more likely to help a person who has fallen. In general, when people experience awe they often say that the experience changed the way they see the world. My guess is that the awe that Keltner studied overlaps with our experience of the plane of possibility. Perhaps the varied sources of awe, such as being in nature, or being with human artifacts, like in a religious cathedral, at the Great Wall of China, or the Wailing Wall of Jerusalem, or the social awe created in interactions with others, may create an access to the open plane of possibility. When we drop beneath the plateaus and peaks of our separate identities and open to the majesty of the world beyond this smaller sense of self, awe is the state created and the experience of self is transformed. The plane enables us to feel something not easily understood at first, something vaster than a private, personal self, something that seems free of the Arrow of Time, something that accesses a perceptual glimpse of our spacetime block universe that feels expansive, if not infinite. This, we can suggest, is the experience of awe.

Awe emerges with a sense of receptivity that we are a part of something vast that may not be readily understood—it is bigger than

the personal me. This may be why awe induces a state of being more open to connection with others.

With our more complex and social cortex, too, we have evolved to possess another way of connecting and helping others. One response to challenging situations, one that Shelly Taylor at UCLA has named "tend-and-befriend" (Taylor, 2006), motivates us to connect with others. Initially found in female subjects, we now know that both genders are capable of activating this social engagement system response to a challenging situation. But the brainstem's more ancient role may still be present when we feel threatened, or completely helpless. Tend-and-befriend seems to be an important pathway mediated by neural regions above our reptilian brainstem. As social creatures, this important way we connect rather than simply react may have its roots, too, in how present we can be. Presence is a learnable skill.

In a recent workshop I taught with Barbara Fredrickson of the University of North Carolina, we discussed her insightful work and powerful book, *Love 2.0* (Fredickson, 2013). She proposes that love comes from the ways, small and large, that we share positive states in something she calls "positivity resonance." It's how we connect with others' positive emotions. I love the idea that love doesn't have to be confined to rigidly defined romantic relationships or attachment relationships, but can be experienced more broadly in our lives. As one participant in the Wheel practice told me years later, "Since the wheel, I now have this frequent experience where I just feel this deep love for people in front of me, people I have just met." "How is that for you?" I asked. Her smile told as much as her words. "Magnificent."

Fredrickson's earlier work on the broaden-and-build theory of positive emotions suggested, too, that when we experience positive states like love, happiness, awe, and gratitude, we are in a state that builds upon itself, connects us to more intricate levels of understanding, and broadens our sense of who we are. I quoted Fredrickson in the second edition of my first book, and was honored to teach with her for the first time.

On the stage, I asked Dr. Fredrickson if she could imagine the notion of integration as being a process potentially relevant to both her broaden-and-build theory, and to her view of love as positivity resonance. She was open to exploring these ideas. In both perspectives, integration, for me, seemed to hold the key to what was going on at the deep micro and macro levels. For love, two differentiated individuals become linked and therefore become integrated. But this resonance need not be limited to sharing positive states that become amplified together, I suggested, but could also take place within connections of compassion. I wondered if what happens in positivity resonance is an increase in integration because positive emotions can be seen as

increases in the level of integration. That statement may feel to you, as I was concerned it would to Dr. Fredrickson, as something coming out of nowhere, so let me explain.

In the 1990s while writing *The Developing Mind,* I became stuck on the chapter on emotions. No one seemed to have a commonly shared view of what emotions really were beyond descriptions of their characteristics. An anthropologist might say that emotion is what links people together in a culture across generations. A sociologist would say that emotions are the glue that holds a group together. A psychologist might say that emotions are what link a person across developmental time, or what links the various processes between appraisal and arousal. Biological researchers, like neuroscientists, might say that emotion is what links the functions of the body-proper to those of the brain. In reading and hearing all of these statements, no one actually used the term *integration.* But for me, these disparate perspectives each seemed to refer to some process of how differentiated things were being linked into a larger whole.

Emotion by itself wasn't always creating more integration, however. Sometimes it was creating less. If we became enraged, our system

became excessively differentiated in reactivity. If we became sad, or frightened, too, excessive differentiation for extended periods could lower integration and actually make us more prone to chaos and rigidity.

The patterns seemed clear: Emotion might be a shift in integration.

I offered to Dr. Fredrickson that broaden-and-build might be re-envisioned as a powerful way to describe increasing states of integration—both within us and our relationships. Those are positive emotions.

The so-called negative emotions could be seen as downshifts in integration, emotional states that often result from threat and can, if prolonged and tenacious, put us into a downward spiral, making us more prone to chaos or rigidity, taking us away from the ease of integrative well-being. All emotion is "good" in that it lets us know how we feel. We can say that feeling what you feel, becoming aware of these feelings, and being open to letting whatever you feel be present and explored, is a deeply important process of living life well. But extended periods of diminished integration, what prolonged "negative emotions" may be from this framework, tend to lead us out of the river of harmony and onto those two banks of chaos and rigidity.

So if integration could usefully be applied to broaden-and-build, could it apply to love? What I suggested was that positivity resonance, as Dr. Fredrickson defined it, seemed to be the amplification of already positively integrated states. We share our joy, excitement, and awe. We are grateful together. Wonderful and connecting, love is indeed a source of positivity resonance.

But what also could be considered a part of love, I felt and feel, is additionally when we connect with someone who is suffering. When we compassionately connect with our closest friends, patients and clients, our children, and even with a stranger, we can experience love. When someone in distress receives our care and concern, when we reach out to that person with a way of understanding his or her suffering and generate an intention and action to help—when we are compassionate—we also are creating an increase in integration.

How?

Even though someone suffering is in a lowered state of integration (by our definitions of mental health and mental un-health), when we connect with that person, the isolated lower-integration state and state of being alone are now shifted into one of being joined. In that joining of two differentiated individuals, even if one is suffering, the overall effect is an increase in the state of integration for both. Compassionate action makes two separate beings linked together into a larger whole. This is how integration makes the whole greater than the sum of its parts.

Compassion can be viewed, as wonderfully articulated by the researcher Paul Gilbert, as the ways we sense another's suffering, make sense of that suffering, and then imagine how we can help and then carry out the intention to help reduce that suffering (Gilbert, 2009; 2015). Empathy, in contrast, can be viewed as sensing or understanding another's experiences, not necessarily a drive to help. We can have empathic concern, though, which may be a precursor to compassion, perhaps for some even a synonym, cognitive empathy that is intellectual, and emotional empathy, when we feel another's feelings. In these ways we can see that empathy may be an important gateway toward compassion.

And what of kindness?

In my own conceptual world, the word *kindness* feels very relevant to all of these important ideas of how we connect, experience love, and how our minds work. One way to view kindness is simply acting on behalf of someone without any expectation for something to be given back in return. So a compassionate act can be an act of kindness for sure.

But kindness for me is also a state of being, a way of approaching one another, and even ourselves, with a certain intention, attitude, and care promoting internal respect, of positive regard. Kindness is a texture of our mental state.

I view kindness as honoring and supporting one another's vulnerability. Being kind is having a state of mind that recognizes that we each have many layers of our mental lives. We have an external adaptive way of being and presenting ourselves to the outside world that can be quite distant from our inner truths. We can have inner states of needs and disappointments, fears and concerns, which can simply be called vulnerabilities, that may lay hidden from view but are nevertheless ever present in our inner mindscape. In fact, we may have many aspects of our "self," our "self-states," that are really many layers of a multifaceted collection of internal states. These would be seen in the framework of the plane of possibility as clusters of recurrent plateaus and their propensities for certain peaks of activation. Integration would involve a kind regard for them all, embracing the full spectrum of emotion and needs, memory and strategies of living in the world. Being aware of needs unfulfilled, of hurts unhealed, brings us to a state of being open and vulnerable.

It is in these vulnerabilities that kindness emerges as a way of being that can be a state of receptivity to this deeper level of our reality, a drive that shapes how we communicate both with words and in the non-verbal meanings between and before words, and which can motivate us to create random or planned acts of kindness. Kindness is a state of mind we can cultivate to bring love into our lives.

If we hypothesize that positive emotional states are increased integration, we can see how acts of kindness create happiness. If we broaden and build on Fredrickson's notion of love to include not just positivity resonance but also all resonance, even when someone else is suffering, that too can reveal how love itself is a state of mind with increased integration. Compassion can cultivate happiness, and since joining happens at the level of vulnerability, reaching out to help someone reduce their suffering is both an act of kindness and a defining feature of compassion.

The Systems of a Plural Self and Integration of Identity

Systems function at many levels, as we've discussed throughout our journey. Molecules within cells of the body are configured in ways that carry out certain cellular functions, such as metabolism or membrane maintenance. Cells within the body cluster together as they differentiate and then link to form organs, such as the heart, liver, or lungs. These organs work in concert to create systems, enabling the immune, cardiovascular, musculoskeletal, and gastrointestinal systems to work

efficiently and effectively at their tasks. The nervous system is but one aspect of the collection of many systems of the body. Comprised of neurons that have microtubules and interconnections enabling a range of energy patterns to flow within and between these neural cells, supported by trillions of glial cells that carry out other important functions, the nervous system serves as a conduit and constructor of energy into information, part of the bottom-up and top-down flow of our mental lives. These are all aspects of the systems of the body. What do these organ systems do? They work together to create the system of the body as a whole.

But where is the system of the whole person? How do we go from molecules to mind? How does the synergy of enfolding layers of our reality, our 'wholeness and implicate order' discussed earlier as Bohm described (1980/1995), emerge from the interactions of our component parts?

If we microscope inward, we see molecules underlying our physical structure. Go even more micro, and we see atoms comprising the molecules. An even more fine-grained focus gets us to the now accepted reality that the vast majority of an atom itself is empty space. If we move even further in, we come to see that particles are comprised in various wondrous ways of forces that can be conceptualized ultimately by the general term, energy. As we've seen, knowing Einstein's discovery of the notion that energy equals mass times the speed of light squared (E=mc2), we can see that even what we think of as the physical nature of the world, the world of things comprised of mass, is actually made up of very dense energy.

If we now telescope outward from molecules to cells to organs to the body, where do we end? We go further out in a macro view and see bodies interacting with each other in what we call relationships based on communication patterns. That's the "social" of our social lives, the mindsphere that envelops us. These ways we connect with each other,

as seen from this wide-angled view, involve the exchange of energy and information. Information can be seen simply as energy patterns with meaning beyond the energy flow itself. As we move macro and outward, we come to where we arrive when we move micro and inward: energy flow. At this macroscopic level, for example, we are learning that much of the universe is comprised of dark matter and dark energy—mass and forces that we cannot directly detect with our eyes or our instruments, though the shadows of their effects, such as dark holes, can be deduced and their existence all around us ascertained. These are the many powerful yet often-invisible ways energy manifests itself.

Living in these bodies, it's sometimes hard to go in either direction to see clearly the micro or the macro. We tend to stay at the "midcro" between micro and macro, as we stay at the bodily level. But even at this level, if we think about it a bit, it's amazing we are even here. Even more, it's astonishing that we know we are here, that we can be aware of how astonishing this all is.

When we embrace the reality of both macro and micro, when we perceive both outer and inner, we come to see energy at the heart of being alive. If we stay at the size of our bodies, resting in our perceptions at the midcro somatic level, we may naturally come to believe that our mental lives are somehow at that size too, some outcome of this bodily level of life, something material, something concrete, like the hand you feel as you hold this book, like the hand you use to shake my hand if or when we meet in the future.

The somatic level is real, but it's not the whole deal.

Our eyesight enables the somatic level to be readily seen.

Our mindsight enables the micro and macro levels of energy flow of the mind to be perceived.

With mindsight we can sense that a separate self is not the entire story of the mind. What are visible to the eye as separate parts may be perceived by mindsight as indivisible facets of one whole. Mindsight makes the indivisible visible.

Our interpreting constructor can create a top-down filter that has us perceive the self living only in the body, a me of everyday life. That generated perception of a private, personal, bodily-based, single-skull view of the mind's self keeps us isolated from one another. There may be a sense of insufficiency and aloneness, a deep longing that never quite becomes fulfilled.

We do live in a body, a body with a nervous system with synaptic connections shaped by genes and experience. The brain's structure is molded by the messages we receive, from parents, teachers, peers, and society. Some of those messages in modern times are that the *self* is separate—and we often believe that misinformation.

The brain is also shaped by genes and the epigenetic regulators that control those genes' expression. We've inherited genetic information that molds how the brain influences our experience of self as it guides the emergence of embodied energy and information flow. That's the embodied aspect of mind. Our evolutionary history gives us a bias to focus on the negative, to pay attention to the things that could harm us. The survival value in that bias is clear, but the result is a tendency toward depression and anxiety, worry and despair. Mindsight may be needed to rise above this natural proclivity of the brain to focus on the negative. It is now clear that we can use the mind to change the way the brain functions and is structured.

Another brain proclivity we've inherited is to focus on the distinction between those in our "in-group" and those not, those in the "out-group." A wide range of studies reveal that we will, especially under conditions of threat, tend to care more kindly for those in the in-group, and treat those in the out-group with more hostility (MacGregor, et al. 1998; Banaji and Greenwald, 2013; Choudhury, 2015) . While we may indeed have a history of collaboration with alloparenting, and while we may have evolved with collaboration with the in-group as our innate nature, confronting others who are not like us, others who our evolutionarily shaped body's brains declare as out-group members, will spontaneously give rise to potentially harsh and destructive behavior. Instead of kindness and compassion, we shut off empathy and can even treat such individuals as non-humans. We shut off our mindsight circuitry and perception. We don't apply our mindsight skills to those in the out-group, especially when we feel threatened.

Here is our challenge. We have been born into a body with an evolutionary history of millions of years of establishing hierarchies, of which individuals, families, and groups are "in" and which are "out," who is alpha and who is beta. At our annual Interpersonal Neurobiology conference, as we were walking to dinner one night primatologist Steve Suomi described to me that in Rhesus monkeys, for example, when a dominant family becomes vulnerable, the next family down in the hierarchy will seize the opportunity to kill all the members of that family and achieve their long-awaited alpha status. I asked Steve how long that species had existed, and therefore how long such a primate history may have been in our past: 25 million years.

Another speaker at our conference, cultural psychologist Shelly Harrell, powerfully expressed some views from the field of multicultural psychology, suggesting that, "Culture is what everybody knows that everybody else knows." In her view, the aspects of our lives subsumed under the notion of culture include our experiences of being, believing, bonding, belonging, behaving, and becoming. These are insightful ways of illuminating the mindsphere that shapes our relationships with one another in our society. Culture is a fundamental facet of who we are (Baumeister, 2005). Many layers of self-identity shape our experience within culture. As Harrell went on to reframe the views from anthropologist Clyde Kluckhohn and psychologist Henry Murray (1948), an individual experiences layers of tension in identity paraphrased this way: "We are at the same time like all others, like some others, and like no others."

The question we can ask, then, is how do we take into account our evolutionary proclivities for both collaboration and competition? We have our individual identity, as we are unique, like no others; we have our group identity as we are like some others; and we have our collective identity, as we are like all others—including not just our human family, but all living beings.

To achieve such integration—of competition and collaboration, of individual, group, and collective identity—we can suggest a deep dive into compassion. As clinical psychologist Paul Gilbert expressed at our meeting, "Compassion is the courage to descend into the reality of human experience." Descending into our reality entails delving within as well as between.

In our inner life, if the brainstem's reactive state of fight or flight, or perhaps even freeze and faint, are activated when we are near those in the out-group, we shut off the limbic and cortical capacity to trust and to tend-and-befriend. We shut off our love for others, and we do not empathically sense the inner mind of the other. The out-group becomes a non-group, non-human, and we set our sights on only those we view with a very narrow definition of who is "like us."

How can we rise above these biases of evolution that our bodies have inherited? How can we address these often hidden implicit biases, these internal filters, that narrow our circle of compassion? In many ways, one of the goals of this journey is to address these questions by examining the underlying mechanisms of mind in a deep and hopefully useful way. The mind can change the way the brain functions in the moment. The mind can change the structure of the brain in the long-run to alter how the brain then begins to naturally function. That's the power and promise of a mindsight approach. But how? Through the portal of awareness, an awakened mind that uses the plane of possibility to sense these imprisoning patterns and frees itself from their hold. One example of such an approach is in the cultivation of mindfulness and compassion that has been demonstrated to decrease in-group and out-group biases and enhance our sense of connection to a wider group (Lueke & Gibson, 2014; and see Gilbert, 2009; 2015).

We have inherited a negativity bias and an out-group hostility. These are a given of our evolution. Modern culture teaches us that the self is a private affair. These proclivities can be seen as vulnerabilities that do not have to be our destiny.

When we relax this privatized view of a separate self, generated by our own top-down filtering constructive minds initiated and reinforced by lessons taught from our social world, we may become more open to a broader, deeper, reality. When we move beyond the level of the body, even with its evolutionarily shaped social brain, when we receive both a macro and micro view, we can come to see the deeply interconnected nature of not only reality "out there," but the reality of our minds. Yes, the personal self is real, lives in our body, and can be called me. There is no one else like me, you are unique. Yes, we have some people "like me" whom we see as members of a small in-group. This personal self and personal group are real and really important. These often shape our identity. We also have a wider, collective relational self that is real, lives in our connections with each other and the planet, and can be called we. But this is a we beyond the limitations of our personal in-group affiliations; this is a more universal we that enables us to sense our membership in a larger humanity, one that is fundamentally interwoven with all living beings.

To integrate our identity, to embrace not only the differentiated me with its personal in-group, and the differentiated broader we as a wider relational self, but also have both, together, we can call this *MWe*.

MWe can be viewed as our integrated identity, the linkage of a differentiated *me* with a differentiated *we*, all in one integrated and integrating self.

Our conscious sense of identity is created as we interpret the sensory stream of the conduit of the mind, generating a sense of who we are in the world. This sense of self arises from the mind, and if we take it in through *s*ensory bottom-up, *o*bserve, and also open our top-down *c*oncepts and *k*nowing, our interpreting constructive narrating minds may begin to open to the possibility that our self is truly a plural verb.

If we open to this possibility of an integrated and integrating self, then we can perhaps embrace a larger sense of purpose in life, become a part of that larger whole that so many studies and ancient wisdom traditions suggest is essential for our well-being. Whether we call this spirituality with meaning and connection, or simply how to live a flourishing and thriving life, cultivating an integrated identity as MWe synergistically combines our embodied and our relational nature. The whole of a life of MWe is greater than the sum of its parts.

As we've seen, the self may be viewed as a node that is a part of a wider system comprised of interacting nodes. For us, that node is the body, and the "self" is often equated in modern times with our bodily node. But what if this view of the self as body-node is limiting our sense of what is real? What if the self is also the whole system? Yes, we have a bodily-self, a me. And yes, we have small group identity, our in-group clan. But we also have a system-self, our membership in a wider whole, the whole of other humans, the whole of our living planet. The self is as much the individual as it is this collective self. Differentiating and linking these layers of self cultivates an integrated identity that emerges as energy and information flow emerges within and between. That is our integrated self as MWe.

As a plural verb rather than a singular noun, we are forever unfolding in how MWe take in and receive, observe and narrate, send out and connect, as we become not merely a set of neural responses to stimuli, but a fully embodied part of the deeply interconnected relational world in which we are embedded.

Reflections and Invitations: MWe, an Integrating Self, and a Kind Mind

On this journey we are on, exploring the inner and inter, diving into subjective seas and scientific concepts, complex systems and self-organization, we've come to a time of saying goodbye, for now, for the now of this moment. We've taken on some compelling questioning and wrestled with some potential preliminary answering, all with the mind in our minds. These inquiries and responses have guided us, together, along this journey to discover and uncover the many layers of who we are. Within these questions are invitations for further exploration, windows into clarity, and calls to action.

Imagine what in your own life might be inviting you to take this journey and open your mind to perhaps a new way of experiencing the world. If you reflect on the studies of awe and gratitude, for example, you may choose to find ways of being with nature or other people that allow the vastness of life to fill you with a sense of respect and reverence for this sacred journey through the moments of our lives. In many ways, we can see this awe and gratitude as a part of developing presence. What does presence do for us? Our body's immune system thrives, inflammatory response is soothed, epigenetic regulators are optimized, cardiovascular profile is improved, and even the telomerase enzyme that repairs and maintains the ends of our chromosomes will be elevated.

Photo by Alexander Siegel

Photo by Lee Freedman

Presence is the portal for integration, and integration is the basis of health and well-being.

We come to embrace the indivisible yet often invisible reality of the interconnected nature of our place in the world. With presence, we come to be open to our own history, the peaks and plateaus of our personal, separate identity. With presence we've learned to harness the hub of our Wheel of Awareness, the metaphoric practice that deepens and broadens our sense of self by making the mechanism of entering a plane of possibility part of our daily life.

Moment by moment, having access to this sea of potential frees us from the prisons of preoccupations with the past and frets over the future.

Yet presence does more than free our minds and bring health to our bodies.

The opportunity to transform identity is an invitation to bring health to our individual selves, yes. But the self is not bound by skull or skin. The mind, and the self that comes from it, is embodied, yes, but is also fully relational. Honoring the personal self of me and the interconnected self of we, we link these two with the integrated self of MWe.

Can you imagine if science embraced the reality of the mind's integrating potential? Can you imagine if schools focused on collaboration instead of individual competition, or if society was infused with this

Photo by Madeleine Siegel

integrating sense of self in our mindsphere world? If competing is to be encouraged, why not support the competition of student innovation to collaboratively create possible solutions to some of the world's most challenging problems? When the competition we engage with becomes a problem challenging us all, then when someone wins the competition and the problem is solved, everyone benefits. Whether it is the shortage of healthy food, water, and air, violence, the loss of animal and plant species, or climate change issues, there is plenty to compete with. We are an inherently collaborative species (Keltner, Marsh, & Adams-Smith, 2010), and this is the true human narrative that ought to be taught by science, schools, and society as a whole, including in messages from the media and parents who are raising the next generation. This collaboration is at the heart of what MWe can do in this world, in this life.

One of the biggest problems with the lie of the separate self is that if we believe it, we will experience, deep beneath the surface or even at the front of our minds, a sense of disconnection, isolation, and despair. Our modern digital technology seems to only reinforce this experience of a separate self (Turkle, 2011). Another problem with this delusion of separateness is we come to treat the planet as

a trash can. Instead of being in love with nature, we treat Earth like a dumpster.

When we open our minds with presence, we come to experience the deeply interconnected nature of our lives. We feel Earth is a part of us, an extended mental body that is as much a part of who we are as these somatic bodies we live in. If we are to cultivate the motivation to protect and preserve our planet, we need to be in love with our natural world.

We are built to connect. We are waiting to become liberated from the destructive implicit messages that the self lives in the body alone, or only in the head of that body. The self is an emergent property of the mind, and the mind is itself an emergent, embodied, and relational process. MWe can find a way to create a new generation that knows, from the inside out, the power and potential of realizing this truth about our integrated identity and integrating minds.

The mind is a verb we all share, not a noun that we only individually own in our isolated despair.

Transforming identity is about liberating the mind from the peaks and plateaus of culturally reinforced beliefs that an isolated, separate self is the whole truth of our existence. As we gain access to the plane of possibility, the potential source of consciousness, we are empowered to experience directly the interconnected nature of our lives. This enables not merely a concept to be thought, but an experience in which MWe can be immersed. Transformation is not learning facts—transformation is shifting perspective by expanding consciousness.

In your own travels, within yourself on this journey into mind, or in exploring various geographical locations, have you noticed that people seem to be aching to find a new way to live? In my own journeying, this has become evident. Something in our human family seems to be waiting to be opened up, transformed, liberated. Not only does this need seem to emanate from economic strife and uncertainty, not only from global concerns about our future, but from the very nature of the mindsphere that surrounds us with messages reinforcing our isolation. What are the ways that you might take these observations, of others, of yourself, and create a more integrated way to awaken and

live, create a more integrated identity, an integrating self, an integrating mind? Can we be part of an integrating team together, to transform, explore, and awaken the mind?

These reflections and invitations have been a chance for you and me to examine where we've come, entry-by-entry, experience-by-experience. I hope the invitation to continue reflecting and exploring will be helpful in your life as your journey unfolds.

For me, this journey brings us to where many wisdom traditions have come. Though they may not use the idea of integration, their teachings are consilient with the notion that differentiating and linking are the sources of well-being. There are simple truths that transcend our various cultures, transcend the unfoldings of life we call time, transcend our sense of separateness. These truths may come down to this: When we drop down into presence, we come to realize how deeply interdependent and interconnected we all are in this world. When we drop down from the peaks and plateaus of learned top-down lessons of a separate self, we come to sense our integrated self.

What arises from embracing the reality of both this we self and me self is another simple truth: Kindness and compassion, toward the self of the body and the distributed self of our interconnected lives, is the natural way of integration. We honor differences and promote linkages. As Arthur Zajonc recently stated in a gathering on contemplative and ethical leadership, we can promote a "pervasive leadership" in which each individual is empowered to lead from a position of contemplative inner knowledge and ethical responsibility for the benefit of the whole. When we realize what Dacher Keltner and colleagues have explored, that awe and gratitude inspire such a devotion to the greater good, we can be filled with deep reverence for what kindness and compassion are all about (Keltner et al., 2010). Contemplation means bringing in the sacred. Having reverence, honoring the sanctity of life, embracing each other with love and care, these are all the sacred ways of an integrating mind.

Can you feel deep within you, from the hub of your mind, from the presence of the plane of possibility, right here, right now, a sense of empowerment and clarity, woven within the fundamental idea that arises from this journey, from this exploration we have been on, this simple truth: Integration made visible is kindness and compassion?

Can you imagine a world in which we can not only explore a definition of the mind, but also share in the view of what may be fundamentally needed to cultivate healthy minds and a healthy world? This is where our journey has taken us. Kindness and compassion are integration made visible. With presence as our portal for integration, kindness for the mind is as natural as breathing for the body.

Together, MWe can make the potential of these simple truths the actual reality of our shared lives.

Photo by Madeleine Siegel

The Beginning

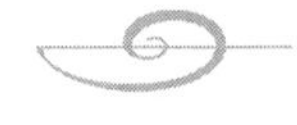

REFERENCES

Ackerman, D. (2014). *The human age: The world shaped by us.* New York, NY: W. W. Norton & Company.

Ackerman, D. (2011). *One hundred names for love: A memoir.* New York, NY: W.W. Norton & Company.

Albom, M. (1997). *Tuesdays with Morrie.* New York, NY: Doubelday.

American Psychiatric Association. (2014). *Diagnostic and statistical manual of mental disorders, 5th edition.* Washington, DC: APA Press.

Anderson, M.L. (2003). Embodied cognition: a field guide. *Artificial Intelligence 149*(1): 91-130.

Banaji, M.R. & Greenwald, A.G. (2013). *Blindspot: Hidden biases in good people.* New York, NY: Random House.

Barbour, J. (2000). *The end of time.* New York, NY: Oxford University Press.

Barbour, J. (2008). The nature of time. First juried prize essay for FQXi contest on the NATURE OF TIME, www.FQXi.org.

Barks, C. & Moyne, J. (Trans.). (1995). *The Essential Rumi.* New York, NY: Harper Collins.

Bateson, G. (1972). *Steps to an ecology of mind.* Chicago, IL: University of Chicago Press.

Bauer, P. V., Hamr, S. C., & Duca, F. A. (2015). Regulation of energy balance by a gut–brain axis and involvement of the gut microbiota. *Cellular and Molecular Life Sciences, 73*(4), 737-55. doi:10.1007/s00018-015-2083-z.

Baumeister, R. (2005). *The cultural animal: Human nature, meaning, and social life.* New York, NY: Oxford University Press.

Bharwani, A., Mian, M. F., Foster, J. A., Surette, M. G., Bienenstock, J., & Forsythe, P. (2016). Structural & functional consequences of chronic psychosocial stress on the microbiome & host. *Psychoneuroendocrinology, 63,* 217-227.

Bird-David. N. (1999). 'Animism' revisited: personhood, environment, and relational epistemology. *Current Anthropology*, 40: 67-91.

Bohm, D. (1980/1995). *Wholeness and the implicate order.* London: Routledge.

Brooks, R.A. (1999). *Cambrian intelligence: The early history of the new.* Cambridge, MA: MIT Press.

Bruner, J. (2003). *Making stories: Law, literature, life.* New York, NY: Farrar, Straus, and Giroux.

Cacioppo, J. T. & Freberg, L. A. (2013). *Discovering psychology: The science of mind* Belmont, CA: Wadsworth.

Cahill, T. (1998). *The gifts of the Jews: How a tribe of desert nomads changed the way everyone thinks and feels.* New York, NY: Anchor/Doubleday.

Callender, C. (2008). What makes time special? Essay for Fqxi contest on THE NATURE OF TIME. wwwFqxi.org.

Cameron, W.B. (1963). *Informal Sociology: A casual introduction to sociological thinking.* New York, NY: Random House.

Carroll, S. (2010). *From here to eternity: The quest for the ultimate theory of time.* New York, NY: Dutton.

Chopra, D. & Tanzi, R.E. (2012). *SuperBrain: Unleashing the explosive power of your mind to maximize health, happiness, and spiritual well-being.* New York, NY: Three Rivers/Penguin Random House.

Choudury, S. (2015). *Deep diversity: Overcoming us vs. them.* Toronto, Canada: Between the Lines.

Christakis, N. & Fowler, J. (2009). *Connected: The surprising power of our social networks and how they shape our lives.* New York, NY: Little Brown.

Clark, A. (1997). *Being there: Bringing brain, body and world together again.* Cambridge, MA: MIT Press.

Clark, A. (2011). *Supersizing the mind: Embodiment, action, and cognitive extension.* New York, NY: Oxford University Press.

Clark, A. & Chalmers, D. (1998). The extended mind. *Analysis* 58, no. 1: 7-19.

Cook, J. (2013). *Ice age art: The arrival of the modern mind.* London, England: The British Museum Press.

Cozolino, L. (2006). *The neuroscience of human relationships: Attachment and the developing social brain.* New York, NY: W.W. Norton & Company.

Creswell, J. D., Taren, A. A., Lindsay, E. K., Greco, C. M., Gianaros, P. J., Fairgrieve, A., Marsland, A. L., Warren Brown, K., May, B. M., Rosen, R. K., & Ferris, J. L. (2016). Alterations in resting-state functional connectivity link mindfulness meditation with reduced interleukin-6: A randomized control trial. *Biological Psychiatry*. doi: http://dx.doi.org/10.1016/j.biopsych.2016.01.008

Crooks, G.E. (2008). Whither time's arrow. Essay for Fqxi contest on THE NATURE OF TIME, www.Fqxi.org.

Csikszentmihalyi, M. (2008). *Flow: The psychology of optimal experience*. New York, NY: Harper.

Damasio, A.R. (2000). *The feeling of what happens: Body and emotion in the making of consciousness*. Orlando, FL: Harcourt Brace.

Damasio, A.R. (2005). *Descartes' error: Emotion, reason, and the human brain*. New York, NY: Penguin.

Davidson, R. J. & Begley, S. (2012). *The emotional life of your brain: How its unique patterns shape how you feel, think and live—and how you can change them*. New York, NY: Plume/Penguin Random House.

Dinan, T. G., Stilling, R. M., Stanton, C., & Cryan, J. F. (2015). Collective unconscious: how gut microbes shape human behavior. *Journal of Psychiatric Research, 63*, 1-9.

Doidge, N. (2015). *The brain's way of healing: Remarkable discoveries and recoveries from the frontiers of neuroplasticity*. New York, NY: Viking.

Doll, A., Hölzel, B. K., Boucard, C. C., Wohlschläger, A. M., & Sorg, C. (2015). Mindfulness is associated with intrinsic functional connectivity between default mode and salience networks. *Frontiers in Human Neuroscience, 9*, 461.

Dorling, J. (1970). The dimensionality of time. *Am J. Phys.*, 38:539-540.

Dossey, L. (2014). *One mind: How our individual mind is part of a greater consciousness and why it matters*. New York, NY: Hay House.

Dunbar, R., Gamble, C., Gowlett, J. (Eds.). (2010). *Social brain, distributed mind*, New York, NY: Oxford University Press.

Dweck, C. (2006). *Mindset: The new psychology of success*. New York: Ballantine/Random House.

Edelman, G. M. (1993). *Bright air, brilliant fire: On the matter of the mind*. New York, NY: Basic Books.

Edelman, G.M., & Tononi, G. (2000). *A Universe of consciousness: How matter becomes imagination*. New York, NY: Basic Books.

Edwards, B. (2012). *Drawing on the right side of the brain*. New York, NY: Tarcher.

Einstein, E. (1950). Letter in the New York Times (29 March, 1972) and the New York Post (28 November, 1972).

Ellis, G.F.R. (2008). On the flow of time. Essay for Fqxi contest on THE NATURE OF TIME. www.Fqxi.org.

Engel, A.K., Fries, P., & Singer, W. (2001). Dynamic predictions: Oscillations and synchrony in top-down processing. *Nature Neuroscience*, 2, 704–716.

Erneling, C.E. & Johnson, D.M. (Eds.). (2005). *The mind as a scientific object*, New York, NY: Oxford University Press.

Fair, D.A., Cohen, A.L., Dosenbach, N.U.F., Church, J.A., Miesen, F.M., March, D.M., Raichle, M.A., Petersen, A.E., & Schlagger, B.A. (2008). The maturing architecture of the brain's default mode circuitry. *PNAS, 105*(10) 4028-4032.

Fair, D.A., Dosenbach, N. U. F., Church, J. A., Cohen, A. L., Brahmbhatt, S., Miezin, F. M., Barch, D. M., Raichle, M. E., Peterson, S. E., & Schlagger, B. L. (2007). Development of distinct control networks through segregation and integration. *PNAS*, 104 (33), 13507-13512.

Faraday, M. (1860). Course of six lectures on the various forces of matter, and their relations to each other. Delivered before a juvenile auditory at the Royal Institute of Great Britain during the Christmas holidays of 1859-60, William Crookes (Ed.). New York, NY: Harper and Brothers Publishers.

Farb, N.A.S., Segal, Z.V., Mayberg, H., Bean, J., McKeon, D., Fatima, Z., & Anderson, A.K. (2007). Attending to the present: Mindfulness meditation reveals distinct neural modes of self-reference. *Social Cognitive and Affective Neuroscience, 2*(4), 313-322.

Felitti, V., Anda, R.F., Nordenberg, D, Williamson, D.F., Spitz, A.M., Edwards, V., Koss, M.P., & Marks, J.S. (1998). Relationship of childhood abuse and household dysfunction to many of the leading causes of death in adults. *American Journal of Preventive Medicine*, Volume 14 , Issue 4 , 245 – 258.

Feng, E.H. & Crooks, G.E. (2008). Length of time's arrow. *Phys. Rev Lett.* 101 (9): 090602.

Fredickson, B. (2013). *Love 2.0: Finding happiness and health in moments of connection*. New York, NY: Hudson/Penguin.

Freud, S. (1955). *The standard edition of the complete psychoanalytical works of Sigmund Freud*, (edited and translated by James Strachey). London, England: Hogarth Press.

Gambini, R. & Pullin, J. (2008). Free will, undecidability, and the problem of time in quantum gravity. Essay for Fqxi contest on THE NATURE OF TIME. www.Fqxi.org.

Gazzaniga, M. (2004). *The cognitive neurosciences III.* Cambridge, MA: MIT Press.

Gilbert, P. (2009). *The compassionate mind: A new approach to life's challenges.* Oakland, CA: New Harbinger.

Gilbert, P. (2015). The evolution and social dynamics of compassion. *Social and Personality Psychology Compass, 9,* 239–254. doi:10.1111/spc3.12176.

Goldstein, B. & Siegel, D.J. (2013): The Mindful Group: Using mind-body-brain interactions in group therapy to foster resilience and integration. In D.J. Siegel & M.F. Solomon (Eds.), *Healing Moments in Psychotherapy,* (pp. 217-242). New York, NY: W.W. Norton & Company.

Gottman, J.D. (2011). *The science of trust: Emotional attunement for couples.* New York, NY: W.W. Norton & Company.

Graziano, M. (2014). *Consciousness and the social brain.* New York, NY: Oxford University Press.

Harrell, S.P. (2000). A multidimensional conceptualization of racism-related stress: Implications for the well-being of people of color. *American Journal of Orthopsychiatry,* 70 (1): 42-57.

Hassin, R.R., Uleman, J.S., & Baragh, J.A. (2005). *The new unconscious.* New York, NY: Oxford University Press.

Hattiangadi, J. (2005). The emergence of minds in space and time. In Erneling, C.E. & Johnson, D.M. (Eds.), *The mind as a scientific object* (pp. 79-100). New York, NY: Oxford University Press.

Hawking, S. & Ellis, G. (1973). *The large-scale structure of space-time.* Cambridge: Cambridge University Press.

Hensen, B., Bernien, H., Dreau, A.E., Reiserer, A., Kalb, N., Blok, M.S., Ruitenberg, J., Vermeulen, R.F.L., Schouten, R.N., Abellan, C., Amaya, W., Pruneri, V., Mitchell, M.W., Markham, M., Twitchen, D.J., Elkouss, D., Wehner, S., Taminau, T.H., & Hanson, R. (2015). Experimental loophole-free violation of a Bell inequality using entangled electron spins separated by 1.3 km. *Nature,* 526, 682–686 (29 October 2015) doi:10.1038/nature15759.

Hrdy, S.B. (2009). *Mothers and others: The evolutionary origins of mutual understanding.* Cambridge, MA: Harvard University Press.

Hubel, D & Wiesel, T.N. (1970): The period of susceptibility to the physiological effects of unilateral eye closure in kittens, *Journal of Physiology, 206*(2): 419–436

Hutchins, E. (1995). *Cognition in the wild.* Cambridge, MA: MIT Press.

James, W. (1890). *Principles of psychology.* New York: H. Holt and Company.

Johnson, D.M. (2005). Mind, brain, and the Upper Paleolithic, in: Erneling, C.E. & Johnson, D.M. (Eds.). *The mind as a scientific object,* pages 499-510. New York, NY: Oxford University Press

Johnson, P. (1987). *A history of the Jews.* New York, NY: Harper and Row.

Johnson, R. L. (2005) *Gandhi's experiments with truth: Essential writing by and about Mahatma Gandhi*, Lanham, MD: Lexington Books

Kabat-Zinn, J. (2005). *Coming to our senses: Healing ourselves and the world through mindfulness.* New York, NY: Hyperion.

Kabat-Zinn, J. (2013). *Full catastrophe living (revised edition): Using the wisdom of your body and mind to face stress, pain, and illness.* New York, NY: Bantam Random House.

Kafatos, M. & Siegel, D.J. (2015). Quantum physics, consciousness, and psychotherapy. Audio recordings of a professional workshop. Santa Monica, CA: Mindsight Institute.

Keller, H. (1903). *The story of my life.* New York, NY: Double Day, Page & Co.

Keltner, D. (2009). *Born to be good: The science of a meaningful life.* New York, NY: W.W. Norton & Company.

Keltner, D. , Marsh, J., & Adams-Smith, J. (Eds.). (2010). *The compassionate instinct: The science of human goodness.* New York, NY: W.W. Norton & Company.

Kimov, P.V., Falk, A.L., Christie, D.J., Dobrivitski, V.V., & Awschalom, D.D. (2015). Quantum entanglement at ambient conditions in a macroscopic solid-state spin ensemble. *Science Advances, 1*(10) doi:10.1126/sciadv.1501015.

Kluckhohn, C. & Murray, H.A. (Eds.). (1948). *Personality in nature, society, and culture.* New York, NY: Alfred A. Knopf.

Kornfield, J. (2011). *Bringing home the dharma: Awakening right where you are.* Boston, MA: Shambhala Publications.

Kornfield, J. (2008). *The wise heart: A guide to the universal teachings of Buddhist psychology.* New York, NY: Bantam RandomHouse.

Krasner, M., Epstein, R., Beckman, H., Suchman, A., Chapman, B., Mooney, C., & Quill, T. (2009). Association of an educational program in mindful communication with burnout, empathy, and attitudes among primary care physicians. *JAMA*, September 23-30, 1284-1293.

Lakoff, G. & Johnson, M. (1999). *Philosophy in the flesh: The embodied mind and its challenge to Western thought.* New York, NY: Basic Books.

Langer, E. (1989/2014). *Mindfulness.* New York, NY: Da Capo.

Levit, G. S. (2000). The biosphere and the noosphere theories of V.I. Vernaksy and P. Teilhard de Chardin, a methodological essay. *International Archives on the History of Science/Archives Internatinales D'Histoire des Sciences, 50*(144), 160-176.

Lewis-Williams, D. (2002). *The mind in the cave: Consciousness and the origins of art.* London: Thames and Hudson.

Llinás, R. R. (2014). Intrinsic electrical properties of mammalian neurons and CNS function: A historical perspective. *Frontiers in Cellular Neuroscience, 8*, 320.

Lueke, A. & Gibson, B. (2014). Mindfulness meditation reduces implicit age and race bias: The role of reduced automaticity of responding. *Social Psychological and Personality Science,* 6: 284-291. doi:10.1177/1948550614559651.

Maleeh, R. (2015). Minds, brains and programs: An information-theoretic approach. *Mind and Matter* 13(1), 71-103.

Mayer, E. A. (2011). Gut feelings: The emerging biology of gut-brain communication. *Nature Reviews Neuroscience, 12*, 453-466.

McGilchrist, I. (2009). *The master and his emissary: The divided brain and the making of the western world.* New Haven, CT: Yale University Press.

McGonigal, K. (2015). *The upside of stress: Why stress is good for you.* New York, NY: Penguin/RandomHouse.

McGregor, H.A., Lieberman, J.D., Greenberg, J., Solomon, S., Arndt, J., & Simon, L. et al. (1998). Terror management and aggression: Evidence that mortality salience motivates aggression against world-view-threatening others. *Journal of Personality and Social Psychology,* 74(3), 590-605.

Meaney, M.J. (2010). Epigenetics and the biological definition of gene x environment interaction. *Child Development, 81*(1), 41-79.

Mesquita, B., Barrett, L.F., & Smith, E.R., (Eds.) (2010). *The mind in context.* New York: The Guilford Press.

Miles, L., Nind L., & Macrae C. N. (2010). Moving through time. *Psychological Science.* 21, 222-223.

Moloney, R. D., Desbonnet, L., Clarke, G., Dinan, T. G., & Cryan, J. F. (2014). The microbiome: stress, health and disease. *Mammalian Genome, 25*(1-2), 49-74.

Mountcasle, V. (1979). The columnar organization of the neocortex. *Brain, 120,* 701-722.

O'Donohue, J. (1997). *Anam cara: A book of celtic wisdom.* New York, NY: HarperCollins.

O'Donohue, J. (2008). *To bless the space between us: A book of blessings.* New York, NY: Doubleday Random House.

Palmer, P, & Zajonc, A. (2010). *The heart of higher education: A call to renewal.* San Francisco, CA: Jossey-Bass.

Pattakos, A. (2010). *Prisoners of our thoughts: Victor Frankl's principles for discovering meaning in life and work.* (2nd ed.). San Francisco, CA: Berrett-Koehler Publishers.

Perlmutter, D. A. (2015). *The brain maker: The power of gut-microbes to heal and protect your brain—for life.* New York, NY: Little Brown.

Pinker, S. (1999). *How the mind works.* New York, NY: W. W. Norton & Company.

Porges, S. (2011). *The polyvagal theory: Neurophysiological foundations of emotion, attachment, communication, and self-regulation.* New York, NY: W. W. Norton & Company.

Prigogine, I. (1996). *The end of certainty: Time, chaos, and the new laws of nature.* New York, NY: The Free Press.

Rakel, D., Barrett, B., Zhang, Z., Hoeft, T., Chewning, B., Marchand, L., & Scheder, J. (2011). Perception of empathy in the therapeutic encounter: Effects on the common cold. *Patient Education and Counseling, 85*(3), 390-397.

Reader, J. (1999). *Africa: A biography of the continent,* New York, NY: Vintage Books.

Rosenblum, B. & Kuttner, F. (2011): *Quantum enigma: Physics encounters conciousness.* (2nd ed.). Oxford: Oxford University Press

Rupert, R.D. (2009). *Cognitive systems and the extended mind.* New York, NY: Oxford University Press.

Sato, W., & Aoki, S. (2006). Right hemispheric dominance in processing of unconscious negative emotion. *Brain and Cognition*, 62 (3), 261-266.

Schacter, D., Addis, D., & Buckner, R. (2007). Remembering the past to imagine the future: the prospective brain. *Nature Reviews Neuroscience*, 8(9), 657-661.

Scharmer, C. O. (2009). *Theory u: Leading from the future as it emerges*. San Francisco, CA: Berrett-Kohler Publishers.

Schore, A.N. (2012): *The science and art of psychotherapy*. New York, NY: W.W. Norton & Company.

Scott, D. (2004). *Conscripts of modernity: The tragedy of colonial enlightenment*. Durham, NC: Duke University Press.

Scott, D. (2014). *Omens of adversity: Tragedy, time, memory, and justice*. Durham, NC: Duke University Press.

Semendeferi, K., Lu, A. Schenker, N., & Damasio, H. (2002). Human and great apes share a large frontal cortex. *Nature Neuroscience, 5*, 272-276.

Senge, P. (1990). *The fifth discipline: The art & practice of the learning organization*. New York: Doubleday/Random House.

Shapiro, S., Astin, J., Bishop, S., & Cordova, M. (2005). Mindfulness-based stress reduction for health care professionals: Results from a randomized trial. *International Journal of Stress Management. 12*(2), 164-176.

Shapiro, S. & Carlson, L. (2013). *The art and science of mindfulness: Integrating mindfulness into psychology and the healing professions*. Washington, DC: American Psychological Association.

Shiota, M.N., Keltner, D., & Mossman, A. (2007). The nature of awe: Elicitors, appraisals, and effects on self-concept. *Cognition and Emotion, 21*(5), 944-963.

Siegel, A.W. (2015). Good Leg. Song with lyrics and music © Alex Siegel, 2015.

Siegel, D.J. (2006). An interpersonal neurobiology approach to psychotherapy: How awareness, mirror neurons and neural plasticity contribute to the development of well-being. *Psychiatric Annals, 36(4), 248-258.*

Siegel, D.J. (2007). *The mindful brain: Reflection and attunement in the cultivation of well-being*. New York, NY: W.W. Norton & Company.

Siegel, D.J. (2009): Mindful awareness, mindsight, and neural integration. *Journal of Humanistic Psychology, 37(2), 137-158.*

Siegel, D.J. (2010a): *Mindsight: The new science of personal transformation*. New York, NY: Bantam/Random House.

Siegel, D.J. (2010b). *The mindful therapist: A clinician's guide to mindsight and neural integration*. New York, NY: W.W. Norton & Company.

Siegel, D.J. (2012a). *The developing mind: How relationships and the brain interact to shape who we are*. (2nd ed.). New York, NY: Guilford Press.

Siegel, D.J. (2012b). *Pocket guide to interpersonal neurobiology: An integrative handbook of the mind*. New York, NY: W.W. Norton & Company.

Siegel, D.J. (2014). *Brainstorm: The power and purpose of the teenage brain*. New York, NY: Tarcher/Penguin.

Siegel, D.J., & Bryson, T.P. (2012). *The whole-brain child: 12 revolutionary strategies to nurture your child's developing mind*. New York, NY: Bantam/Random House.

Siegel, D.J., & Bryson, T.P. (2014). *No-drama discipline: The whole-brain way to calm the chaos and nurture your child's developing mind*. New York, NY: Bantam/Random House.

Siegel, D.J., & Hartzell, M. (2003). *Parenting from the inside out: How a deeper self-understanding can help you raise children who thrive*. New York, NY: Tarcher/Penguin.

Siegel, D.J. & Siegel, M.W. (2014). Thriving with uncertainty. In Le, A., Ngnoumen, C.E., & Langer, E.J. (Eds.), *The Wiley Blackwell handbook of mindfulness* (Vol. 1, Ch. 2, pp. 21–47). Malden, MA: Wiley Blackwell.

Smith, S.M., Nichols, T. E., Vidaurre, D., Winkler, A. M., Behrens, T. E. J., Glasser, M. F., Ugurbil, K., Barch, D. M., Van Essen, D. C., & Miller, K. L. (2015). A positive-negative mode of population co-variation links brain connectivity, demographics, and behavior. *Nature Neuroscience, 18*(11), 1567-71.

Solomon, M.F. &. Siegel, D.J. (Eds.). (2003). *Healing trauma: Attachment, mind, body and brain*. New York, NY: W.W. Norton & Company.

Sperry, R. (1980). Mind-brain interaction: Mentalism, yes; Dualism, no. *Neuroscience* 5 195-206.

Spreng, R. N., Mar R. A., & Kim A. S. N. (2009). The common neural basis of autobiographical memory, prospection, navigation, theory of mind and the default mode: A quantitative meta-analysis. *Journal of Cognitive Neuroscience*. 21, 489-510.

Stapp, H. (2011). *Mindful universe: Quantum mechanics and the participating observer* (2nd ed.). New York, NY: Springer.

Stoller, R..J. (1985). *Observing the erotic imagination*. New Haven, CT: Yale University Press.

Strathern, M. (1988). *The gender of the gift: Problems with women and problems with society in Melanesia*. Berkeley, CA: University of California Press.

Taylor, S. (2006). Tend and befriend: Biobehavioral bases of affiliation under stress. *Current Directions in Psychological Science, 15*(6), 273-277.

Teicher, M.H., Andersen, S.L., Polcari, A., Anderson, C.M., Navalta, C.P., & Kim, D.M. (2003). The neurobiological consequences of early stress and childhood maltreatment. *Neurosci Biobehav Rev*, 27(1-2), 33-44.

Teicher, M.H., Dumont, N.L., Ito, Y., Vaituzis, C., Giedd, J.N., & Andersen, S.L. (2004). Childhood neglect is associated with reduced corpus callosum area. *Biol Psychiatry*, 56(2), 80-5.

Thompson, E. (2014). *Waking, dreaming, being: Self and consciousness in meditation, neuroscience, and philosophy*. New York, NY: Columbia University Press.

Tononi, G., & Koch, C. (2015). Consciousness: Here, there and everywhere? *Philosophical Transactions of the Royal Society B: Biological Sciences, 370*(1668), 20140167.

Tulving, E. (2005). Episodic memory and autonoesis: Uniquely human?. In H.S. Terrace & J. Metcalfe (Eds.), *The missing link in cognition: origins of self-reflective consciousness* (pp. 3-56). New York, NY: Oxford University Press.

Turkle, S. (2011). *Alone together: Why we expect more from technology and less from each other.* New York, NY: Basic Books.

Varela, F., Lachaux, J., Rodriguez, E. & Martinerie, J. (2001). The brainweb: Phase synchronization and large-scale integration. *Nature Reviews Neuroscience*, 2, 229-239.

Varela, F., Thompson, E., & Rosch, E. (1991). *The embodied mind: Cognitive science and human experience.* Cambridge, MA: MIT Press.

Vieten, C. & Scammell, S. (2015). *Spiritual & religious competencies in clinical practice: Guidelines for psychotherapists & mental health professionals.* Oakland, CA: New Harbinger.

Vygotsky, L. (1986). *Thought and language,* Cambridge, MA: MIT Press.

Weinstein, S. (2008). *Many times.* Essay for Fqxi contest on THE NATURE OF TIME. www.Fqxi.org.

Wilson, E.O. (1998). *Consilience: The unity of knowledge.* New York, NY: Vintage/Penguin.

Yehuda, R., Daskalakis, N. P., Lehrner, A., Desarnaud, F., Bader, H. N., Makotkine, I., Meaney, M. J. (2014). Influences of maternal and paternal PTSD on epigenetic regulation of the glucocorticoid receptor gene in Holocaust survivor offspring. *The American Journal of Psychiatry, 171*(8), 872–880.

Youngson, N.A. & Whitelaw, E. (2012): Transgenerational epigenetic effects. *Annual Review of Genomics and Human Genetics,* 9:233-257.

Zajonc, A. (Ed). (2006). *We speak as one: Twelve Nobel laureates share their vision for peace.* Portland, OR: Peacejam Foundation.

Zajonc, A. (2009). *Meditation as contemplative inquiry: When knowing becomes love.* Great Barrington, MA: Lindisfarne Books.

Zhang, T. & Raichle, M.E. (2010). Disease and the brain's dark energy. *Nature Reviews Neurology. 6*(1) 15-28.

INDEX

Note: Italicized page locators indicate figures or photographs.

abuse, impediments to brain integration and, 80–81, 223
academic publishing world, *69*
acceptance, grief, forgiveness, and, 71
Ackerman, D., 216, 223
action potential, 44, 45
actuality, possibility transformed into, 31
Adams, A., 19
adaptation, epigenetic shifts and impact on, 193
adolescence, ESSENCE of, 119–20
adolescent brain remodeling, early twenties and, 102, 120
adolescents, teaching mindsight ideas to, 119
adrenaline, trauma and excess secretion of, 169
Africa, human origins and, 194
Albom, M., 105
alloparenting, 194, 308, 320
All Saints Church, Pasadena, gathering at, 263, 264, 265–66, 268, 271–75
American Psychiatric Association, honorary fellowship with, 148
amnesia, horse accident, suspended identity, and, 125–27, 133, 160, 168, 193
amplitude, of energy, 41, 55, 56
ancient wisdom traditions, leading a wise and kind life as taught by, 204
anguish, manifesting as chaos and rigidity, 138
anterior insula, 131
anxiety, mindfulness meditation and, 224
ape brain, cortex and human brain *vs.*, 136
Arrow of Time, 241, 243, 244, 270, 285, 306, 310
  journey into mind and, 246
  quantum microstates, classical macrostates, and, 245
artifacts, 181
  definition of, 180
  of digital culture, 201
atoms, 242, 318
attachment, *66*
  close figures in adulthood and, *65*, *66*, *70*
  mindfulness meditation and, 221, 222

attachment (*continued*)
parents' sense of their own past and dynamics of, 28, 175, 251
selective, 195
see also secure attachment
attachment research, *where* and *what* of mind and, 175
attention, 284
brain structure shaped by, 225
definition of, 225
energy and information flow shaped by, 179
focal, 171, 172, 284
interpersonal shaping of, 171
mind and, 58
nonfocal, 167
attention deficit, mindfulness meditation and, 224
attitudes, 2
attuned communication, prefrontal cortex and, 204
attunement
definition of, 227
feeling felt, healing, and, 167
internal, 227–28, 231, 233
mindsight and, 112
secure attachment and, 227–28
term "relational" and, 154
autism, 9, 217
brain regions, lack of integration, and, 80
MEG studies on, 218, 219
autobiographical memory, 93, 169
explicit processing by hippocampus and, 172
grief and, 96
autonomic nervous system, 46
awakening the mind, meaning of, 141–42
awareness, 2
cultivating intentional states within, 284
knowns of, 278
magic within, 305
mindful, 283
potential origin of, 285
probability position in moment in near-zero plane and, 296
reflections and invitations on time and, 235–56
streams of, 224, 229–31, 232, 272
subjective experiences and, 34
wide-open, flow *vs.*, 292–93
see also consciousness; perception; subjective experience; Wheel of Awareness
awe, 301, 312, 329
experiencing, studies of, 309–10
openness to communal connection and, 309, 310
positivity resonance and, 314
presence and, 324

Bacon, E., 264, 271, 272, 273, 274, 275
ballroom dance team, 124, 125
Barnes, B. V., 174
Bateson, G., 12
"beginner's mind," open plane of possibility and, 269
behavior, 3
being, self-organization as recursive feature of, 50
beliefs, 2, 3, 278
belonging, Wheel of Awareness and sense of, 267

betweenness of mind
being present to, 157
opening to sense of connection and, 186
presence in psychotherapy and, 166–68, 170, 171
see also within and between
Biamonte, J., 247
Big Bang Theory, 243–44
bilateral integration
brain hemispheres and, 91
narrative integration and, 93
bilateral tapping technique, grief and, 87–88
binge eating, mindfulness meditation and, 224
"Biology of Compassion, The" (Siegel), 189, 236
biopsychosocial views, 53
bipolar disorder, 9, 198
chaos of mania and rigidity of depression in, 77
impaired integration and, 80, 218
Bitbol, M., 33
block universe, four-dimensional, 240, 263
blurry state, awareness without flow and presence and, 293
body based therapies
interoception and, 92
"body consciousness," 155
body(ies)
epigenetic legacy, brain function and structure, and, 193
as open system, 43
prefrontal cortex and regulation of, 204
vertical integration and, 92
Bohm, D., 83, 84, 317
Bohr, N., 248, 249
Bottom-Up Conduit, mind as, 131
bottom-up information processing, 132, 133, 134, 135, 136, 137, 138
description of, 127–31
schematic of, *136*
sensory bottom-up conduit and, 135–36, 141, 273
see also top-down information processing
boundaries of the self, 60–61
brain, 34, 43
adulthood and types of changes in, 177
as anticipation machine, 305
consciousness and, 9
deep interdependence between mind and, 39
defining, 27
embodied, 8, 34, 39
experience and shaping of, 176, 177, 192
exploring relationship between mind and, 27
focused attention and physical structure of, 225
growth of, throughout lifespan, 177
integrated relationships and integration in, 82
micro-systems in, 44
mind and transformation of, 7, 184, 185
mindfulness meditation and integration in, 221
neuroplasticity of, 167, 177–82, 226
rewiring toward integration, 199
as source of mind, 9
see also hemispheres of the brain; mind

brain activity, 29
  contemporary neuroscience perspective on, 47
  electrochemical energy and, 45
  energy flow and, 45
  use of term, 15
brain-activity=mind perspective, 10
"brainbound" model, 10
brain-centric view of mind, support for, 8–10
brain hemispheres
  bilateral integration and, 91–92
  Sperry's research on, 108, 161
brain proclivities, evolutionary, in-groups/out-groups and, 320–22
brainstem, 309, 311
Brainstorm (Siegel), 119
"Breeze at Dawn, The" (Rumi), 121
broaden-and-build theory
  integration applied to, 312, 314
  love and, 314
  of positive emotions, 312
Bruner, J., 13, 28, 29, 73
burnout, reducing with mindsight, 110

Callender, C., 246, 250
Cameron, W. B., 149
Cantwell, D., 28
cardiovascular system, 43, 316
Carroll, S., 239, 244
categorical diagnoses, limits of, 151
Catholic faith, 190, 191
causal influences, 48
causality, relaxing search for, 50
cell body, 45
cells, organ formation and, 316
cell systems, 44
Chalmers, D., 181
change, 48
  across probability curve, 32
  "being the change we wish to see," 209, 210, 211
  consciousness and, 266
  ever-flowing experience of, 18
  flow and, 40–41, 56
  mind and capacity for, 6
  presence and, 295
  time and, 281–82
chaos, 96, 117, 254, 284
  acute grief and, 97
  anguish manifested as, 138
  assessments of, in clinical setting, 88
  clinging to sense of control and, 280
  entering state of flourishing from, 200
  excessive differentiation and, 314
  grief and, 83
  impaired integration and, 86, 87, 198
  negative emotions and, 114
  origins of, gaining sense of, 89–90
  patients and pattern of, 76, 77
  reflecting on, 85, 86, 210
  relieving distressing symptoms of, 206
  stress, shift to right brain hemisphere, and, 92
  transforming into harmony of integration, 298
  see also rigidity

chaos-capable energy and information flow, 35, 36
child development, parent's inner self-understanding and, 69–70
child-rearing, alloparenting and, 194, 308, 309
children, being present with and secure attachment in, 224
choice, 48, 208
  consciousness and, 266
  mind and capacity for, 6
  mind shaped by, 187
  presence and, 295
Christakis, N., 162
Christian faith, history behind, 191
cingulate cortex, 133
circadian rhythms, 303
clarity, 287
  body's wisdom and call for, 120
  mindsight and renewed sense of, 117
Clark, A., 10, 181
Classical (or Newtonian) physics, 48, 57, 58, 164, 240, 244, 247, 270
CLIFF of energy, monitoring and modulating, 56, 57
climate change, notions of integration, mindsight, and, 236
clocks, 245, 303
  bodily, 239, 240
  passage of time and, 239
  see also time
cloud computing, 38
cloud of construction, awakening from slumber of, 305
clouds
  as complex systems, 49
  as open systems, 42
COAL, mindfulness and, 225
cognition, continuity of consciousness, community, and, 299
cognitive empathy, 315
coherence
  features of resilience in, 203
  integration and, 78
COHERENCE, 80, 81
  features of integrated flow, 203
  mental health emerging from integration and, 89
collaboration, 325
  evolutionary proclivities for both competition and, 320–21
  at heart of MWe, 326
collaborative connections, clarity and promotion of, 38
collective intelligence, wisdom and, 209
collective mental emergence, sense of, 160–61
collective self, 323
color spectrum, 55
community
  continuity of consciousness, cognition, and, 299
  deep interconnections within, 300
comparative ape neuroanatomy, 136–37
compassion, 109, 190, 250, 304, 312
  Dalai Lama as symbol of, 220
  decrease in in-group/out-group bias and, 322
  defining features of, 316
  descent into reality of human experience and, 321
  empathy *vs.*, 315

compassion (*continued*)
integration as source of, 196
integration made visible and, 329, 330
interpersonal integration and, 94, 95
love, integration, and, 314–15
mind as within and between and, 186
mindfulness and being present with, 233
mindsight and, 214
optimal self-organization and, 112
for self, others, and planet, integration and, 98
subjective sense approached with, 122
see also empathy; kindness; love
compassionate world, integration and seeds of, 236
competition
evolutionary proclivities for both collaboration and, 320–21
mindsphere, MWe, and, 325
complexity
energy and information flow and, 35
entropy and, 244
optimizing, 79
complexity science, 112
complexity theory, 77, 78
complex systems
emergent properties of, 36
emergent, self-organizing aspect of, 36
mind as part of, 202
self-organization and, 49, 50
conceptualization, as construction, 255
conceptual stream
defining, within SOCK, 272
mindfulness meditation and, 230, 231
conduit, free flow and, 130
conduition, 157
bottom-up, 135–36, 137, 141, 273
differentiating construction from, mindfulness and, 224
flow, wide-open state, and, 293
honoring both construction and, 138, 139, 142, 143–44
regaining presence and, 231–32
sensation as core of, 255
who of who we are and, 140
see also construction
conduit sensory stream, vivid present and, 138
connection, 210
betweenness of mind and opening to sense of, 186
integration and, 122
see also compassion; empathy; kindness; relationships
connectome, 44
diffusion tensor imaging of, 234
life traits, integration, and, 81
conscious intention, 55
consciousness, 6, 13, 38, 39, 53, 57, 75, 76, 152, 253
brain and, 9
brain firing and, 45
choice and change catalyzed by, 266, 283
continuity of cognition, community, and, 299

head-brain property and contrasting views of, 155
identity transformation and expansion of, 327–28
as inherent part of the universe, 269
integration of, 90–91, 206, 297
internal integration and, 232
knowing of, describing, 270
mind and, 1, 2
neural correlates of, 8
neural processing and, 7
"optical delusion" of, 307
plane of possibility, probability shifts, and, 291–92
as possible prime of the plane of possibility, 276
potential overlaps of self-organization and, 59
pure, 289, 290
source of, questions related to, 156, 158, 161
as a waking dream, 263
within and between, 159
see also awareness; mind
consilience, 29, 30
construction, 157
conceptualization as, 255
conceptualization stream and, 230
differentiating conduition from, mindfulness and, 224
flow, wide-open state, and, 293
honoring both conduition and, 138, 139, 142, 143
knowns of, 278
personal self notion and, 141
present moment and, 232
top-down layers in, 133
who of who we are and, 140
see also conduition
contemplation, 329
contemplative neuroscience, 223
contour, of energy, 56
corpus callosum
abuse and neglect, lack of integration, and, 81
bilateral integration and, 91
developmental trauma and growth impediments in, 223
mindfulness meditation and, 221, 234
cortex
bidirectional flow of energy in, 135
human brain *vs.* ape brain and, 136–37
cortical columns, six-layered, schematic of, 135, *136*, 137
cortisol, trauma, hippocampus, and, 169
coupling, term "relational" and, 154
Covey, S., 208
Cozolino, L., 28
Csikszentmihalyi, M., 292
cultural evolution, distributed mind and, 181
cultural systems, neuroplasticity and, 177–82
culture, 179
collective story, and creation of future of, 251
deep interconnections within, 300
digital, 201
mindsphere and, 321
social nature of, 28
curanderos (folk healers), 125

Dalai Lama, 250
connection in meeting with janitors and, 219–20, 224
keynote address, Mind and Life Institute, 220
participating on panels with, 235–36
dance, 108
dark energy, 319
dark holes, 319
dark matter, 319
Davidson, R., 225
day vision, perception, betweenness, and, 157, 162
"Decade of the Brain" (1990s), 26, 37, 41, 63, 77, 84, 88, 90, 97, 145, 173, 232, 255
decision-making, 3
default mode circuitry in brain, 133
mindfulness meditation and integration of, 234
OATS circuit and, 65, 132
dehumanizing the patient, medical socialization process and, 104
delusion of separateness, problems with, 326–27
delusions, definition of, 307
dendrites, 45
density, of energy, 41, 56
depression, mindfulness meditation and, 224
despair, 67, 326
Developing Mind, The (Siegel), 68, 69, 74, 78, 82, 114, 188, 214, 218, 234, 313
developmental psychology, 251
developmental trauma, impediments to brain integration and, 80–81, 223
diabetes, 193
Diagnostic and Statistical Manual of Mental Disorders (DSM), 30, 77, 150, 198
Diaspora, 191, 192
differentiation, 123
cultivating, 207
diminished, therapeutic intervention, and, 89
domains of integration and, 90
grieving process and, 97
integration and, 286
lack of, impaired integration and, 87
mind as within and between and, 186
mindsight and, 118
optimal self-organization and, 78
well-being and, 329
see also linkage
diffusion tensor imaging, connectome revealed with, 234
digestive system, 43, 46
digital culture, 201
digital processing, 158
discernment, 223, 232
disconnection, lie of the separate self and, 326
dis-ease
active grief and, 71, 72
impaired integration and, 84
dis-order, impaired integration and, 84
disorganized attachment, unresolved trauma and intergenerational transfer of, 176
distributed mind, 153
cultural evolution and, 181
what it is to be human and, 154
diurnal pattern, unfolding of now and, 303

domains of integration, 90–95, 198, 199, 210
bilateral integration, 91–92
identity integration, 95, 206
integration of consciousness, 90–91, 206
interpersonal integration, 94–95, 206
memory integration, 92–93
MWe integrated identity and, 96
narrative integration, 93–94
state integration, 94
temporal integration, 95, 206
vertical integration, 92, 206
dorsolateral prefrontal cortex, 131
Drawing on the Right Side of the Brain (Edwards), 108
dreams, 2
DSM. see Diagnostic and Statistical Manual of Mental Disorders (DSM)
DSM-III, 148
dualism, 113
Dunbar, R., 153
Dweck, C., 184, 185
dynamical system, subjective reality of lived life and, 202
dysregulation, 62

Earth
delusion of separateness and impact on, 326–27
plane of possibility and deep connection to, 286
time unfolding and, 303
ecosystems, 174
ecphoric sensation, 172
education, spirituality and, 236–37
Edwards, B., 107–8
eighth sense
eighth sense, Wheel of Awareness and, 90, *91,* 186, 267, 278
Einstein, A., 149, 209, 238, 240, 307, 318
electrochemical energy, brain activity and, 45
electroencephalogram studies, integration and, 81
electrolysis, discovery of, 31
electromagnetic fields, 162
electromagnetism, discovery of, 31
electrons, 163
Ellis, G., 246
embedded cognition, 153
embodied brain
meaning of term, 8–9
triangle of human experience and, 39–40, *40*
embodied cognition, 153
embodied language, 21
embodied mind, 9, 10, 46, 152, 320, 325, 327
emergence, 247
present moment and, 249
self-organization and, 49, 55
"Emergence of Minds in Space and Time, The" (Hattiangadi), 247–48, 249
emergent cognition, 153
emergent events, uncertainty and, 251
emergent flow of energy, symbolic meaning and, 31–32
emergent process, feeling of, 54
emergent synergy, 80
emotional balance, prefrontal cortex and, 204

emotional communication
energy, information, and, 112–13
suicide prevention and, 30, 105, 124, 183
emotional empathy, 315
emotions, 2, 3, 6, 34, 278
disparate perspectives on, 313
enzymes and, 30, 107, 124
as shift in integration, 114
SIFTing the mind and, 122
wide distribution of, in brain, 177
empathic joy, 210
empathy, 147, 304
compassion *vs.*, 315
health and healing and, 112
mindsight and, 109, 110, 214
optimal self-organization and, 112
prefrontal cortex and, 204
see also attunement; compassion; kindness; love
empowerment
how you live your life and, 209
integration and, 89, 200, 211
energy
betweenness and, 163
boundaries of, 60
contour of, 56
defining, 238
forms and description of, 31
at heart of being alive, 319
information and, 31–32
potentialities of, 57, 237–38
probability curve and, 237–38, 239, 268, 281
regulation of, 56–57
size of, 59
time and, 237
various forms of, 55
energy and information flow, 77, 131, 145, 280
attention and, 171, 179, 225
characteristics of, 35
domains of integration and, 96
embodied mechanism of, 39
emergent mind and, 32
human brain and impaired integration in, 138–39
information processing and, 53
integration and, 82
mind as potential emergent property of, 52, 53, 118, 152
mindsight and, 116–17
nonlinear, 35, 36
relationships and direction and nature of, 70
self-organizational aspect of emergence and, 55
self-organization of, 51–61
subjective experience and, 113
triangle of human experience and, 39–40, *40*
working definition of one aspect of the mind and, 37
energy-as-information, moment by moment emergence of, 52
energy flow
brain activity and, 45
defining, 238
emergent properties of, 52
meaning of, 253
physicists on process of, 55–59
social and neural connections and, 47
energy patterns, perceptual capacities and, 132

energy probability curve
  change without space and, 245
  chaos and rigidity and movement of, 254
energy probability distributions, mapping, plane of possibility, and, 291–95
entanglement, 48, 163, 164, 168, 180
entrainment, term "relational" and, 154
entropy
  complexity and, 244
  more complete definition of, 242
  "Past Hypothesis" and, 243
enzymes, emotions and, 30, 107, 124
epigenetic changes, experience and, 179
epigenetic regulation, of gene expression, 180, 184, 192
epigenetic regulators
  brain shaped by, 320
  mindfulness practice and optimization of, 225
epistemic unfolding, 241
"epoch entries," description of, 18–19
Erneling, C., 11
Esalen Institute, 258
eternal imprint, 262, 263, 264
ethics, integration and, 208
events, 226
  culture and, 179–80
  mindsphere and, 181
evolutionary brain proclivities, in-groups/out-groups and, 320–22
executive function, 234. *see also* self–regulation
experience
  brain shaped by, 176, 177, 192
  epigenetic changes and, 179
expertise
  awakening from slumber of, 305
  top-down dominance and, 130
explanation, description *vs.*, 230
explicit memory
  PART and achieving integrated state of, 172
  trauma resolution and integration into, 93
extended cognition, 153

FACES (flexible, adaptive, coherent, energised, and stable) flow of integration, 81, 88, 118, 198
  integrating self-organizing system and, 203
  self-organization and, 78, 79
  transient states moving away from, 85–86
  widening window of tolerance and, 172
FACES transformation, vitality and well-being and, 89
factual memory, explicit processing by hippocampus and, 172
famine, epigenetic changes in utero and, 192–93
Faraday, M., 31, 162, 164, 168, 180
feeling felt, 188
  healing and, 166–67
  subjective sense of joining me and we and, 228–29
feelings, elusiveness at core of, 34
fibromyalgia, mindfulness meditation and, 224
fight, flee, freeze, or faint, reactivity and, 309, 321
Fine, P., 174

fixed mindset, 184
flashbacks, PTSD and, 169
flow
  within and between, 33, 35, 36
  change and, 40–41, 56
  energy shifts, time, and, 239
  origins of peaks, ratio of plane to non-plane positions, and, 292
  presence and, 294–95
  wide-open awareness *vs.*, 292–93
  see also energy and information flow
flying, gravity and example of, 48–49, 57–58
fMRI studies, integration and, 81
focal attention, 171, 172, 284
foreign travel, mindfulness and, 130
forgiveness, acceptance and, 71
form, of energy, 41, 55, 56
four-dimensional block universe, 240, 263
Fowler, J., 162
FQXi 2008 contest, on nature of time, 239
Frankl, V., 208, 209
Fredrickson, B., 311, 312, 313, 314, 316
Freedman, L., *146, 325*
freedom
  meaning and, 208
  plane of possibility and, 285
free will, 2, 48, 54, 55
frequency, of energy, 41, 55, 56
Freud, S., 279
fruit juice concoction, directionality of time for macrostates and, 241, 242
future, 252
  fretting about, 231, 243, 251, 303, 305, 325
  open nature of, 241
future events, uncertainty and, 251

Galileo, Pope John Paul II and pardoning of, 195
Gambini, R., 246
Gamble, C., 153
gamma waves, diminished in autism, MEG studies on, 218, 219
gastrointestinal system, 316
gene expression, epigenetic regulation of, 180, 184, 192
General Relativity Theory, 240
genes, brain shaped by, 320
genome, mindfulness and optimized epigenetic regulators in, 225
Gilbert, P., 315, 321
glial cells, 317
Goldstein, B., 28
Gowlett, J., 153
gratitude, 263, 301, 312, 329
  presence and, 324
  Wheel of Awareness and sense of, 267
gravity, 48, 49, 244
  flying and, 48–49, 57–58
  relative nature of time and, 49
  shape of spacetime block and, 240
Graziano, M., 9, 155
grief, 66, 67
  acceptance and, 71
  challenged domains of integration and, 96
  father-in-law's death and, 87

moving through, after loss of mentor, 67, 68, 70–73, 74, 83, 88, 96–97
state of chaos and rigidity in, 83
groups, in-group/out-group distinctions, 206–7
growth mindset, 184, 185
gut-brain, 153, 158

habits of action, 180
habits of thought, 180
hallucinogenic drugs, altered perception and, 127
happiness, kindness and, 316
harmony, 253, 285
FACES flow of integration and, 78
of integration, 211
optimizing complexity and, 79
reflecting on, 210
self-organizational movement toward, 203
Harrell, S., 321
Hartzell, M., 69, 214, 215
Hattiangadi, J., 247, 248, 249
head-brain, 44, 153
head trauma, identity and, 125–27
healing, 213
feeling felt and, 166–67
as integration, 83
mindsight and, 110, 115
photons and, 164
plane of possibility and experiences of, 271, 283
health, 213
integration and, 80, 81, 85, 98, 99, 117, 190, 197
mindsight and, 110, 115
presence and, 308, 324
well-being and, 308
see also well-being
healthy mind
cultivating, essence of mind and, 6
defining, 76, 77
FACES and working definition of, 80
fundamental questions about, 72
hearing, 132
heart, parallel distributed processing systems and, 153
heart-brain, 153, 158
hemispheres of the brain
bilateral integration and, 91–92
mindfulness meditation and, 221
Sperry's "split-brain patients" research and, 108, 161
"higher consciousness," 155
hippocampus, 175, 176, 177
abuse and neglect, lack of integration, and, 81
developmental trauma and growth impediments in, 223
explicit processing by, 172
focal attention and activation of, 171
memory integration and, 93
mindfulness meditation and, 221, 234
PTSD and, 169, 176
Hippocrates, 9, 152
homo sapiens sapiens, 5, 277
hopes, 2
horse accident, suspended sense of self, and, 125–27, 133, 160, 168, 193
Hrdy, S., 308
Hubel, D., 175, 176

"Human Age," 6
human brain, cortex and ape brain *vs.*, 136–37. *see also* brain
human connectome project, 81
human experience, triangle of, 39–40, *40*
human family, awaiting transformation, 328
human potential movement, 258
humans, as interpretative constructors, 303, 304
hypertension, mindfulness meditation and, 224

ideas, fully felt experience and, 17
identity, 140, 141, 142
  accident and suspension of, 125–27, 160, 168, 193
  expanding sense of self and, 60–61
  exploring layers of experience beneath, 124–27
  integration within and between, 95
  internal and relational origin of mind and, 12–13
  new sense of, moving toward, 206
  transforming, expansion of consciousness, and, 327–28
  see also self
images, SIFTing the mind and, 121, 122
immune function
  empathy and, 112
  mindfulness meditation and, 224
immune system, 43, 316
implicate order, 83–84
implicit memory, 172
  encoding of, 93
  hippocampus, PTSD symptoms, and, 169
  mental suffering of PTSD and, 173
  PART and guiding patient's focal attention to, 172
inductive reasoning, 36
inferior parietal cortex, 133
infinity, plane of possibility and, 274
information, 45, 281
  energy patterns and, 319
  as energy patterns with symbolic value, 56
  energy transformations and, 31–32
  networks of energy patterns and, 153
  see also energy and information flow
information flow
  within and between, 158, 159, 161
  neuroplastic changes, mindsphere, and, 180
  regulation of, 57
  social and neural connections and, 47
information processing, 38, 39, 52, 75, 118, 152, 155
  differentiated modes of, 161
  energy and information flow and, 53
  source of, questions related to, 156, 158, 161
  top-down and bottom-up, 127–40
"information processor," mind as, 2, 3, 4, 152

in-groups
  in-group/out-group distinctions and, 206–7
  threat conditions, and tending to those in, 320
inside-out parenting approach, 69, 70
insight
  mindsight and, 109
  optimal self-organization and, 112
  prefrontal cortex and, 204
insight meditation practice, 217
insula, mindfulness meditation and, 221, 234
integrated consciousness, self-organization and, 297
integrated self, presence and, 329
integration, 123, 285
  connection and, 122
  definition of, 253
  differentiation and, 286
  empathic connections and, 112
  empowering nature of, 89, 200
  empowerment and, 211
  ethics and, 208
  FACES flow of harmony across time and, 80
  harmony of, 211
  healing as, 83
  health and, 80, 81, 85, 98, 99, 117, 190, 197
  hypothesis on mind and, supportive empirical findings on, 218–19, 234
  kindness and, 302
  love, compassion, and, 312, 314–15
  mental health and, 81
  mindsight and, 109, 122, 214
  natural drive toward, 199
  open potential of, 211
  optimal self-organization and, 78
  presence as portal for, 256, 280, 281, 283, 295, 298, 325, 330
  promoting, in community, 299
  psychotherapy, transformation, and, 197
  as "purpose of life?," 200–205, 210
  relationships, and rewiring our brains toward, 199
  in relationships, study of, 82
  river of, 78, *79*
  self-organization and, 82–83, 145
  as source of compassion and kindness, 196
  time and, 306
  transpirational, 95, 206
  unending journey of, 211
  unimpeded unfolding of, 97
  use of term, 80
  visible, kindness and compassion as, 329, 330
  well-being and, 115, 255, 306
  world's wisdom traditions and, 205
  see also domains of integration; self-organization
integration of consciousness, 90–91
intensity, of energy, 56
intentions, 2, 55, 257, 281, 283
inter-life, emergence of mind from, 116
internal attunement, 227, 231, 233

internal compass, placing value on, 117
internal neural integration, interpersonal integration and, 234
internal subjective experience
  aligning, mindsight and, 112
  SIFTING the mind and, 105
Internet, 158, 200
interoception, cortical brain and, 155
interoceptive sense, 132
interpersonal attunement, secure attachment and, 233
interpersonal integration, 94–95, 206
  compassionate communication and, 229
  focus on subjectivity as portal to, 111
  grief and, 96
  internal neural integration promoted by, 234
interpersonal neurobiology (IPNB), 188, 194, 196, 197, 218
  as health-based view of human life, 209
  here and now framework offered by, 219
  meditative practices and, 217
  mindsight built upon consilient view of, 198, 206
  parents and principles of, 214
intestines, parallel distributed processing systems and, 153
intrinsic nervous system, 46
introspection, James on, 8
intuition, prefrontal cortex and, 204
Inuit tribes, Alaska, leading a wise and kind life as taught by, 204
ions, 44
IPNB. *see* interpersonal neurobiology (IPNB)
isolation, lie of the separate self and, 326

James, W., 8, 150
Jesus Christ, 191
Jewish identity, Diaspora and, 191, 192
Jewish Synagogue, Rome, 195
Jews, basic history of, 191
John Paul II (pope)
  Galileo pardoned by, 195
  invitation from, 188–89
joy
  empathic, 210
  positivity resonance and, 314
Judaism, 191
judgments, discerning, 224

Kabat-Zinn, J., 215, 216, 217, 222, 223, 225, 233
Kandel, E., 225
Keller, H., 22
Keltner, D., 309, 310, 329
Khayyam, O., 247
kindness, 190, 236
  approaching subjective sense with, 122
  integration and, 109, 196, 302
  integration made visible and, 329, 330
  mindfulness and being present with, 233
  mindsight and, 214
  optimal self-organization and, 112
  random or planned acts of, 316

for self, others, and planet, integration and, 98
vulnerabilities and emergence of, 315–16
see also compassion; empathy; love
Kluckhohn, C., 321
knowing, Wheel of Awareness and, 232
knowing of consciousness, describing, 270
knowing of wisdom and truth stream, mindfulness meditation and, 231
knowing stream, defining, within SOCK, 272
Koch, C., 263
Kornfield, J., 217
Krishnamurti, J., 117

Langer, E., 130, 139
language
embodied, 21
neuroplastic changes, mindsphere, and, 181
observational flow and, 134
social nature of, 28
lateral parietal cortex, 133
Lazar, S., 221, 223
leadership, within each of us, 210
learning
mindful, Langer's notion of, 130, 139
trust and, 167
lesions in brain, brain-centric view of mind and studies of, 10
life stories, "canonical violation" and, 73
life well lived, consilient view of integration and, 307
light, 56
ultraviolet, 164
visible, spectrum of colors in, 55
Likert Scale, 151
linkage, 123
of differentiated parts, maximal complexity, and, 80
diminished, therapeutic intervention, and, 89
domains of integration and, 90
grieving process and, 97
integration in relationships and, 82
interpersonal attunement and, 228
lack of, impaired integration and, 87
mind as within and between and, 186
mindsight and, 118
optimal self-organization and, 78
promoting, 207
well-being and, 329
see also differentiation
Lithuania, *shtetls* in, 192
Llinás, R., 263, 289
Lloyd, S., 247
logical reasoning, left brain hemisphere and, 161
longings, 2
loss
of mentor, moving through grief after, 67, 68, 70–73, 74, 83, 88, 96–97
rim of Wheel of Awareness and, 96
love, 329
attachment and, 222
compassion, integration, and, 314–15

love (*continued*)
kindness and, 316
mindful awareness and, 295
positivity resonance and, 311, 312
see also compassion; empathy; kindness
Love 2.0 (Fredrickson), 311
lunar cycle, 303

macrostates, 243, 250, 270, 305
directionality of time for, 241
living creatures and, 244
time and, directionality to unfoldings of, 245
magnetoencephalogram (MEG) study, on diminished gamma waves and autism, 218, 219
Mahatma Gandhi, 195, 209
making sense, integration and, 83
Malamud, D., 208, 209
Massachusetts General Hospital, lecture at Ether Dome, 213, 214, 221
mass-energy equivalence formula (Einstein), 318
mathematics, integration in, 80
matter
betweenness and, 163
mind and shaping of, 173
maximal complexity
interpersonal integration and, 111–12
linkage of differentiated parts and, 80
self-organization and, 78, 79, 203, 253
MBSR program. *see* Mindfulness Based Stress Reduction (MBSR) program
meaning
freedom and, 208
in the mind, interpreting and generating, 264
mind and, 196
reflections and invitations on, 206–11
medial lobes, 133
medical knowledge, blending with human communication, 103
medical model, 148
medical school
acceptance to, 127
drive for objectivity in, 148
experiences in, 100–107, 117, 118
leave of absence from, 106
post-graduate training in pediatrics, 146–47
medical socialization process, 171, 193
dehumanizing the patient and, 104
mindsight and, 109–10, 111, 114
medical students, teaching empathy to, 110
meditation, 215
Mind and Life Institute and focus on, 250
"nothingness of non-dual reality" question and, 295–96, 298
see also mindfulness meditation
MEG study. *see* magnetoencephalogram (MEG) study
membrane maintenance, molecular configurations and, 316
memory(ies), 2, 3, 6, 66, 278
neural circuitry and mediation of, 169
prospective, 251

wide distribution of, in brain, 177
see also explicit memory; implicit memory
memory integration, 92–93
grief and, 96
narrative integration and, 93
mental health
brain as source of mind and challenges to, 9
energy and its boundaries and, 60
integration and, 81
mental illness, push for objective stance toward, 148
"Mentalism, Yes; Dualism, No" (Sperry), 113
mental representation, 41, 45
mental sea, mindsight and, 115–16
mental suffering, impaired integration and, 138
"mental time travel" (Tulving), 18
metabolism, molecular configurations and, 316
microstates, 245, 250, 270, 285
composition of, 242
physics of, 241
microtubules, 277, 317
"midcro" somatic level, 319
mind
activities of, 7
with "a mind of its own," 54, 203, 273
attention and, 58
awakening, 141–42
being beyond our interior notion, 154
as being what the brain does, 7–11
betweenness of, 157
brain as source of, 9
choices and shaping of, 187
classical physics notions and, 58
combining social and neural views of, 47
consciousness and, 1, 2
creating working definition of, 3
deep interdependence between brain and, 39
deeply interconnected between and within of, 171
defining, personal to planetary importance of, 6
"distributed," 153
embodied aspect of, 9, 10, 46, 152, 320, 325, 327
energy, time, and *when* of, 237
energy and information flow and, 32
exploring fundamental nature of, 24
exploring relationship between brain and, 27
finding fuller definition of, 12
as fully embodied and relational process, 11, 13, 48, 53, 59, 70, 77, 85, 88, 131, 158, 168, 197, 202, 218, 233, 236, 306, 325, 327
as heart of being human, 194, 213
hypothesis on integration and, supportive empirical findings on, 218–19, 234
identity, and internal and relational origin of, 12–13
as "information processor," 2, 3, 4, 152

mind (*continued*)
inner and inter nature of, 16, 17, 260
integration as *why* of, 201
interpreting and generating meaning in, 264
loss of loved ones and relational aspect of, 262
matter and shaping of, 173
meaning and purpose and, 196
meaning of, 1
nature of, open issue of, 14
as a noun, 3, 4
as part of complex system, 202
perennial curiosity about, 5
personal solo-self view of, 158, 160
as potential emergent property of energy and information flow, 52, 53, 118, 257
"relational" term and, 154
relaxing top-down conceptualizations of, 186
removing from focus of medicine, 104, 106
science of energy and, 163
self-organization and, 37–38, 50, 54, 156
sharing and birth of, 22
shift in probability state and flow of, 245
SIFTing, 143
"single skull" or "enskulled" view of, 10
subjectivity and, 1, 2
transformation of brain and, 7, 184, 185
triangle of human experience and, 39–40, *40*
use of term, 15–16
as verb, 3, 4, 327
when of, and emergent property of now, 226
working on a working definition of, 26–40
see also awareness; brain; consciousness
Mind and Life Institute, 220, 250
Mind and Life Summer Research Institute, 223
"Mind and Moment" conference, 216, 222–23
mindful awareness, 283, 298
mindful awareness training, brain structure changed toward integration with, 308
Mindful Brain, The (Siegel), 216, 223
mindful learning, Langer's notion of, 130, 139
mindfulness
decrease in in-group/out-group bias and, 322
foreign travel and, 130
parents and, 215
possible overlaps with secure attachment and, 215
Mindfulness Based Stress Reduction (MBSR) program, 222, 225
"Mindfulness is Relationality" (Kabat-Zinn), 233
mindfulness meditation
corpus callosum growth and, 92
fixed sense of identity and perception altered with, 127
goal of, 222
OATS system and, 133
streams of awareness and, 224, 229–31

mindfulness practices, interoception and, 92
"Mindful Revolution," 221
mindscape, 175, 188
  energy and information flow and, 202
  feelings connected to sensation of, 255
  inner and inter mind and, 260, 306
  intentional shaping of, 184
  neuroplasticity, culture, and, 182
  receptive state and, 309
mindscape work, weaving with mindsphere around you, 299
mindsets, 171, 184
mindsight, *110*, 146, 183
  betweenness and, 157
  capabilities embraced with, 109
  differentiation and, 118
  health and healing and, 110–15
  indivisible made visible with, 319
  integration and, 122, 214
  linkage and, 118
  medical socialization system and, 109–10, 111
  mental sea and, 115–16
  resilience and, 146
  well-being and, 122
Mindsight Institute, 216
Mindsight (Siegel), 106, 200, 204
mindsight tripod, 281, *282*
mindsphere, 173, 174, 175, 180, 181, 188
  bringing integration into the world and, 300
  collective, relational, 182
  culture and, 321
  energy and information flow and, 186, 202
  feelings connected to sensation of, 255
  inner and inter mind and, 260
  recursive, self-reinforcing nature of, 181
  well-being, and, 186
mind-system, 43
mind-therapy, 176
mode, 54, 65, 84, 91, 92, 96, 130, 132, 133, 161, 162, 183, 264, 299
modification
  mind's reliance on, 280
  strengthening, 281
moment-by-moment living, meaning embedded in, 207
monism, current mind-brain theory and, 113
monitoring
  mind's reliance on, 280
  strengthening, 281
mood, 257, 284
moon, 303
morality, prefrontal cortex and, 204
mortality, time and, 196
Mothers and Others (Hrdy), 308
motor activation, wide distribution of, in brain, 177
Mountcastle, V., 135
movement, well-being and, 124
multiple sclerosis, mindfulness meditation and, 224
Murray, H., 321
musculoskeletal system, 316
MWe
  collaboration at heart of, 326
  continual unfolding of, 323

MWe (*continued*)
domains of integration and, 96
as integrated identity, 322–23, 325, 327
our shared lives and full potential of, 330
transformation and, 328
myelin growth, 177, 180

"naming it to tame it," 87
narrative integration, 93–94, 96
narrative(s), 66
descriptive aspect of, 230
landscape of action and consciousness in, 73
as social process, 28, 29, 73
see also stories
National Institute of Mental Health (NIMH), Research Fellowship through, 175
natural world, presence and love for, 327
NCC. *see* neural correlates of consciousness (NCC)
near-zero (near-zero probability), 59, 238, 252, 268, 270, 274, 276, 285, 289, 292, 296
negative emotions, diminished integration and, 114, 314
negative life traits, less integrated connectome and, 81
neglect, impediments to brain integration and, 80–81, 223
neocortex, 138, 155
nervous system, 317, 519–20
embodied energy flow patterns and, 131
interactive components of, 43–44
neural activity, current understanding of, 44–45
neural behavior, subjective mental life and, 113
neural correlates of consciousness (NCC), 263, 277, 289
neural firing, 44, 45, 177
neural growth, 177, 180
neural integration
enhanced, 234
healthy self-regulation and, 234
mindfulness training and, 221
self-regulation and, 82
neural "maps" of world, cortex and, 135
neural processing, consciousness and, 7
neural representation, 45
neural system, 46
neurobiology, 47
neuroception, 193
neuronal firing, attention, energy and information flow and, 225–26
neurons, 44, 317
neuroplasticity of brain
attention, neuronal firing, and, 226
cultural systems and, 177–82
trust, learning, and, 167
neuroscience, 152
neurotransmitters, 44, 46
"New Narrative Project, The," 271–72
Newtonian (or Classical) physics, 48, 57, 58, 164, 240, 244, 247, 270
night vision, perception, betweenness, and, 157, 162
NIMH. *see* National Institute of Mental Health (NIMH)
"noble silence," mindfulness meditation and, 222

non-conscious energy flow, 279, 280
non-conscious mind, 119, 159
non-declarative memory, hippocampus, PTSD symptoms, and, 169
"non-dual" view, 295–96, 297
non-focal attention, 167
non-linear energy and information flow, 35, 36
non-locality, 48, 163, 164
non-verbal communication, 103
non-worded world, silence and, 23
noosphere, 174
"no," reactive state and, 309
now, *when* of mind and emergent property of, 226. *see also* present moment
now-here, becoming no-where, 303
nuclei, 44

OATS (others and the self) circuitry system, 138
  horse accident and temporary disabling of, 133
  mindfulness meditation and, 234
  worry and, 65
obesity, 193
objectivity
  mindsight tripod and, 281, *282*
  presence and, 287
observation, 287
  as bridge between conduition and construction, 255
  mindsight tripod and, 281, *282*
  nature of energy shifted through, 268–69, 276, 283
observational stream
  defining, within SOCK, 272
  mindfulness meditation and, 224, 229, 230, 231, 232
obsessive-compulsive disorder, mindfulness meditation and, 224
O'Donohue, J., 71, 216, 223, 256, 261, 262, 263, 264, 272, 273
Ohlers, L., *27*
On the Sacred Disease (Hippocrates), 9
openness
  mindsight tripod and, 281, *282*
  presence and, 287
open systems, 42
organ formation, cells and, 316
out-group
  seen as non-group, non-human, 321
  threat conditions, and hostile treatment of, 320
OWN (observing, witnessing, and narrating), 134, 230

pain, plane of possibility and abatement of, 271, 283
parallel-distributed-processor (PDP) system
  in brain, 46
  in intrinsic nervous system of heart and intestines, 153
parasympathetic nervous system, 46
parental presence, at heart of secure attachment, 228
Parenting from the Inside Out (Siegel & Hartzell), 69, 214
parents, mindfulness for, 215
particles, 268, 269, 283

PART (presence, attunement, resonance, and trust)
feeling felt and, 167
interpersonal integration, internal neural integration, and, 234
PTSD treatment and, 170
secure attachments and, 229
trust and, 172
past, 252
certainty and events in, 251
fixed nature of, 241
preoccupation with, 231, 243, 251, 270, 303, 305, 325
Past Hypothesis, 243, 244, 270
past-present-future, embracing integration of, 18, 19
patients, caring for by caring about, 102, 111
patterns, culture and, 179, 180
Paul (apostle), 191
PDP system. *see* parallel-distributed-processor (PDP) system
perception, 3, 6, 132
altered, hallucinogenic drugs and, 127
reflecting on meaning of, 133
right- and left-mode of brain and, 162
subjective nature of, 121
top-down learning and, 142
training, 162
see also awareness
perceptual capacities, energy patterns and, 132
personal identity, levels of, 206
personal self, construction and notion of, 141
"pervasive leadership," Arthur Zajonc, promoting, 329
pharmaceutical interventions, 30
photographic images, 23
photons, 164, 239, 244, 268, 283
physicians, mindful awareness training and, 110
physics
entanglement and, 48
process of energy flow in, 55–59
see also Newtonian (or Classical) physics; quantum physics
piloerection, 309
placebo effect, 10
plane of possibility, *254*, 270, 283, 284
awareness and open plane of, 269
awe and overlap with, 310
consciousness as possible prime of, 276
creating cultural waves of positive influence and, 300
deep resonance and, 286
emptiness or fullness experiences with, 295–97, 298
examination of, flow of energy, and, 291–95
feelings reported by participants about, 271
hub of Wheel of Awareness and, 274
increasing access to, presence, and, 298
integrated awareness and, 299
in-the-plane or not-in-the-plane, energy probability and, 291, 292
knowing of awareness emerging from, 289
receptive state and, 309
"self-states" and, 315

as source of consciousness?, 268
sweep ratio, quality of consciousness, and, 294, 295
see also Wheel of Awareness
planet
delusion of separateness and impact on, 326–27
human mind and shaping of, 6
integration and kindness and compassion for, 98
planetary well-being, expanding sense of self and, 61
Porges, S., 193, 309
positive emotions, increased levels of integration and, 114
positive life traits, more integrated connectome and, 81
positive psychology, lens of integration and, 114
positivity resonance, love and, 311, 312, 314
possibility, transforming into actuality, 31
postsynaptic neurons, 45
posttraumatic stress disorder (PTSD), 176, 198
chaos and rigidity in, 77
controlled hallucinogen use and benefits for, 127
hippocampus and, 169
implicit memory and mental suffering of, 173
PART and treatment of, 170
Vietnam war veteran with, 168
potentials, energy, 57
preconscious, 279
precuneus, 133
prefrontal cortex, 133
developmental trauma and growth impediments in, 223
integrative fibers of, 215
mindfulness meditation and, 221
nine functions arising in, 204
prefrontal regions of brain, abuse and neglect, lack of integration, and, 81
presence, 185, 298
being in love with natural world and, 327
choice and change and, 295
cultivating, 287–301
flow and, 294–95
health and, 324
integrated self and, 329
"non-judgmental" as a term and, 224
as portal for integration, 256, 280, 281, 283, 295, 298, 325, 330
regaining, conduition and, 231–32
repair of ruptures and, 228
"tend-and-befriend" response and, 311
well-being cultivated with, 308
pre-sence, opening to, 157
present moment, 252
embracing full potential of, 61
emergence and, 249
losing touching with, 231
opening to, integration and, 235
past, present, and future linked in, 18
resting in, 256
primates, 308
prime, 201, 252
priming, future events and, 173
Principles of Psychology, The (James), 8

probability distribution curve, 31
change across, 32
energy, Wheel practice, and, 274
energy and, 237–38, 239, 268, 281
integrated consciousness and, 297
probability distribution function, of energy, awareness and alteration of, 269
probability(ies)
directionality of time and, 243
nature of time and role of, 250
proprioceptive sense of motion, 132
prospective memory, 251
psoriasis, mindfulness meditation and, 224
psychiatric disorders, chaos and rigidity in, 77–78
psychiatric medications, 9–10
psychiatry field
conflict and turf battles in, 146
division of the objective from the subjective and, 148–52
drive for objectivity in, 148
transfer into, 147–48
psychotherapy, 149
interpersonal neurobiology applied to, 197
within and between mind and, 176
Psychotherapy Networker magazine, 215
PTSD. *see* posttraumatic stress disorder (PTSD)
Pullin, J., 246
pure consciousness, 289, 290
purpose
mind and, 196
reflections and invitations on, 206–11
purpose of life, integration as?, 200–205, 210

Quakers, 190
quantum complex networks theory, 247
quantum effects, obscuring of, by classical physics' properties, 164
quantum mechanics, 48, 248
orthodox Copenhagen interpretation of, 269
possibilities, probabilities, and, 57
quantum physics, 163, 268
plane of possibility consistent with, 271
time and, 240
quantum theory, 244

Raichle, M., 80
reactive state, "no" and, 309
reactivity, 309, 314, 321
Reader, J., 194
reality, quantum or probability nature of, 58
reasoning, 3
receptive state, "yes" and, 309
receptive stillness, wide-open awareness and, 293
recursive feedback, 203
reductionism, 248, 249
reflections and invitations
within and between, 182–87
awareness and time, 235–56
centrality of subjectivity, 115–22
cultivating presence, 287–301
identity, self, and mind, 140–44

integration and well-being, 85–98
MWe, an integrating self, and a kind mind, 323–30
purpose and meaning, 206–11
self-organization of energy and information flow, 51–61
relational, contextual perceptions, right brain hemisphere and, 161
relational mental lives, 13
"relational," use of term, in reference to the mind, 154
relationships
energy and information shared in, 39
integration in, study of, 82
interpersonal integration and, 94–95
lives unfolding within, 28
science of mind and, 70
supportive, well-being and, 307
transformative power of, 216
triangle of human experience and, 39–40, *40*
see also attunement; compassion; empathy; kindness; MWe
relationship science, 175
religion, science and, 195, 236
resilience, 78
of coherence, features of, 203
integration and, 98
integration in brain and self-regulation at heart of, 82
mindsight and, 147
resonance
attunement, trust, and, 229
feeling felt and, 167
plane of possibility and, 286
term "relational" and, 154
see also positivity resonance
respect, interpersonal integration and, 94, 95
respiratory system, 43
response flexibility, prefrontal cortex and, 204
reverence, 329
Rhesus monkeys, killing by, and gain in alpha status, 320
rigidity, 96, 117, 254, 284
acute grief and, 97
anguish manifested as, 138
assessments of, in clinical setting, 88
clinging to sense of control and, 280
entering state of flourishing from, 200
excessive differentiation and, 314
grief and, 83
impaired integration and, 86, 87, 198
negative emotions and, 114
origins of, gaining sense of, 89–90
patients and pattern of, 76
reflecting on, 85, 86, 210
relieving distressing symptoms of, 206
stress, shift to left brain hemisphere, and, 92
transforming into harmony of integration, 298
see also chaos
river of integration, 78, *79*
Roman Pantheon, 195
romantic relationships, integration in, 82
Rumi, 121

Rupert, R., 154
ruptures, presence and repair of, 228
Russia, *shtetls* in, 192, 193

sadness, 74. *see also* grief
Saint Peter's Square, Rome, 195, 196
salmon, fresh water to salt water switch and, 30, 107, 115, 124, 183
sanctity of life, honoring, 329
Scharmer, O., 162
schema, 128, 172
schizophrenia, 9, 198
  brain regions, lack of integration, and, 80
  chaos of hallucinations and rigidity of delusions in, 77
  impaired integration and, 218
Schore, A., 28
science, 5
  religion and, 195, 236
  spirituality and, 195, 236–37, 272, 273
  world's wisdom traditions, spirituality, and, 205
science of mind
  relationships and, 70
  subjective mental experiences and, 69–70
Scott, D., 251
seasons, 304
Second Law of Thermodynamics, 270
  directionality of Arrow of Time and, 242
  expansion of the universe and, 244
  "Past Hypothesis" and, 243, 244
secure attachment, 28
  attunement and, 227–28
  being present with your children and, 224
  as integrated relationship, 82
  integration, within and between, and, 223
  interpersonal attunement and, 233
  mindfulness and, 215
  mindshpere in family and, 175
  trauma resolution and development of, 176
selective attachments, 195
self, 140, 141, 142
  betweenness and sense of, 160
  expanding sense of, 60–61
  illusions of time and, 61
  see also identity
self-care, integration and, 211
self-discovery, 120
self-identity, culture and, 321
"self-object," loss of, 96–97
self-organization, 38, 84, 99, 123, 152, 190, 284
  complex systems and, 49, 50, 98
  defined, 36
  domains of integration and, 97
  drive toward integration and, 82–83
  of energy and information flow, 51–61
  FACES flow and, 78, 79
  health and, 62
  integration and, 78, 111, 145
  living struggle of, 75
  maximal complexity and, 78, 79, 111–12, 203, 253

mind and, 156
natural essence of unfolding with, 51
optimal, 61, 76, 78, 199
potential overlaps of consciousness and, 59
probability distribution curve, consciousness, and, 297–98
recursive feature of being and, 50, 54, 60, 203
stories and, 72–73
within and between, 158–59
see also integration
self-organizing aspect of mind, defining, 37
self-reflection, cortical brain and, 155
self-regulation
integration in the brain and, 221
neural integration and, 82, 234, 235
"self-states," integration and, 315
self-transformation, awe and, 310
Senge, P., 162, 179
sensations
bottom-up, 132
ecphoric, 172
reflecting on meaning of, 133
SIFTing the mind and, 121, 122
senses, in Wheel of Awareness, 90, *91*
sensing stream, mindfulness meditation and, 224, 229, 231, 232
sensory conduit stream, vivid present and, 138
sensory stream, defining within SOCK, 272
separate self
digital technology and reinforced experience of, 326
family and societal messages about, 320, 322
serotonin, 46
seventh sense
of mental abilities, 278
in Wheel of Awareness, 90, *91*
Shapiro, S., 225
shared attention, subjectivity honored in focus of, 111
sharing, birth of the mind and, 22
shtetls, 192, 193
sickle cell trait, case of patient with, 103, 104
Siegel, A., *141*, 258, *324*
Siegel, M., *42*, *190*, *200*, *227*, *326*, *330*
SIFTing the mind, 105, 106, 143
focusing on subjective reality of your life and, 121–22
reflections on plane of possibility and, 300
sight, 132
silence
importance of, 22
truth illuminated in, 23
situated cognition, 153
sixth sense (interoceptive ability), 278
vertical integration and, 92
in Wheel of Awareness, 90, *91*
smartphones, 162, 240
smell, 132
social awe, 310
social brain, 47
social brain hypothesis, distributed mind and, 153

social engagement system
  "tend-and-befriend" response
    and, 311
  trust and, 309
social field, 115, 116, 117, 162,
    168, 170–71, 180–81
  of acceptance and receptivity,
    172
  invisible neuroplastic changes
    and, 181
  mind as within and between
    and, 186
social intelligence, promoting,
    236
social mind, 119
social neuroscience, 158
socio-cognitive relationships,
    distributed mind and, 154
SOCK (sensation, observation,
    concept, knowing), 255, 272
  of mindful awareness, 231
  power and possibility of pres-
    ence and, 256
Solomon, M., 28
soothing fear, prefrontal cortex
    and, 204
"Soul and Synapse" event, 264
sound, 56
sound waves, 55
spacetime, 239, 240, 243
  eternal imprint in, 263
  unfoldings in, 245
speed
  relative nature of time and, 49
  shape of spacetime block
    and, 240
Sperry, R., 108, 113, 161
spirituality
  science, world's wisdom tra-
    ditions, and, 205
  science and, 195, 236–37, 272,
    273
"split-brain patients" research
    (Sperry), 108
starlight, betweenness of mind
    and, 157
state integration, 94, 96
Stoller, R., 24, 28, 65, 67, 68, 71
stories, 18, 20, 29
  humans as social creatures
    and, 196
  self-organization and, 72–73
  social nature of, 28
  see also narrative(s)
stress, new interpretations of, 185
string theory, 240
structure, culture and, 179, 180
subconscious, 279
subjective experience, 38, 39,
    52, 74, 151, 152, 284
  attending to, respecting, and
    sharing, 111
  awakening and paying atten-
    tion to, 121
  awareness and, 34
  definition of, 119
  describing, 253
  as gateway for interpersonal
    connection, 112
  honoring, integrated life, and,
    120–21
  importance of, 111, 112
  source of, questions related
    to, 155, 156, 158
subjectivity, 13, 20, 123
  centrality of, 115–22
  definition of, 119
  expanding source of, 155
  mind and, 1, 2
  prime of energy and informa-
    tion flow and, 52–53
  as "prime" of the mind, 33
substance abuse, mindfulness
    meditation and, 224

suffering
reducing, through love and compassion, 314–15
source of, in Buddhist philosophy, 222, 231
suicide prevention, emotional communication and, 30, 105, 124, 183
Sullivan, A., 22
sun, passage of Earth around, 303
Suomi, S., 320, 321
survival, distrust and, 193
Suzaki, K., *116*
sweep ratios, 294, 295
sympathetic nervous system, 46
synapse, 44
synaptic architecture, energy flow and, 304–5
synaptic growth, 177, 180
synergy, emergent, 80
systems
composition of, 42
types and sizes of, 42–43
system-self, 323
systems science, 77

Tai' Chi Chu'an, 124, 125
Taoism, 124
taste, 132
Taylor, S., 311
Teicher, M., 80
telomerase, 224, 308, 324
temporal integration, 95, 206, 251
temporal lobes, 133
"tend-and-befriend" response, presence and, 311
theory of mind, 153
therapeutic experience, changing patient's brain in positive way and, 171
therapeutic intervention, promotion of well-being, and, 89
therapy, basis of mindsight approach to, 199
Theravada Buddhist tradition, insight meditation practice of, 217
thermal equilibrium, 244
thought, 2, 3, 4, 6, 34, 122, 278. *see also* mind
Tibetan Buddhism, 220
time
academic discussions on nature of, 246–49
bottom-up perceiving and, 130
calendar of nows and, 304
change and, 41, 281–82
defining, 239, 245
energy and, 237
exploring nature of, 23
flow and, 32
flow in relation to energy shifts and, 239
illusions of self and, 61
integration and, 306
macrostates and directionality of, 241
mortality in, 196
reflecting on deep nature of, 212
reflections and invitations on awareness and, 235–56
risk in believing mind's illusion of, 303
seeming irreversibility of, 241
velocity, gravity, and nature of, 49
time-asymmetry, 247
timeless reality, 220

Time magazine, "Mindful Revolution" cover story, 221
To Bless the Space Between Us (O'Donohue), 216
Tononi, G., 263, 289
top-down categorical construction, clinician, diagnostic categories, and, 151–52
Top-Down Constructor, mind as, 131
top-down information processing, 132, 135, 136, 139
  description of, 127–31
  OATS system and, 133, 134
  schematic of, *136*
  top-down construction and, 137, 140, 273
  see also bottom-up information processing
touch, 132
traffic jams, inner sanctuary and dealing with, 299–300
transpirational integration, 95, 206
trauma
  chaos and rigidity in, 78
  corpus callosum impairment and, 91–92
  resolving, integration into explicit memory and, 93
  unresolved, brain function and, 66
  unresolved, intergenerational transfer of disorganized attachment and, 176
traumatic childhood history, Wheel of Awareness and individuals with, 266
traumatic loss, mind's struggle with, 67
travel, bottom-up perceiving and, 130
true human narrative, teaching, 326
trust, 308
  attunement, resonation, and, 229
  learning facilitated by, 167
  PART and, 172
  social engagement system and, 309
truth, 22, 23, 210
Tuesday Sunday manuscript, 67, 68, 73, 105
Tuesdays with Morrie, (Albom) 105
Tulving, E., 18
turning points, 140, 142

Ukraine, *shtetls* in, 192
ultraviolet light, 164
unconscious, 279
Unitarian Church, 190, 194, 195
universe
  consciousness as inherent part of, 269
  energy potentials and nature of, 57
University of California--Berkeley campus, grove of trees on, 309–10, *311*, *312*
University of Freiburg, Germany, 550th birthday conference at, 236
Upper Paleolithic period, cultural evolution in, 181

Varela, F., 250
Vatican
  Pontifical Council on the Family, 188
  public talk in Synod of Bishops at, 194
  visit to, 190, 191, 193–96, 207

velocity, relative nature of time and, 49
ventrolateral prefrontal cortex, 131
Verely, A., 208
vertebrate nervous system, regulatory nature of, 308
vertical integration, 92, 206
  grief and, 96
  narrative integration and, 93
Vietnam War, 190
visual images, 23
vulnerabilities, emergence of kindness and, 315–16
Vygotsky, L., 12

Wailing Wall of Jerusalem, 310
wave function, 31
waves, range of values or locations for, 268
we-identity, longing for, 160. *see also* MWe
Welch, C., 215, *258*
Welch, N., 86, 87
well-being, 62, 236, 250, 266
  within and between and, 97
  compassion, kindness, and, 196
  differentiation, linking, and, 329
  empathy and, 112
  FACES transformation and, 89
  focus on subjective reality and, 111
  integration and, 78, 81, 98, 115, 255, 306
  integration in brain and self-regulation at heart of, 82
  interpersonal integration and, 95, 96
  mindfulness and, 223
  mindsight and, 122
  movement and, 124
  presence and cultivation of, 308
  temporal integration and, 251
  see also health
Wheel of Awareness, 59, 186, 255, 256, 259, 264, 269, 277, 284
  bending of the spoke and, 265–66, 267, 268, 274, 298
  consistency in numerous responses to, 267, 268, 299
  finding commonality in hub of, 287
  four segments on rim of, 232–33
  function of, 232
  hub, plane of possibility, and, 274, 298, 299
  integration of consciousness and, 90, *91*
  living on rim of, 270
  loss and, 96
  love and, 311
  miraculous mechanisms of mind and, 281
  plane of possibility and, 271
  presence and harnessing hub of, 325
  reflecting on experiences with, 299
  shift in perspective with, 265, 266
  sweep ratio and, 294
  see also plane of possibility
Whitfield, T., 28, 68, 76, 105, 146
  cancer diagnosis, 63
  death of, 67
  initial contact with, 102
  mentoring relationship with, 63, 64–65
  moving through grief after loss of, 67, 68, 70–73, 74, 83, 88, 96–97

wholeness, 84, 86, 124
Wholeness and the Implicate Other (Bohm), 83
whole person, seeking system of, 317–19
wide-open awareness, flow *vs.*, 292–93
Wiesel, T., 176
window of tolerance, widening of, trusting relationship and, 172
wisdom, collective intelligence and, 209
within and between
  attention shaping energy and information flow and, 179
  brain hemispheres and perception of, 161–62
  compassion and, 321
  consciousness and, 159, 161
  FACES flow of integration, therapy, and, 172
  finding way toward harmony and, 211
  identity integration and, 95
  information flow and, 158, 159, 161
  mind's emergence from energy flows in, 306
  mindsphere, mindscape, and, 175
  reality of mind and, 175, 180
  reflections and invitations, 182–87
  secure attachment, integration, and, 223
  self-organization and, 158–59
  subjective experience and, 159
  well-being and, 97
witnessing, top-down mode and, 134
worded thoughts, non-worded thoughts and, 23
words, innate limiting and limited nature of, 21
World Health Organization, 125

"yes," receptive state and, 309

Zajonc, A., 250, 253, 329
"zero percent probability" state, ways of experiencing, 296–97